CELLULAR AND MOLECULAR MECHANISMS IN ALZHEIMER'S DISEASE THERAPEUTICS

Volume 2

Molecular Mechanisms

Cellular and Molecular Mechanisms in Alzheimer's Disease Therapeutics, 2-volume set
ISBN: 978-1-77964-176-2 (hbk)

Cellular and Molecular Mechanisms in Alzheimer's Disease Therapeutics, Volume 1: Cellular Mechanisms
ISBN: 978-1-77964-178-6 (hbk)
ISBN: 978-1-77964-179-3 (ebk)

Cellular and Molecular Mechanisms in Alzheimer's Disease Therapeutics, Volume 2: Molecular Mechanisms
ISBN: 978-1-77964-180-9 (hbk)
ISBN: 978-1-77964-181-6 (ebk)

CELLULAR AND MOLECULAR MECHANISMS IN ALZHEIMER'S DISEASE THERAPEUTICS

Volume 2

Molecular Mechanisms

Prashant Tiwari, PhD
Sunil Kumar Kadiri, PhD
Kunjbihari Sulakhiya, PhD
Editors

AAP | APPLE ACADEMIC PRESS

First edition published 2027

Apple Academic Press Inc.
1265 Goldenrod Circle, NE,
Palm Bay, FL 32905 USA

760 Laurentian Drive, Unit 19,
Burlington, ON L7N 0A4, Canada

CRC Press
2385 NW Executive Center Drive,
Suite 320, Boca Raton FL 33431

4 Park Square, Milton Park,
Abingdon, Oxon, OX14 4RN UK

Apple Academic Press exclusively co-publishes with CRC Press, an imprint of Taylor & Francis Group, LLC

Library and Archives Canada Cataloguing in Publication

Title: Cellular and Molecular Mechanisms in Alzheimer's Disease Therapeutics / Prashant Tiwari, PhD, Sunil Kumar Kadiri, PhD, Kunjbihari Sulakhiya, PhD, editors.
Names: Tiwari, Prashant, editor. | Kadiri, Sunil Kumar, 1981- editor. | Sulakhiya, Kunjbihari, editor.
Description: First edition. | Includes bibliographical references and indexes. | Contents: Volume 2: Molecular mechanisms.
Identifiers: Canadiana (print) 20260143553 | Canadiana (ebook) 2026014360X | ISBN 9781779641762 (set) | ISBN 9781779641809 (v. 2 ; hardcover) | ISBN 9781779641816 (v. 2 ; ebook)
Subjects: LCSH: Alzheimer's disease—Cytopathology. | LCSH: Alzheimer's disease—Molecular aspects. | LCSH: Alzheimer's disease—Pathophysiology.
Classification: LCC RC523 .C45 2026 | DDC 616.8/3107—dc23

Library of Congress Cataloging-in-Publication Data

CIP data on file with US Library of Congress

ISBN: 978-1-77964-180-9 (hbk)
ISBN: 978-1-77964-181-6 (ebk)
DOI: 10.1201/9781779641816

About the Editors

Prashant Tiwari, PhD
Associate Professor, Department of Pharmacology,
College of Pharmaceutical Sciences, Dayananda Sagar University,
Bengaluru, India

Prashant Tiwari, PhD, is an Associate Professor in the Department of Pharmacology at the College of Pharmaceutical Sciences, Dayananda Sagar University, Bengaluru, Karnataka, India. He holds a PhD from Siksha O Anusandhan University, Bhubaneswar, and was a fellow of the Indian Council of Medical Research (ICMR), New Delhi. With over 13 years of experience in teaching and research, Dr. Tiwari's expertise includes drug screening, method validation for brain-targeted drug delivery, drug-induced dyskinesia revalidation, and the neuroprotective effects of benzimidazole derivatives. His research has explored the dopaminergic pathway, prolactin inhibition, and neuroprotection, with notable studies on nanoemulsions containing *Lepidium sativum* oil for Parkinson's disease and *Muntingia calabura* leaf extract for Alzheimer's disease. For his outstanding work, Dr. Tiwari received the Dr. P. D. Patil National Award in 2022 for Best Thesis in Pharmaceutical Sciences.

In addition to his research, Dr. Tiwari has authored five books, edited five books, and co-authored six book chapters. He has filed seven patents, two of which have been granted. He organized a national e-seminar on targeted drug delivery against neurodegenerative diseases, sponsored by the Department of Biotechnology (DBT), Government of India. Dr. Tiwari has been honored with various awards, including the *Indian Council of Medical Research (ICMR)* Senior Research Fellowship, Young Scientist Award, and Best Teacher Award. He supervises five PhD students and has guided nine MPharm students. He is also an active member of professional bodies and serves as a peer reviewer for international journals. His research interests span neurodegenerative disorders, drug interactions, pharmacokinetics/pharmacodynamics, and metabolic disorders.

Sunil Kumar Kadiri, PhD
Professor, College of Pharmaceutical Sciences,
Dayananda Sagar University, Bengaluru, Karnataka, India

Sunil Kumar Kadiri, PhD, is a Professor at the College of Pharmaceutical Sciences, Dayananda Sagar University, Bengaluru, Karnataka, India, with 16 years of teaching experience. His research interests encompass neuropharmacology, metabolic disorders, organ-protective activities, and drug interactions. He holds a PhD in Pharmaceutical Sciences from JNTU, Hyderabad and has authored 65 research and review articles. In addition to publishing five books and four book chapters, Dr. Kadiri holds six patents. An active member of Association of Pharmaceutical Teachers of India (APTI), his recent research includes drug interaction studies involving Alprazolam and Pomegranate, Carbamazepine and Losartan, Naproxen and St. John's Wort, as well as investigations into the antiparkinsonian effects of Zinda Tilismath.

Kunjbihari Sulakhiya, PhD
Assistant Professor, Department of Pharmacy, Indira Gandhi National Tribal University, Amarkantak, Madhya Pradesh, India

Kunjbihari Sulakhiya, PhD, is working as an Assistant Professor in the Department of Pharmacy, Indira Gandhi National Tribal University, Amarkantak, Madhya Pradesh, India. He obtained his MS (Pharm) and PhD in Pharmacology and Toxicology from the National Institute of Pharmaceutical Education and Research (NIPER), Guwahati, Assam, India. He has published 22 research papers, seven review articles, and 11 book chapters in reputed peer-reviewed national and international journals and books. Dr. Sulakhiya has received 13 awards/fellowships including six travel awards to present his research papers at international conferences in the United States, Canada, Japan, France, and Macau. He is a member of various professional societies, including the Indian Academy of Neurosciences (IAN), the International Society for Neurochemistry (ISN), Asia Pacific Society for Neurochemistry (APSN), APTI, and so forth. Dr. Sulakhiya is serving as an editorial member and reviewer for various reputed journals. His area of research is pre-clinical research in the field of neuropharmacology. He has completed three research projects funded by DBT, UGC, and ISN.

Contents

Contributors

Arshi
S.D. College of Pharmacy, Patel Nagar, Muzaffarnagar, Uttar Pradesh, India

Abhishek Agarwal
H&H Healthcare and Cosmetics Private Limited, Indore, Madhya Pradesh, India

Ankur Agrawal
Jai Institute of Pharmaceutical Sciences and Research, Gwalior, Madhya Pradesh, India

Mohit Agrawal
School of Medical & Allied Sciences, K.R. Mangalam University, Gurugram, Haryana, India

Shashi Alok
Institute of Pharmacy, Bundelkhand University, Jhansi, Uttar Pradesh, India

Meenakshi Attri
School of Medical & Allied Sciences, K.R. Mangalam University, Gurugram, Haryana, India

P. Veeresh Babu
Department of Pharmacology, Gokaraju Rangaraju College of Pharmacy, Bachupally, Hyderabad, Telangana, India

Lakhyajit Borah
Arunachal University of Studies, Namsai, Arunachal Pradesh, India

Hema Chaudhary
School of Medical & Allied Sciences, K.R. Mangalam University, Gurugram, Haryana, India

Udit Chaube
Department of Pharmaceutical Chemistry, Institute of Pharmacy, Nirma University, Ahmedabad, Gujarat, India

Lalit Pratap Chandrvanshi
Department of Forensic Science, Sharda University, Greater Noida, Uttar Pradesh, India

Anup Kumar Chakraborty
IES Institute of Pharmacy, IES University, Bhopal, Madhya Pradesh, India

Diksha Diwakar
Sahu Onkar Saran School of Pharmacy, Faculty of Pharmacy, IFTM University, Moradabad, Uttar Pradesh, India

Pawan Kumar Dubey
Department of Pharmacology, Swami Vivekananda College of Pharmacy, Indore, Madhya Pradesh, India

Biswajit Dash
Amity Institute of Pharmacy, Amity University, Kolkata, West Bengal, India

Ekta
Department of Pharmaceutics, School of Health Sciences, Sushant University, Gurugram, Haryana, India

Mohd. Asad Farooqi
Institute of Pharmaceutical Research, GLA University, Mathura, Uttar Pradesh, India

Reena Gupta
Institute of Pharmaceutical Research, GLA University, Mathura, Uttar Pradesh, India

Jitendra Gupta
Institute of Pharmaceutical Research, GLA University, Mathura, Uttar Pradesh, India

Pradeep Golani
Gyan Ganga Institute of Technology and Sciences, Jabalpur, Madhya Pradesh, India

Dishant Gupta
Department of Pharmacology, Swami Vivekananda College of Pharmacy, Indore, Madhya Pradesh, India

Deepak Kumar Gupta
Bengal College of Pharmaceutical Technology, Dubrajpur, West Bengal, India

Richa Gupta
Department of Pharmacology, Institute of Pharmacy, Nirma University, Ahmedabad, Gujarat, India

Radha Goel
Lloyd Institute of Management and Technology, Greater Noida, Uttar Pradesh, India

Praveen Kumar Gaur
Metro College of Health Sciences and Research, Greater Noida, Uttar Pradesh, India

Ramza Rahat Hashmi
Metro College of Health Sciences and Research, Greater Noida, Uttar Pradesh, India

Ittishree
Institute of Pharmaceutical Research, GLA University, Mathura, Uttar Pradesh, India
SRM Modinagar College of Pharmacy, SRMIST Delhi-NCR Campus, Modinagar, Ghaziabad, Uttar Pradesh, India

Ashok Jangra
Department of Pharmaceutical Sciences, School of Interdisciplinary and Applied Sciences, Central University of Haryana, Mahendragarh, Haryana, India

Deenanath Jahde
School of Pharmacy, Mody University of Science and Technology, Sikar, Rajasthan, India

Sanjay Khattri
Derpartment of Pharmacology, King George'e Medical University, Lucknow, Uttar Pradesh, India

L. Karpagavalli
Saveetha College of Pharmacy, Saveetha Institute of Medical and Technical Sciences, SIMATS, Tamil Nadu, India

Shahbaz Khan
Pharmacy Academy, Faculty of Pharmacy, IFTM University, Moradabad, Uttar Pradesh, India

Virendra Kushwaha
Institute of Pharmaceutical Research, GLA University, Mathura, Uttar Pradesh, India

Sunil Kumar Kadiri
College of Pharmaceutical Sciences, Dayananda Sagar University, Bengaluru, Karnataka, India

Nitin Kumar
Department of Pharmacy, Meerut Institute of Technology, Meerut, Uttar Pradesh, India

Devendra Singh Lodhi
Gyan Ganga Institute of Technology and Sciences, Jabalpur, Madhya Pradesh, India

Alka Lohani
Amity Institute of Pharmacy, Amity University, Noida, Uttar Pradesh, India

Arun Kumar Mishra
SOS School of Pharmacy, IFTM University, Moradabad, Uttar Pradesh, India

Rosaline Mishra
Metro College of Health Sciences and Research, Greater Noida, Uttar Pradesh, India

Niharika
Metro College of Health Sciences and Research, Greater Noida, Uttar Pradesh, India

Sanjay Nagdev
Shri. Prakashchand Jain College of Pharmacy and Research, Palaskheda (Bk.), Jamner, Maharashtra, India

Akhilesh Patel
NIMS Institute of Pharmacy, NIMS University, Jaipur, Rajasthan, India

Shruti Poriya
Department of Pharmacology, Institute of Pharmacy, Nirma University, Ahmedabad, Gujarat, India

Mohammed Nazish Quasmi
Department of Pharmaceutical Sciences, School of Interdisciplinary and Applied Sciences, Central University of Haryana, Mahendragarh, Haryana, India

Asha Raghav
Department of Pharmaceutics, School of Health Sciences, Sushant University, Gurugram, Haryana, India

Harpreet Singh
School of Pharmaceutical Sciences (Faculty of Pharmacy), IFTM University, Moradabad, Uttar Pradesh, India

M. Sri Ramachandra
Dr. K. V. Subba Reddy Institute of Pharmacy, Dupapdu, Kurnool, Andhra Pradesh, India

Jaya Sharma
Department of Pharmacology, Swami Vivekananda College of Pharmacy, Indore, Madhya Pradesh, India

Rajendra K. Shukla
Department of Biochemistry, MSD Autonomous State Medical College & Maharishi Balark Hospitals, Bahraich, Uttar Pradesh, India

Kangkan Jyoti Saud
Arunachal University of Studies, Namsai, Arunachal Pradesh, India

Rajib Singh
Shakti College of Pharmacy, Balrampur, Uttar Pradesh, India

Brij Raj Singh
Malti Memorial Trust CSM Group of Institutions, Faculty of B. Pharmacy, Prayagraj, Uttar Pradesh, India

Shriram Sisodia
Institute of Pharmaceutical Research, GLA University, Mathura, Uttar Pradesh, India

Kantrol Kumar Sahu
Institute of Pharmaceutical Research, GLA University, Mathura, Uttar Pradesh, India

Junaid Tantray
NIMS Institute of Pharmacy, NIMS University, Jaipur, Rajasthan, India

Prashant Tiwari
College of Pharmaceutical Sciences, Dayananda Sagar University, Bengaluru, Karnataka, India

Himangi Vig
PSIT-Pranveer Singh Institute of Technology (Pharmacy), Kanpur, Uttar Pradesh, India

Pranay Wal
PSIT-Pranveer Singh Institute of Technology (Pharmacy), Kanpur, Uttar Pradesh, India

Yashraj Yadav
Department of Pharmacology, Swami Vivekananda College of Pharmacy, Indore, Madhya Pradesh, India

Mohd. Yaasir
Arsi University, Asella, Oromia, Ethiopia

Preface

Alzheimer's disease (AD) stands as a profound and growing global health concern, characterized by progressive neurodegeneration and cognitive decline. As research into the underlying causes of AD advances, it becomes clear that a deeper understanding of its molecular mechanisms is crucial for the development of effective therapeutic interventions.

This book, *Cellular and Molecular Mechanisms in Alzheimer's Disease Therapeutics,* presents a detailed exploration of the molecular processes that drive the onset and progression of AD. Edited by Dr. Prashant Tiwari, Dr. Sunil Kumar Kadiri, and Dr. Kunjbihari Sulakhiya, this volume brings together leading experts who offer insights into the intricate molecular pathways involved in AD pathology. From the misfolding and aggregation of amyloid-beta and tau proteins to the disruptions in cellular homeostasis, mitochondrial dysfunction, and neuroinflammation, the chapters provide a comprehensive understanding of the disease at a molecular level.

The primary objective of this book is to bridge the gap between molecular research and therapeutic applications. By shedding light on the latest findings related to disease mechanisms, such as oxidative stress, synaptic alterations, and the role of neuroinflammatory pathways, we hope to inspire new strategies that target the root causes of AD. The contributors to this volume explore both well-established and novel therapeutic avenues, including small molecules, immunotherapies, gene editing, and precision medicine approaches.

As editors, we are privileged to work alongside a diverse group of researchers and clinicians whose expertise has shaped this volume. The insights gathered here reflect the current state of the field and point to future directions in the ongoing quest to develop effective treatments for AD. We believe this book will serve as a valuable resource for scientists, healthcare professionals, and students alike, all of whom are working to unravel the molecular complexities of AD and translate these insights into meaningful therapeutic outcomes.

We extend our heartfelt thanks to the contributors for their dedication and to the team at Apple Academic Press for their support in bringing this book to publication. It is our hope that *Cellular and Molecular Mechanisms in Alzheimer's Disease Therapeutics* will serve as an important reference for those committed to combating the devastating effects of AD through a molecular lens.

—Editors

CHAPTER 1

Investigating the Role of Amyloid Beta and Tau Peptides in Alzheimer's Histopathological Brain Lesions

ASHA RAGHAV[1], EKTA[1], MOHIT AGRAWAL[2], MEENAKSHI ATTRI[2], HARPREET SINGH[3*], ARUN KUMAR MISHRA[4], and HEMA CHAUDHARY[2]

[1]*Department of Pharmaceutics, School of Health Sciences, Sushant University, Gurugram, Haryana, India*

[2]*School of Medical & Allied Sciences, K.R. Mangalam University, Gurugram, Haryana, India*

[3]*School of Pharmaceutical Sciences (Faculty of Pharmacy), IFTM University, Moradabad, Uttar Pradesh, India*

[4]*SOS School of Pharmacy, IFTM University, Moradabad, Uttar Pradesh, India*

*Corresponding author

ABSTRACT

Alzheimer's disease (AD) is a neurological condition that worsens with time. Amyloidogenic peptides and tau proteins have been shown to aggregate and accumulate in the brains of people with neurodegenerative diseases such as AD, and these results are considered pathogenic characteristics. In addition to a person's genetic makeup, various conditions and an unhealthy lifestyle are recognized risk factors for AD development. The critical role that membrane-bound Aβ assemblies play in neurotoxicity has been highlighted in several recent investigations.

Cellular and Molecular Mechanisms in Alzheimer's Disease Therapeutics, Volume 2: Molecular Mechanisms. Prashant Tiwari, Sunil Kumar Kadiri, and Kunjbihari Sulakhiya (Eds)

DOI: 10.1201/9781779641816-1

Additionally, AD is further aggravated by neuroinflammation and the production of free radicals. Through direct and indirect interactions, metal ions can influence their profiles of aggregation and cytotoxicity. Thus, it would be helpful to clarify the complex pathogenic nature of AD and create efficient diagnostics and treatments if we had a better knowledge of the interactions between metal ions and amyloidogenic peptides and proteins. As of right now, scientific methods for stopping the advancement of AD are currently lacking. This chapter is based on our understanding of AD pathology with biophysical, biochemical, and biological processes involved in modeling techniques for AD and developing an effective treatment strategy for AD.

1.1 INTRODUCTION

The hallmark of Alzheimer's disease (AD), a neurodegenerative condition, is functional decline without a shift in consciousness. An impairment in memory plus at least one other cognitive area (language, praxis, gnosis, executive function, judgement, and abstract thought) is referred to as dementia. According to estimates, there are 46.8 million instances of AD globally, with as many as 74.7 million cases in 2030 and 131.5 million cases in 2050; the disease has drawn increasing attention in recent decades due to its increased incidence in aging populations. In addition to being a major health and social issue, the condition drives up expenses for the healthcare system, which, in just five years, saw a 35.4% growth to $604 billion in predictions [1, 2]. There have been many attempts to combat the illness, but no medication has shown promise in stopping the onset or progression of AD yet.

The pathophysiology of AD has been determined to be primarily driven by tau protein and amyloid-β peptide (Aβ), with information from thousands of basic, pre-clinical, and clinical research supporting this theory. This is mainly because of their accumulation in the typical histopathological brain lesions, such as tau's neurofibrillary tangles (NFTs) and Aβ's senile plaques, as well as because AD patients' brains have more of their soluble forms. Nevertheless, attempts at therapy targeted at lowering Aβ levels have not succeeded thus far. In a similar vein, positive results from tau-based clinical trials have not yet been obtained. The main objective of this chapter is to offer a comprehensive evaluation of the literature on the mechanisms driving the onset and course of disease. In particular, we will re-examine research on tau and how they interact, throwing doubt on the amyloid theory, which for the last 25 years has dominated research on AD. According to this theory, tau comes after Aβ in a series of pathological events where Aβ is established as the *primum movens*. This theory states that treatments targeting Aβ would cease to be effective once tau pathology has developed since the illness would then proceed on its own [3]. In order to allow the peptides to carry out their harmful actions, we suggest rearranging the complex puzzle of AD pathogenesis by positioning tau and Aβ soluble forms in parallel and upstream of the amyloid-β precursor protein (AβPP). According to this perspective, the failure of anti-Aβ therapies should be reevaluated and should no longer be linked to the fact that tau pathology was triggered before the therapy was started. This will necessitate a change in the methodology used in studies of the etiopathogenesis of AD and pave the way for novel therapeutic strategies [4].

1.2 HISTORY OF PEOPLE WITH ALZHEIMER'S DISEASE

When German psychiatrist Alois Alzheimer examined Auguste Deter, a patient displaying severe memory loss and cognitive decline, the disease's hallmark pathological features—amyloid plaques and NFTs—were discovered in Auguste Deter's brain. This study brought AD to a wider audience. When AD was first identified as uncommon and mostly affecting those under 65, it was sometimes confused with "senile dementia" in older folks. Advances in biochemical research and neuroimaging helped to distinguish Alzheimer's from other dementias in the 1960s and 1970s. With the discovery that NFTs and amyloid plaques are essential to the pathophysiology of the illness, the 1980s saw tremendous advancements. Understanding of its hereditary components was furthered by genetic findings in the 1990s, including mutations on chromosome 21 and the APOE ε4 allele. Early detection was enhanced by advances in diagnostics, including biomarker research and PET and MRI scans. Therapeutic advancement was signaled by the 2003 approval of memantine for the treatment of symptoms. Renowned individuals like Terry Pratchett, Glen Campbell, and Ronald Reagan all candidly discussed their personal experiences with AD, bringing attention to the continued need for activism and research [5–9].

1.3 CLINICAL AND PATHOLOGICAL FEATURES IN AD

1.3.1 ROLE OF AMYLOID BETA PEPTIDE

Important pathological hallmarks of AD include significant NFTs in medial temporal-lobe structures, plaques in the brain containing β-amyloid peptide (Aβ), and dystrophic neurites in the terminal fields of the neocortex. Congophilic (amyloid) angiopathy, inflammation, oxidative damage, and loss of neurons and white matter are also observed. Aβ peptides, which are made up of 36–43 amino acids, are naturally occurring products of metabolism. Compared with the harmful and aggregation-prone Aβ 42 species, monomers of Aβ 40 are far more common. Two enzymes, γ-secretase (a protein complex with presenilin 1 at its catalytic core4) and beta-site amyloid precursor protein—cleaving enzyme 1 (*BACE-1*), a β-secretase, sequentially break down the amyloid precursor protein (APP) to produce β-amyloid peptides. Aβ builds up due to an imbalance in peptide production, clearance, and aggregation; this excess may be the first sign of AD [10]. Based on studies on hereditary variations of AD, like Down syndrome [11], and evidence that Aβ 42 is toxic to cells [12, 13], the "amyloid hypothesis" was proposed. Aβ self-aggregates into multiple coexisting physical forms on its own. Oligomers, which consist of two to six peptides, can form intermediate assemblies in one form [14, 15]. Furthermore, fibrils that assemble into β-pleated sheets to form the insoluble fibers present in mature amyloid plaques are produced by β-amyloid. The types of Aβ that are most harmful are soluble oligomers and intermediate amyloids [16]. Trimers and dimers of Aβ are harmful to synapses in brain-slice preparations [17, 18]. Not the overall Aβ load, but rather the amount of oligomers in the brain, correlates with the degree of the cognitive impairment in AD [19]. Aβ secretion at the synapse is quickly increased by neural activation; this process is linked to the regular release of vesicles containing neurotransmitters. Neuronal hyperactivity may be avoided and excitatory transmission dampened by physiological levels of synaptic A [20]. Steady-state Aβ levels are controlled by the proteases insulin-degrading enzyme and neprilysin. Aβ monomers and oligomers are broken down by the zinc endopeptidase nemilysin, which is membrane-anchored [21]. Brain Aβ builds up when neprilysin levels are lower. 16 A thiol metalloendopeptidase enzyme called insulin-degrading enzyme breaks down tiny peptides, such as monomeric Aβ and insulin [22]. Deletion of the insulin-degrading enzyme in mice results in a reduction in Aβ degradation of over 50% [23]. On the other hand, plaque formation is inhibited by overexpressing neprilysin or an enzyme that breaks down insulin [24].

1.3.2 TAU PEPTIDES IN ALZHEIMER'S DISEASE

Tauopathies are neurodegenerative illnesses that include AD, are associated with filamentous inclusions in pyramidal neurons called NFTs [25]. One pathologic indicator of the severity of AD is the quantity of NFTs. Tau that has been excessively hyperphosphorylated and aggregated makes up the majority of the tangles. Tau, a soluble protein that is often found in large amounts in axons, aids in microtubule construction, stability, and vesicle transit. Hyperphosphorylated tau self-assembles into paired helical filament formations, is insoluble, and has no affinity for microtubules. The amount of tau phosphorylation is controlled by enzymes that add and remove phosphate residues [26]. Similar to Aβ oligomers, intermediate aggregation of abnormal tau molecules is cytotoxic [27] and impairs cognition [28, 29]. Nonetheless, as reductions in neuronal population and axonal transit occur regardless of the presence of NFTs, insoluble helical filaments might be inactive [30]. These helical filaments may provide protection by sequestering harmful intermediate tau species [31]. Frontotemporal dementia with Parkinsonism has been linked to over 30 tau mutations on chromosome 17 [32]. On the other hand, neuron loss in AD is disproportionate to the amount of NFTs, and tau mutations are not associated with the disease [33]. However, there is a correlation between greater levels of total and phosphorylated tau in the cerebrospinal fluid (CSF) with worse scores on cognitive tests [34]. Elevated levels of the phosphotau amino acids T181, T231, and total tau in the CSF together form a biomarker test that can reliably predict the start of AD in patients with moderate cognitive impairment [35]. Tau aggregation is driven by and preceded by Aβ buildup, according to experimental findings [36–38]. Moreover, Aβ-induced degeneration of cultured neurons and cognitive abnormalities in mice afflicted with a condition resembling AD require endogenous tau [39, 40]. Increased oxidative stress, decreased endoplasmic reticulum protein-folding function, and insufficient proteasome- and autophagic-mediated removal of damaged proteins all contribute to the buildup of amyloid and tau proteins in AD. These factors are also linked to aging [41, 42]. There are currently no agents that can reverse these alterations, however studies are being conducted on small-molecule inhibitors of aggregation and tau oxidation (e.g., methylene blue) (NCT00568776) and β-amyloid (e.g., scylloinositol) (NCT00568776) [43]. Additionally promising as therapeutic agents are polyphenolic extracts from grape seeds, such as resveratrol, which trigger genes that inhibit aging [44].

1.4 MULTIFACTORIAL ETIOLOGY

It is commonly acknowledged that AD has a multifactorial etiology, which means that a variety of genetic, environmental, and lifestyle variables interact to cause the ailment. The disease's complex nature makes diagnosis and treatment more difficult, but it also makes a variety of preventive measures possible.

1.4.1 GENETIC ELEMENTS

1. *APOE* gene: The apolipoprotein E *(ApoE) ε4* allele is the primary genetic risk factor for late-onset AD. Though not everyone with the gene develops AD, people with one copy of the ε4 allele are at higher risk, and those with two copies are much more so.
2. AD in families: In rare cases of early onset, there is often a mutation in the genes encoding the *APP*, presenilin 1 (*PSEN1*), and presenilin 2 (*PSEN2*).This results in elevated synthesis of amyloid-beta, a major pathological characteristic of AD.

1.4.2 ENVIRONMENTAL ELEMENTS

1. Toxin exposure: Chronic exposure to pesticides, heavy metals, and pollutants has been linked to the onset of AD, possibly via processes involving inflammation and oxidative stress.
2. Head trauma: Traumatic brain injury (TBI), particularly when severe or recurring, has been connected to a higher risk of AD. Amyloid-beta and tau protein buildup in the brain may be accelerated by TBI.

1.4.3 LIFESTYLE ELEMENTS

1. Diet: There is a link between a heightened risk and a diet that is heavy in saturated fats and deficient in antioxidants. On the

other hand, diets like the Mediterranean diet that are high in fruits, vegetables, whole grains, fish, and healthy fats have been associated with a decreased risk of AD.

2. Physical activity: Engaging in regular physical activity is good for the brain. It promotes the production of growth factors that enhance neurogenesis and synaptic plasticity, lowers inflammation, and improves blood flow.
3. Cognitive engagement: Research has demonstrated that lifelong participation in cognitively stimulating activities, such as reading, picking up new skills, and socializing, lowers the risk of cognitive decline and AD. Both neuroplasticity and cognitive reserve are enhanced by these exercises.

1.4.4 INTERPLAY OF FACTORS

1. AD development is greatly influenced by the interplay of genetic predisposition, environmental variables, and lifestyle choices. For example, people with a genetic predisposition may be more susceptible to the negative consequences of a high-fat diet or physical inactivity. On the other hand, making good lifestyle decisions may reduce some of the inherited risks [45–53].

1.5 ROLE OF MEDIATORS AND MODULATORS IN AD

Numerous mediators and modulators have an impact on AD, and they are essential to the disease's etiology. These variables interact with important cellular processes, which can either worsen or lessen the disease's course. We now examine the functions of modulators and mediators in AD.

1.5.1 CYTOKINES AND CHEMOKINES

1. Pro-inflammatory cytokines: AD is associated with increased levels of tumor necrosis factor-α (TNF-α), interleukin-6 (IL-6), and of interleukin-1β (IL-1β), all of which contribute to neuroinflammation and neuronal degeneration [54].
2. Cytokines that reduce inflammation: Interleukin-10 (IL-10) and TGF-β both have anti-inflammatory properties and can protect neurons from inflammation [55].

1.5.2 REACTIVE OXYGEN SPECIES (ROS) AND ANTIOXIDANTS

1. ROS: Increased ROS levels in AD cause oxidative neuronal injury [56].
2. Antioxidants: Materials that scavenge ROS and lessen oxidative stress include polyphenols, vitamin E, and coenzyme Q10 [57].

1.5.3 CHOLESTEROL AND LIPID METABOLISM MODULATORS

1. Apolipoprotein E (ApoE): The APOE ε4 allele affects amyloid deposition and clearance and is a major genetic risk factor for AD [45].
2. Drugs that lower cholesterol: Statins have been shown to reduce amyloid formation and cholesterol levels, which may have a protective impact [58].

1.5.4 NEUROTRANSMITTERS AND THEIR MODULATORS

1. Acetylcholine: Cognitive impairment is a common result of deficits in cholinergic signaling in AD [59].
2. Cholinesterase inhibitors: By blocking acetylcholinesterase, medications such as galantamine, rivastigmine, and donepezil raise acetylcholine levels and alleviate symptoms [60].

1.5.5 MITOCHONDRIAL FUNCTION MODULATORS

1. Mitochondrial dysfunction: Reduced energy generation and increased apoptosis are two ways that impair mitochondrial function which leads to AD [61].
2. Mitochondrial protectants: The ability of substances like creatine and coenzyme Q10 to safeguard mitochondria and enhance cellular energy metabolism is being investigated [62].

1.5.6 INSULIN SIGNALING MODULATORS

1. Insulin resistance: The pathophysiology of AD, often known as "type 3 diabetes," is associated with impaired insulin signaling in the brain [63].

2. Insulin sensitizers: The ability of medications, such as metformin and intranasal insulin to enhance insulin signaling and cognitive performance is being studied [64].

1.5.7 PROTEOSTASIS REGULATORS

1. Autophagy modulators: Misfolded proteins like tau and Aβ can be cleared more easily when autophagy is increased [65].
2. Proteasome inhibitors: By altering the ubiquitin-proteasome system, damaged protein breakdown is influenced, which may have an effect on the advancement of AD [66].

1.6 ROLE OF INTRACELLULAR SIGNALING PATHWAY IN AD

1.6.1 AMYLOID PRECURSOR PROTEIN PATHWAY

APP belongs to the class I membrane proteins. In humans, the *APP* gene, which is found on chromosome 21, encodes it [67–69]. The body expresses *APP* extensively, with the brain's neural tissues expressing it at a higher level. Different *APP* isoforms are produced in a tissue-dependent manner via alternative splicing and differential processing [70]. All three of the main variants—APP695, APP751, and APP770—have the ability to produce amyloid beta [71]. While isoform APP695 is primarily found in neurons and thought to be the most hazardous, isoforms APP751 and APP770 are primarily found in non-neuronal organs. The many roles played by *APP*, include cell cycle and adhesion, calcium metabolism, synaptic processes, neurogenesis and differentiation of neuronal cells, and more [72–74]. In the brain regions commonly impacted by the disease, APP-deficient mice show reduced metal homeostasis, a shorter lifespan, and cognitive and learning impairment [75–77].

Multiple cleavage events are involved in the multistep mechanism of *APP* processing, which releases distinct products. The great bulk of research focuses on the last steps of *APP* processing, which involve cleavage events and the insertion of the entire *APP* into the intracellular membrane or other plasma membrane organelles. Two paths exist for the breakdown of membrane-inserted APP: the most prevalent pathway is non-amyloidogenic, but there is also an amyloidogenic one. The late processing of *APP* is primarily mediated by three secretases: α-, β-, and γ-secretase. α-secretase releases the membrane-attached C83 fragment and the N-terminal extracellular soluble APP domain (sAPPα) to initiate the non-amyloidogenic pathway. By cleaving the middle of the Aβ region, the α-secretase stops further formation of Aβ. Extracellular sAPPα can lower the production of amyloid beta by suppressing β-secretase (BACE1), the enzyme responsible for one of the steps in the amyloidogenic pathway of *APP* processing. Consequently, the non-amyloidogenic pathway is stimulated by sAPPα [78, 79]. A normal phenotype is restored with reintroducing sAPPα into APP-depleted animal models, suggesting that the sAPPα fragment plays a critical function in development [80].

In the amyloidogenic process, β-secretase, also known as β-site APP-cleaving enzyme-1 (BACE1), is in charge of the first cleavage. This cleavage results in the membrane-bound C99 domain and the extracellular soluble APP β (sAPPβ) fragment. Numerous studies have shown the significance of BACE1 in the generation of aberrant amyloid beta.

Surprisingly, when BACE1 was silenced, studies using mice models for AD showed that amyloid beta was completely absent [81–84]. Membrane-associated β-secretase is an enzyme with intricate trafficking and a variety of cellular routes. It is produced as a proenzyme in the ER and reaches full activity only after undergoing a number of post-translational changes in the golgi. These changes, which include phosphorylation, acetylation, glycosylation, and palmitoylation, have been demonstrated to be necessary for amyloidogenic events to be triggered by this enzyme [85, 86]. It is rare for β-secretase to process *APP* on the plasma membrane since it can be released extracellularly after being inserted into the lipid rafts of the plasma membrane. This extracellular BACE1, which is then endocytosed in order to enter and function within endosomes or move on to lysosomes for destruction. BACE1 is mostly found on the membranes of endosomes, which house *APP* processing, and the trans-Golgi network (TGN) within cells [87, 88].

1.6.2 WNT SIGNALING'S FUNCTION IN AD AND NEURAL SYNAPTOGENESIS

Numerous critical cellular processes involve glial and neuronal cells—the primary resident cells of the central nervous system (CNS)—that are connected to differentiation, cellular proliferation, adhesion, apoptosis, and survival in diverse anabolic and catabolic cells—are mediated by the Wnt signaling pathway [89, 90]. Named for the Drosophila proteins wingless and Int-1, the Wnt proteins class of glycosylated, secreted, cysteine-rich proteins plays a crucial role in embryonic development, cell proliferation, differentiation, migration, and oncogenesis through complex signaling pathways regulating gene expression and tissue patterning [89]. As of right now, genetic research conducted on populations of humans, mice, Drosophila, Xenopus, and zebrafish has discovered 80 Wnt target genes. Conversely, human identification of 19 of the 24 *Wnt* genes that express Wnt protein highlights the evolutionary conservation and functional diversity of this signaling family, which regulates cell fate, proliferation, migration, and organogenesis across multiple developmental and pathological contexts. When Wnt attaches itself to frizzed (Fz) transmembrane receptors on the cell surface, at least three different downstream signaling pathways are triggered [90]. The first is referred to as the canonical Wnt pathway, or Wnt/β-catenin, and through β-catenin, it regulates gene transcription. The second pathway, called as Wnt/Ca^{2+}, is non-canonical and regulates intracellular Ca2+ release. The third pathway, known as the Wnt/PCP-JNK pathway, deals with Wnt cell polarity and involves the activity of Jun N-terminal kinase (JNK) [91–93].

1.6.3 WNT SIGNALING PROTECTS SYNAPTIC INTEGRITY FROM AB TOXICITY

Some investigations have demonstrated the alteration of Wnt signaling components in AD: (a) Reduced levels of β-catenin were observed in AD patients with presenilin-1(PS1)-inherited mutations [94]; (b) Aβ exposure inhibited canonical Wnt signaling in cultured hippocampus neurons [95, 96]; (c) The Aβ protein induced Dickkoff-1 (*Dkk1*), a Wnt antagonist, in neurons of the hippocampus [97]; it was also higher in brains from transgenic AD models of animals and the postmortem samples from patients with AD [98, 99]; (d) In the CSF and plasma of AD patients, *Dkk3*, which is highly related to *Dkk1*, is elevated [100]; (e) A risk factor for AD, apolipoprotein E (ApoE ε4), inhibits canonical Wnt signaling [101]; (f) The development of the disease is caused by a common genetic variation in low-density lipoprotein receptor-related protein 6 (LRP6) [102]; (h) Clustering, a susceptibility factor for late-onset AD, controls the toxicity of Aβ amyloid by inducing the non-canonical Wnt/PCP-JNK pathway via *Dkk1* [103–105]. This process contributes to tau phosphorylation and cognitive impairment. (g) *Dkk1* causes synaptic disintegration at pre- and post-synaptic sites by reversibly lowering the quantity of synaptic proteins and the quantity of active pre-synaptic sites. An early event in AD is synaptic failure, and the synaptic pathology preceding plaque formation and neuronal death is thought to be caused by soluble Aβ oligomers [106]. The non-canonical Wnt-5 ligand inhibits the reduction in excitatory post-synaptic current amplitude caused by Aβ oligomers, suggesting that Aβ does not cause synaptic injury [107]. By blocking the Wnt/PCP-JNK pathway, Wnt-5a stops the PSD-95 post-synaptic clusters from shrinking. Wnt-5a, however, also activates the Wnt/Ca^{2+} pathway. This further encourages the creation of dendritic spines [110], the trafficking of the receptors for NMDA and GABAA to the neurons surface [108, 109], and the protection of neuronal mitochondria against Aβ oligomers [111]. By activating both canonical and non-canonical Wnt signaling *in vivo* using small Wnt ligands, more recent study enhances cognition in adult mice and reverses cognitive deficits and LTP in the APPswe/PS-1 transgenic model of AD [112]. These findings provide credence to the theory that modifications in the Wnt signaling system, which encompasses the canonical Wnt/β-catenin pathway, as well as the non-canonical Wnt/PCP and Wnt/Ca^{2+} pathway, are in charge of influencing synaptic development and AD progression [113]. In the end, the activation of several signaling pathways that interact with the Wnt pathway, such as the nicotine and muscarinic ACh receptors, peroxisome proliferator-activated receptors (PPAR) γ and α, and antioxidants and anti-inflammatory pathways, supports the potential for neurological protection of the Wnt signaling cascade in AD [114–116].

1.6.4 CHOLINERGIC SYSTEM AND WNT SIGNALING CROSS-TALK: ANCIENT AND NEW STRATEGY

According to the "cholinergic hypothesis" of AD, cholinergic neurons in the basal forebrain are lost, and cortical deficits in cholinergic neurotransmission are partly to blame for the cognitive decline and behavioral abnormalities seen in AD patients. This theory was developed after anomalies in acetylcholine metabolism were found in postmortem AD brains, [117]. Acetylcholinesterase (AChE) and cholineacetyl transferase activity are two examples of the proteins in the cholinergic system that are altered when cholinergic neurons are reduced [117]. Prior research has demonstrated that a macromolecule present in synapses combines with Aβ to generate a compound that modifies the typical role of synapses in hippocampus neurons [118, 119]. Furthermore, our team has shown that depending on the quantity of AChE, Aβ-AChE complexes were more harmful than Aβ alone [34], indicating that AChE may be crucial in the neurodegenerative alterations seen in the AD brain. Interestingly, Aβ neurological damage *in vivo* and Aβ-induced spatial memory loss can be avoided by using hyperforin, a phytochemical medication that controls acetylcholine release in the CNS [119–121]. Furthermore, in a double transgenic mouse model of AD, tetrahydrohyperforin (THH), a semi-synthetic derivative of hyperforin, protects cholinergic neurons, lowers levels of cholinergic markers linked to oxidative stress, apoptosis, and amyloid plaques, and restores brain AChE activity [121, 122]. According to a recent study, a subset of acetylcholine receptors (AChRs) may be translocated to synapses by a Wnt signaling pathway, which may be involved in preserving synaptic strength in the CNS [123]. Mutations in *Caenorhabditis elegans* that affect canonical and non-canonical Wnt signaling include those found in the Fz receptor, LIN-17; the DSH-1, DSH cytoplasmic phosphoprotein; the CAM-1, Ror receptor tyrosine kinase; and the Wnt ligand, CWN-2. These mutations result in severe behavioral defects, including a mutant α7−nACh receptor (ACR-16/α7), reduced synaptic function, and synaptic accumulation of the AChR [122, 123]. The findings of this study imply that Wnt signaling mediates synaptic plasticity, at least in part.

1.6.5 REDUCING OXIDATIVE STRESS BY WNT SIGNALING ACTIVATION

A new study suggests that a Wnt signaling pathway may translocate a subset of AChRs to synapses, perhaps contributing to the maintenance of synaptic strength in the CNS [123]. Mutations affecting both non-canonical and canonical processes are observed in *Caenorhabditis elegans* in the Wnt ligand, CWN-2; the Fz receptor, LIN-17; the Ror tyrosine receptor kinase, CAM-1; and the DSH cytoplasm phosphoprotein, DSH-1 Wnt signaling. These mutations result in severe behavioral defects, a mutant α7−nACh receptor (ACR-16/α7), decreased synaptic accumulation of AChR, and reduced synaptic activity [122, 123].The results of this investigation suggest that Wnt signaling, at least partially, mediates synaptic plasticity. In the past, we have demonstrated that in cultured rat hippocampal neurons, peroxisomal proliferation in conjunction with an increase in catalase can protect against the neurotoxicity of Aβ. This leads to significant improvements in spatial memory as well as long-term potentiation (LTP), a drop in Aβ aggregate levels, glial activation, post-synaptic protein accumulation, and tau phosphorylation [127]. In AD, neuroprotection against oxidative stress may also be provided via Wnt signaling. In primary hippocampal murine neurons, it has been shown that overexpressing Wnt1 protects neurons against oxidative DNA damage and Aβ-mediated oxidative stress. Neuronal cells may be more susceptible to oxidative damage when Wnt activity is reduced [128]. GSK-3β, a serine/threonine protein kinase that phosphorylates β-catenin and ultimately causes its depletion, can be produced in AD as a result of Aβ toxicity. Additionally, decreased Aβ production may be the outcome of increased PKC activity, which is regulated by the Wnt pathway. It has been demonstrated that overexpression of *DSH-1* and *DSH-2* prevents tau protein from being phosphorylated by GSK-3β; hence, it enhances neuroprotection and avoids the production of NFTs, one of the primary pathological hallmarks of AD [129].

1.6.6 CROSS-TALK BETWEEN AMPK AND MTOR PATHWAY

AMPK, a heterotrimeric kinase protein complex, is widely expressed in most types of cells and tissues. AMPK is primarily involved in the phosphorylation of downstream substrates of ATP-producing pathways and serves as a sensor of intracellular ATP levels [130]. Two upstream key enzymes are involved in the regulation of AMPK liver kinase B1 (*LKB1*), also known as serine/threonine kinase 11 (*STK11*), and the Ca^{2+}/CaM-dependent protein kinase kinase β (CaMKKβ), as well as other stimuli, such as nitric oxide (NO) [131]. Dephosphorylation of AMPK, which is facilitated by certain phosphatases (PPase), returns it to its inactive state [132]. Enzymes linked to lipolysis and glycolytic pathways are among the downstream targets of AMPK, which also includes "master energy regulators" [133], like the PPARγ coactivator-1α (PGC-1α), which, when AMPK is activated, de-acetylates mitochondrial biogenesis via *Sirt1* [134]. Additionally, the transcription factor Forkhead box O3 (FOXO3) is directly phosphorylated at many locations by AMPK, which activates the transcription of multiple genes, some of which are linked to oxidative stress resistance [135]. Furthermore, AMPK suppresses the process of directly phosphorylating ULK1, a novel serine/threonine kinase, and raptor to produce new proteins and mTORC1 complex subunit. This process initiates autophagy, which recycles amino acids and other cell components in the event of cellular famine [136, 137]. Enhanced mTOR activity in neuronal cells is triggered by multiple stimuli, including BDNF, leptin, and Ca^{2+} influx. It also plays a role in maintaining synaptic plasticity by controlling the synthesis of proteins needed for LTP [138]. As a result, AMPK serves as a "master of master cell physiology," and its pharmacological regulation offers a compelling therapeutic target for a variety of age-related conditions, including AD and neurodegenerative illnesses.

1.6.7 IDENTIFYING AMPK-MTOR'S ACTIVITY IN AD

As of this moment, little is known about how AMPK and mTOR function in the onset and course of AD. However, new research has shown that both pathways are frequently dysregulated in AD [139, 140]. While some researchers have established that AMPK may phosphorylate tau at multiple locations (including Thr231 and Ser396/404) and disrupt tau's attachment to microtubules, others have presented controversial *in vitro* models that suggest AMPK activation suppresses tau phosphorylation in rat cortical neurons [141, 142]. However, AMPK activation has been demonstrated in multiple studies to suppress amyloidogenesis in neurons [144]. Furthermore, AMPK activation promotes autophagy and lysosomal Aβ breakdown while suppressing mTOR signaling [145–148]. However, a recent study showed that at therapeutic doses, the oral antidiabetic medication metformin, which belongs to the biguanide class, can activate the AMPK and transcriptionally upregulate β-secretase, the enzyme that limits the rate at which Aβ is generated. This results in a significant increase in the production of both intracellular and extracellular Aβ species [149].

1.7 DIAGNOSTIC TECHNIQUES FOR ALZHEIMER'S DISEASE

Prior to the development of new methods, such as brain imaging and fluid biomarkers, to determine the underlying pathology linked to the illness, AD was only ever diagnosed postmortem. But even with encouraging outcomes from single- and multi-center studies, there are still significant national differences in the application and payment of fluid and imaging biomarkers to aid in the analysis of AD [151].

1.7.1 IMAGING BIOMARKERS

Recent developments have made it possible for doctors to use PET scanning to see the proteins tau and Aβ, which are linked to AD. As of right now, only amyloid PET is recommended by the Alzheimer's Association and the Amyloid Imaging Task Force to assist in the diagnosis of AD [150]. In addition to verifying the existence of Aβ plaques in the brain [151, 152] and directly quantifying brain amyloid pathology [153], a positive amyloid PET scan will show increased cortical retention of the tracer in regions of Aβ deposition within the brain [151]. This makes the PET scan a valuable adjunct to a clinical battery for the diagnosis of AD [151]. The tracers, flutemetamol, flutetaben, and flutebetapir are used in amyloid PET because they attach to Aβ within amyloid plaques selectively. Nevertheless, a positive amyloid PET scan by itself does not provide a conclusive diagnosis of clinical AD; other clinical evaluations, such as cognitive testing, must be performed in order to obtain an accurate diagnosis [151]. It is worth noting that amyloid PET is costly and may not always be covered by health insurance [150]. In cases where amyloid PET is not accessible, CSF can be used to evaluate biomarker confirmation.

1.7.2 FLUID BIOMARKERS

The analysis and collection of CSF for the existence of biomarkers linked to AD disease is an additional or substitute method to amyloid PET. It may be advised for those displaying AD-like symptoms to have lumbar punctures in order to check their CSF for specific biomarkers linked to the disease. Protein concentrations in the CSF of the lumbar sac are measured to determine CSF biomarkers, which indicate the rates of protein production and clearance at a given point in time. Spinal punctures can be performed in an outpatient or memory clinic environment safely and often. Nonetheless, a lot of patients still have concerns about the procedure's pain and potential adverse effects, and they might need more information and encouragement from the doctor before having it done [154]. HCPs can utilize appropriate use criteria to determine which patients are a good fit for CSF testing and lumbar punctures. For instance, patients who exhibit symptoms suggestive of AD or chronic, progressive, and unexplained MCI should be referred for CSF testing and lumbar puncture. But for individuals who have already been diagnosed with AD, lumbar puncture and CSF testing are not advised as a means of assessing the severity of the disease or as a substitute for genotyping probable carriers of autosomal dominant mutations [155]. Both amyloid PET and CSF biomarkers have strong concordance, therefore either can be employed to confirm the amount of Aβ [156]. As a result, the AD community generally accepts CSF biomarkers as diagnostically useful. In early-stage AD, brain indicators for the disease can be found in CSF long before obvious clinical symptoms appear. To assess the presence of illness, core AD CSF biomarkers can be evaluated, including Aβ 42 (one of the two main isoforms of Aβ and a significant component of Aβ plaques), total tau (t-tau), and phosphorylated tau (p-tau) [157]. Ultimately, a number of pragmatic considerations may influence the clinical choice between amyloid PET and CSF for confirming tau and amyloid disease [150, 154, 157–161].

1.7.3 EMERGING DIAGNOSTIC TOOLS

Due to the limitations of amyloid PET access, alternative sensitive and specific blood-based and CSF biomarkers that can diagnose AD-associated pathology in its early stages are needed [162]. Many efforts have been made in the last 10 years to identify blood-based biomarkers that can be used to: (1) detect AD pathology; (2) identify individuals who are susceptible to AD in the future; and (3) monitor the disease's course [163].Currently, doctors can only identify AD pathology using a few number of authorized blood-based assays [164]. Nevertheless, a number of innovative assays are being investigated, including those that measure different phosphorylated forms of tau, such as p-tau181 and p-tau217. It has been demonstrated that using plasma p-tau181, an isoform of tau, in an investigational manner can distinguish AD from other neurodegenerative illnesses and predict cognitive deterioration in AD patients. Research is being done on CSF p-tau217, a distinct isoform of tau, as a potential biomarker for preclinical and advanced AD detection [162]. Blood-based biomarkers may eventually be used at the initial stage of a multistage diagnostic procedure because blood testing is now a widely accepted component of clinical routines worldwide and can be conveniently conducted in a range of clinical settings. Clinicians would benefit from this, especially those working in primary care, as it would make it easier to identify patients who need to be sent to a specialist for diagnostic testing [163].

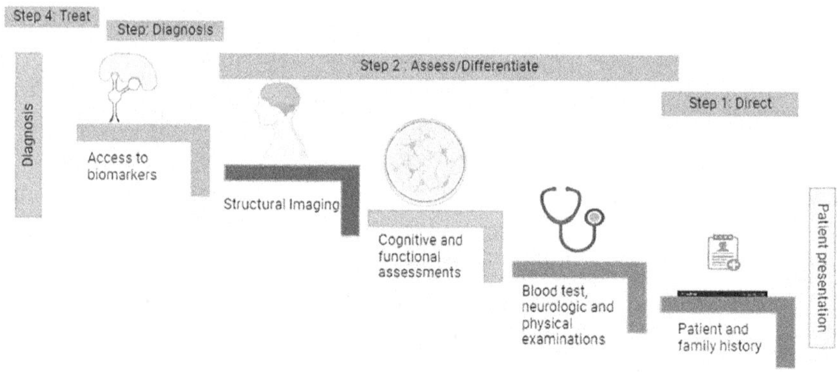

FIGURE 1.1 A stepwise information graphic highlighting key stages within the diagnostic process, along with the recommended tests to support each step.

Case study: Following a lumbar puncture for CSF analysis, J.K.'s diagnosis revealed elevated t-tau and p-tau protein and decreased Aβ 42. The clinician validated that early-stage AD was most likely the origin of the patient's cognitive difficulties conducted on the outcomes of the cognitive evaluations, genotyping, and CSF and MRI biomarkers. This was particularly true in light of a favorable family history of dementia with a comparable onset age [164].

1.8 POSSIBLE THERAPEUTIC INTERVENTION FOR AD

The goals of therapeutic interventions for AD are to reduce symptoms, enhance quality of life, and delay the illness's development. These treatments focus on a number of pathological features of the illness, such as tau tangles, amyloid plaques, neuroinflammation, oxidative stress, and synaptic dysfunction.

1.9 FUTURE PERSPECTIVE AND CONCLUSION

With the development of cutting-edge diagnostic methods and creative therapeutic approaches, the future of AD research and therapy appears bright. Even if there aren't any preventive treatments available yet, research is still running strong and could lead to discoveries that improve patient care. New diagnostic techniques, including sensitive blood-based biomarkers and sophisticated imaging technologies, are opening doors for

TABLE 1.1 List of Some Therapeutic Intervention for Alzheimer's Disease.

Therapies	Classification	Example	Mechanism	References
Amyloid-targeting	β-secretase inhibitors	Verubecestat	By inhibiting the β-secretase enzyme (BACE1), the amount of amyloid-beta (Aβ) peptides produced is decreased.	[165]
	γ-secretase modulators	Semagacestat	Modulates γ-secretase activity to reduce the formation of pathogenic Aβ 42 while sparing the enzyme's other functions.	[166]
	Anti-amyloid antibodies	Aducanumab and Donanemab	Monoclonal antibodies that specifically target Aβ to facilitate its excretion from the brain.	[167]
Tau-targeting therapies	Tau aggregation inhibitors	LMTX (methylthioninium chloride)	Stops tau from clumping together to form NFTs.	[168]
	Anti-tau antibodies	Gosuranemab and Tilavonemab	Monoclonal antibodies that specifically target tau to improve its removal and stop intracellular aggregation.	[169]
Anti-inflammatory approaches	Non-steroidal anti-inflammatory drugs (NSAIDs)	Trials have shown mixed results; long-term use might offer some protective effects.	Using cyclooxygenase enzyme inhibitors can help reduce neuroinflammation.	[170]
	Microglial modulators	*TREM2* agonists that enhance microglial phagocytosis of Aβ.	Reduces chronic inflammation via altering the activation of microglia.	[171]
Antioxidant therapies	Vitamin E	Studies show mixed results regarding its efficacy in slowing cognitive decline.	Reduces oxidative stress by acting as a scavenger of free radicals.	[172]
	Coenzyme Q10	Preclinical studies suggest benefits, but clinical evidence is limited.	Lessens oxidative damage and promotes mitochondrial activity.	[173]
Neurotransmitter modulators	Cholinesterase inhibitors	Rivastigmine, donepezil, and galantamine	Enhances cognitive function by blocking the breakdown of acetylcholine to raise its levels.	[61]
	NMDA receptor antagonists	Memantine, used in moderate to severe AD.	Glutamatergic signaling should be altered to avoid excitotoxicity.	[174]

the earlier and more precise diagnosis of AD. It is anticipated that these advances would enhance patient outcomes and enable prompt intervention. Furthermore, a better comprehension of the pathophysiology of the illness, namely the function of tau and amyloid-β in soluble forms, may result in the creation of more potent treatments. Researchers hope to stop the chain of pathogenic events that advance AD by focusing on these upstream variables. Moreover, novel therapeutic approaches may be made possible by investigating the Wnt signaling pathway and its ability to protect against Aβ toxicity. Additionally, modifying intracellular signaling pathways, such as AMPK and mTOR may have advantageous effects. There could be major developments in the diagnosis and treatment of this debilitating disease in the future as research into the intricate systems underlying AD continues.

This chapter concludes an important worldwide health issue that is becoming more common as a result of aging populations. There are no effective treatments available to stop or prevent the progression of AD, despite intensive research efforts. The disease is typified by the buildup of tau proteins and amyloid-β (Aβ) peptides, which results in cognitive impairment and neurodegeneration. Although these proteins have been the focus of most current treatment approaches, clinical trials have not produced positive results. This shows that other routes implicated in AD development should be investigated in addition to the amyloid hypothesis.

Important modulators and mediators in AD include neurotransmitters, mitochondrial function, ROS, cytokines, insulin signaling, cholesterol metabolism, and proteostasis. These variables affect how the disease develops by interacting intricately with biological processes. In particular, the Wnt signaling pathway has demonstrated promise in preserving synaptic integrity and guarding against Aβ toxicity. Furthermore, by influencing autophagy and protein synthesis, the interaction between the AMPK and mTOR pathways provides possible targets for therapeutic intervention.

Advances in diagnostic techniques have made it possible to analyze biomarkers in CSF and visualize AD-associated proteins using PET scans. New diagnostic techniques, like blood-based biomarkers, are being researched to enhance early AD progression monitoring and diagnosis.

KEYWORDS

- **Alzheimer's disease**
- **amyloidogenic peptides**
- **neurodegenerative diseases**
- **neuroinflammation**
- **neurofibrillary tangles**
- **tau proteins**
- **therapeutic strategies**

REFERENCES

1. Prince, M., Wimo, A., Guerchet, M., Ali, G. C., Wu Y-T, Prina, M., Chan, K. Y., & Xia, Z. (2015). *World Alzheimer Report 2015. The global impact of dementia: An analysis of prevalence, incidence, cost and trends*. Alzheimer Disease International.
2. Wimo, A., Guerchet, M., Ali, G.-C., Wu, Y.-T., Prina, A. M., Winblad, B., Jönsson, L., Liu, Z., & Prince, M. (2017). The world-wide costs of dementia 2015 and comparisons with 2010. *Alzheimers Dement, 13*, 1–7.
3. Morris, M., Maeda, S., Vossel, K., & Mucke, L. (2011). The many faces of tau. *Neuron, 70*, 410–426.
4. Gulisano, W., Maugeri, D., Baltrons, M. A., Fà, M., Amato, A., Palmeri, A., D'Adamio, L., Grassi, C., Devanand, D. P., Honig, L. S., Puzzo, D., & Arancio, O. (2018). Role of amyloid-β and tau proteins in Alzheimer's disease: Confuting the amyloid cascade. *Journal of Alzheimer's Disease, 64*(s1), S611–S631. https://doi.org/10.3233/JAD-179935.
5. Alzheimer, A. (1907). Über eine eigenartige Erkrankung der Hirnrinde [About a peculiar disease of the cerebral cortex]. *Allgemeine Zeitschrift für Psychiatrie und Psychisch-Gerichtlich Medizin, 64*, 146–148.
6. Hardy, J., & Selkoe, D. J. (2002). The amyloid hypothesis of Alzheimer's disease: Progress and problems on the road to therapeutics. *Science, 297*(5580), 353–356.
7. Bertram, L., & Tanzi, R. E. (2005). The genetic epidemiology of neurodegenerative disease. *The Journal of Clinical Investigation, 115*(6), 1449–1457.
8. Alzheimer's Association. (2020). Alzheimer's disease facts and figures. *Alzheimer's & Dementia, 16*(3), 391–460.
9. Glenner, G. G., & Wong, C. W. (1984). Alzheimer's disease: Initial report of the purification and characterization of a novel cerebrovascular amyloid protein. *Biochemical and Biophysical Research Communications, 120*(3), 885–890.
10. Querfurth, H. W., & LaFerla, F. M. (2010). *Alzheimer's Disease. New England Journal of Medicine, 362*(4), 329–344. DOI: 10.1056/nejmra0909142.

11. Busciglio, J., Pelsman, A., Wong, C., et al. (2002). Altered metabolism of the amyloid beta precursor protein is associated with mitochondrial dysfunction in Down's syndrome. *Neuron, 33*, 677–688.
12. Selkoe, D. J. (2001). Alzheimer's disease: Genes, proteins, and therapy. *Physiological Review, 81*, 741–766.
13. Tanzi, R. E., & Bertram, L. (2005). Twenty years of the Alzheimer's disease amyloid hypothesis: A genetic perspective. *Cell, 120*, 545–555.
14. Kayed, R., Head, E., Thompson, J. L., et al. (2003). Common structure of soluble amyloid oligomers implies common mechanism of pathogenesis. *Science, 300*, 486–489.
15. Klein, W. L., Krafft, G. A., & Finch, C. E. (2001). Targeting small Abeta oligomers: The solution to an Alzheimer's disease conundrum? *Trends Neurosciences, 24*, 219–224.
16. Walsh, D. M., & Selkoe, D. J. (2007). A beta oligomers—A decade of discovery. *Journal of Neurochemistry, 101*, 1172–1184.
17. Klyubin, I., Betts, V., Welzel, A. T., et al. (2008). Amyloid beta protein dimer-containing human CSF disrupts synaptic plasticity: Prevention by systemic passive immunization. *Journal of Neuroscience, 28*, 4231–4237.
18. Lue, L. F., Kuo, Y. M., Roher, A. E., et al. (1999). Soluble amyloid beta peptide concentration as a predictor of synaptic change in Alzheimer's disease. *The American Journal of Pathology, 155*, 853–862.
19. Kamenetz, F., Tomita, T., Hsieh, H., et al. (2003). APP processing and synaptic function. *Neuron, 37*, 925–937.
20. Kanemitsu, H., Tomiyama, T., & Mori, H. (2003). Human neprilysin is capable of degrading amyloid beta peptide not only in the monomeric form but also the pathological oligomeric form. *Neurosciences Letters, 350*, 113–116.
21. Iwata, N., Tsubuki, S., Takaki, Y., et al. (2001). Metabolic regulation of brain Abeta by neprilysin. *Science, 292*, 1550–1552.
22. Qiu, W. Q., Walsh, D. M., Ye, Z., et al. (1998). Insulin-degrading enzyme regulates extracellular levels of amyloid beta-protein by degradation. *Journal of Biological Chemistry, 273*, 32730–32738.
23. Farris, W., Mansourian, S., Chang, Y., et al. (2003). Insulin-degrading enzyme regulates the levels of insulin, amyloid beta-protein, and the beta-amyloid precursor protein intracellular domain in vivo. *Proceedings of the National Academy of Sciences USA,* 100, 4162–4167.
24. Leissring, M. A., Farris, W., Chang, A. Y., et al. (2003). Enhanced proteolysis of beta-amyloid in APP transgenic mice prevents plaque formation, secondary pathology, and premature death. *Neuron, 40*, 1087–1093.
25. Lee, V. M., Goedert, M., & Trojanowski, J. Q. (2001). Neurodegenerative tauopathies. *Annual Review of Neuroscience, 24*, 1121–1159.
26. Iqbal, K., Alonso Adel, C., Chen, S., et al. (2005). Tau pathology in Alzheimer disease and other tauopathies. *Biochimica et Biophysica Acta, 1739*, 198–210.
27. Khlistunova, I., Biernat, J., Wang, Y., et al. (2006). Inducible expression of Tau repeat domain in cell models of tauopathy: Aggregation is toxic to cells but can be reversed by inhibitor drugs. *Journal of Biological Chemistry, 281*, 1205–1214.

28. Santacruz, K., Lewis, J., Spires, T., et al. (2005). Tau suppression in a neurodegenerative mouse model improves memory function. *Science, 309*, 476–481.
29. Oddo, S., Vasilevko, V., Caccamo, A., Kitazawa, M., Cribbs, D. H., & LaFerla, F. M. (2006). Reduction of soluble Abeta and tau, but not soluble Abeta alone, ameliorates cognitive decline in transgenic mice with plaques and tangles. *Journal of Biological Chemistry, 281*, 39413–39423.
30. Andorfer, C., Kress, Y., Espinoza, M., et al. (2003). Hyperphosphorylation and aggregation of tau in mice expressing normal human tau isoforms. *Journal of Neurochemistry, 86*, 582–590.
31. Lee, H. G., Perry, G., Moreira, P. I., et al. (2005). Tau phosphorylation in Alzheimer's disease: Pathogen or protector? *Trends in Molecular Medicine, 11*, 164–169.
32. Goedert, M., & Jakes, R. (2005). Mutations causing neurodegenerative tauopathies. *Biochimica et Biophysica Acta, 1739*, 240–250.
33. Gómez-Isla, T., Hollister, R., West, H., et al. (1997). Neuronal loss correlates with but exceeds neurofibrillary tangles in Alzheimer's disease. *Annals of Neurology, 41*, 17–24.
34. Wallin, A. K., Blennow, K., Andreasen, N., & Minthon, L. (2006). CSF biomarkers for Alzheimer's disease: Levels of beta-amyloid, tau, phosphorylated tau relate to clinical symptoms and survival. *Dementia and Geriatric Cognitive Disorders, 21*, 131–138.
35. Mattsson, N., Zetterberg, H., Hansson, O., et al. (2009). CSF biomarkers and incipient Alzheimer disease in patients with mild cognitive impairment. *JAMA, 302*, 385–393.
36. Oddo, S., Caccamo, A., Shepherd, J. D., et al. (2003). Triple-transgenic model of Alzheimer's disease with plaques and tangles: Intracellular Abeta and synaptic dysfunction. *Neuron, 39*, 409–421.
37. Gotz, J., Chen, F., van Dorpe, J., & Nitsch, R. M. (2001). Formation of neurofibrillary tangles in P3011 tau transgenic mice induced by Abeta 42 fibrils. *Science, 293*:1491–1495.
38. Lewis, J., Dickson, D. W., Lin, W. L., et al. (2001). Enhanced neurofibrillary degeneration in transgenic mice expressing mutant tau and APP. *Science, 293*:1487–1491.
39. Roberson, E. D., Scearce-Levie, K., Palop, J. J., et al. (2007). Reducing endogenous tau ameliorates amyloid beta-induced deficits in an Alzheimer's disease mouse model. *Science, 316*, 750–754.
40. Rapoport, M., Dawson, H. N., Binder, L. I., Vitek, M. P., & Ferreira, A. (2002). Tau is essential to beta-amyloid-induced neurotoxicity. *Proceedings of the National Academy of Sciences USA, 99*, 6364–6369.
41. López Salon, M., Morelli, L., Castaño, E. M., Soto, E. F., Pasquini, J. M. (2000). Defective ubiquitination of cerebral proteins in Alzheimer's disease. *Journal of Neuroscience Research, 62*, 302–310.
42. Hoozemans, J. J., Veerhuis, R., Van Haastert, E. S., et al. (2005). The unfolded protein response is activated in Alzheimer's disease. *Acta Neuropathologica, 110*, 165–172.
43. McLaurin, J., Kierstead, M. E., Brown, M. E., et al. (2006). Cyclohexanehexol inhibitors of Abeta aggregation prevent and reverse Alzheimer phenotype in a mouse model. *Nature Medicine, 12*, 801–808.

44. Ono, K., Condron, M. M., Ho, L., et al. Effects of grape seed-derived polyphenols on amyloid beta-protein self-assembly and cytotoxicity. *Journal of Biological Chemistry,* 2008;283:32176–32187.
45. Corder, E. H., Saunders, A. M., Strittmatter, W. J., Schmechel, D. E., Gaskell, P. C., Small, G. W., & Roses, A. D. (1993). Gene dose of apolipoprotein E type 4 allele and the risk of Alzheimer's disease in late onset families. *Science, 261*(5123), 921–923.
46. Mahley, R. W., & Huang, Y. (2006). Apolipoprotein E: From atherosclerosis to Alzheimer's disease and beyond. *Current Opinion in Lipidology, 10*(3), 207–217.
47. Hardy, J. A., & Higgins, G. A. (1992). Alzheimer's disease: The amyloid cascade hypothesis. *Science, 256*(5054), 184–185.
48. Levesque, S., Surace, M. J., McDonald, J., & Block, M. L. (2011). Air pollution & the brain: Subchronic diesel exhaust exposure causes neuroinflammation and elevates early markers of neurodegenerative disease. *Journal of Neuroinflammation, 8*(1), 105.
49. Fleminger, S., Oliver, D. L., Lovestone, S., Rabe-Hesketh, S., & Giora, A. (2003). Head injury as a risk factor for Alzheimer's disease: The evidence 10 years on; A partial replication. *Journal of Neurology, Neurosurgery & Psychiatry, 74*(7), 857–862.
50. Scarmeas, N., Stern, Y., Tang, M. X., Mayeux, R., & Luchsinger, J. A. (2006). Mediterranean diet and risk for Alzheimer's disease. *Annals of Neurology, 59*(6), 912–921.
51. Larson, E. B., Wang, L., Bowen, J. D., McCormick, W. C., Teri, L., Crane, P., & Kukull, W. (2006). Exercise is associated with reduced risk for incident dementia among persons 65 years of age and older. *Annals of Internal Medicine, 144*(2), 73–81.
52. Stern, Y. (2002). What is cognitive reserve? Theory and research application of the reserve concept. *Journal of the International Neuropsychological Society, 8*(3), 448–460.
53. Norton, S., Matthews, F. E., Barnes, D. E., Yaffe, K., & Brayne, C. (2014). Potential for primary prevention of Alzheimer's disease: An analysis of population-based data. *The Lancet Neurology, 13*(8), 788–794.
54. Heneka, M. T., et al. (2015). Neuroinflammation in Alzheimer's disease. *Lancet Neurology, 14*(4), 388–405.
55. Wyss-Coray, T., & Mucke, L. (2002). Inflammation in neurodegenerative disease—A double-edged sword. *Neuron, 35*(3), 419–432.
56. Butterfield, D. A., & Halliwell, B. (2019). Oxidative stress, dysfunctional glucose metabolism and Alzheimer disease. *Nature Reviews Neuroscience, 20*(3), 148–160.
57. Smith, M. A., et al. (2000). Antioxidant properties of cerebrospinal fluid in Alzheimer's disease. *Annals of the New York Academy of Sciences, 899*(1), 245–254.
58. Sparks, D. L., et al. (2006). Induction of Alzheimer's-like β-amyloid immunoreactivity in the brains of rabbits with dietary cholesterol. *Experimental Neurology, 188*(2), 263–271.
59. Francis, P. T., Palmer, A. M., Snape, M., & Wilcock, G. K. (1999). The cholinergic hypothesis of Alzheimer's disease: A review of progress. *Journal of Neurology, Neurosurgery & Psychiatry, 66*(2), 137–147.
60. Birks, J. S., & Harvey, R. J. (2018). Donepezil for dementia due to Alzheimer's disease. *Cochrane Database of Systematic Reviews,* (6).

61. Swerdlow, R. H., & Khan, S. M. (2004). A mitochondrial cascade hypothesis for sporadic Alzheimer's disease. *Medical Hypotheses, 63*(1), 8–20.
62. Beal, M. F. (2004). Mitochondrial dysfunction and oxidative damage in Alzheimer's and Parkinson's diseases and coenzyme Q10 as a potential treatment. *Journal of Bioenergetics and Biomembranes, 36*(4), 381–386.
63. de la Monte, S. M., & Wands, J. R. (2008). Alzheimer's disease is type 3 diabetes—Evidence reviewed. *Journal of Diabetes Science and Technology, 2*(6), 1101–1113.
64. Craft, S., et al. (2012). Intranasal insulin therapy for Alzheimer disease and amnestic mild cognitive impairment: A pilot clinical trial. *Archives of Neurology, 69*(1), 29–38.
65. Nixon, R. A. (2013). The role of autophagy in neurodegenerative disease. *Nature Medicine, 19*(8), 983–997.
66. Ding, Q., & Keller, J. N. (2001). Proteasome inhibition in oxidative stress neurotoxicity: Implications for heat shock proteins. *Journal of Neurochemistry, 77*(4), 1010–1017.
67. Goldgaber, D., Lerman, M. I., McBride, O. W., Saffiotti, U., & Gajdusek, D. C. (1987). Characterization and chromosomal localization of a cDNA encoding brain amyloid of Alzheimer's disease. *Scienc,*. 235:877–880. DOI: 10.1126/science.3810169.
68. Yoshikai, S., Sasaki, H., Doh-ura, K., Furuya, H., Sakaki, Y. (1990). Genomic organization of the human amyloid beta-protein precursor gene. *Gene, 87*, 257–263. DOI: 10.1016/0378–1119(90).90310-N.
69. Lamb, B. T., Sisodia, S. S., Lawler, A. M., Slunt, H. H., Kitt, C. A., Kearns, W. G., Pearson, P. L., Price, D. L., & Gearhart, J. D. (1993). Introduction and expression of the 400 kilobase amyloid precursor protein gene in transgenic mice [corrected] *Nature Genetics, 5*, 22–30. DOI: 10.1038/ng0993-22.
70. Delvaux, E., Bentley, K., Stubbs, V., Sabbagh, M., Coleman, P. D. (2013). Differential processing of amyloid precursor protein in brain and in peripheral blood leukocytes. *Neurobiology Aging, 34*, 1680–1686. DOI: 10.1016/j.neurobiolaging.2012.12.004.
71. Kang, J., & Müller-Hill, B. (1990). Differential splicing of Alzheimer's disease amyloid A4 precursor RNA in rat tissues: PreA4(695) mRNA is predominantly produced in rat and human brain. *Biochemical and Biophysical Research Communications, 166*, 1192–1200. DOI: 10.1016/0006–291X(90).90992-V.
72. Korte, M., Herrmann, U., Zhang, X., & Draguhn, A. (2012). The role of APP and APLP for synaptic transmission, plasticity, and network function: Lessons from genetic mouse models. *Experimental Brain Research, 217*, 435–440. DOI: 10.1007/s00221–011–2894–6.
73. Soba, P., Eggert, S., Wagner, K., Zentgraf, H., Siehl, K., Kreger, S., Löwer, A., Langer, A., Merdes, G., Paro, R., et al. (2005). Homo- and heterodimerization of APP family members promotes intercellular adhesion. *EMBO Journal, 24*, 3624–3634. DOI: 10.1038/sj.emboj.7600824.
74. Yamazaki, T., Koo, E. H., & Selkoe, D. J. (1997). Cell surface amyloid beta-protein precursor colocalizes with beta 1 integrins at substrate contact sites in neural cells. *The Journal of Neuroscience, 17*, 1004–1010. DOI: 10.1523/JNEUROSCI.17-03-01004.1997.
75. Senechal, Y., Kelly, P. H., & Dev, K. K. (2008). Amyloid precursor protein knockout mice show age-dependent deficits in passive avoidance learning. *Behavioural Brain Research, 186*, 126–132. DOI: 10.1016/j.bbr.2007.08.003.

76. White, A. R., Reyes, R., Mercer, J. F., Camakaris, J., Zheng, H., Bush, A. I., Multhaup, G., Beyreuther, K., Masters, C. L., & Cappai, R. (1999). Copper levels are increased in the cerebral cortex and liver of APP and APLP2 knockout mice. *Brain Research, 842*, 439–444. DOI: 10.1016/S0006-8993(99).01861-2.
77. Zhang, X., Zhong, W., Brankačk, J., Weyer, S. W., Müller, U. C., Tort, A. B., & Draguhn, A. (2016). Impaired theta-gamma coupling in APP-deficient mice. *Scientific Report, 6*, 21948. DOI: 10.1038/srep21948.
78. Young-Pearse, T. L., Chen, A. C., Chang, R., Marquez, C., & Selkoe, D. J. (2008). Secreted APP regulates the function of full-length APP in neurite outgrowth through interaction with integrin beta1. *Neural Developement, 3*, 15. DOI: 10.1186/1749-8104-3-15.
79. Obregon, D., Hou, H., Deng, J., Giunta, B., Tian, J., Darlington, D., Shahaduzzaman, M., Zhu, Y., Mori, T., Mattson, M. P., et al. (2012). Soluble amyloid precursor protein-α modulates β-secretase activity and amyloid-β generation. *Nature Communication, 3*, 777. DOI: 10.1038/ncomms1781.
80. Ring, S., Weyer, S. W., Kilian, S. B., Waldron, E., Pietrzik, C. U., Filippov, M. A., Herms, J., Buchholz, C., Eckman, C. B., Korte, M., et al. (2007). The secreted beta-amyloid precursor protein ectodomain APPs alpha is sufficient to rescue the anatomical, behavioral, and electrophysiological abnormalities of APP-deficient mice. *Journal of Neuroscience, 27*, 7817–7826. DOI: 10.1523/JNEUROSCI.1026-07.2007.
81. Farzan, M., Schnitzler, C. E., Vasilieva, N., Leung, D., & Choe, H. (2000). BACE2, a beta -secretase homolog, cleaves at the beta site and within the amyloid-beta region of the amyloid-beta precursor protein. *Proceedings of the National Academy of Sciences USA, 97*, 9712–9717. DOI: 10.1073/pnas.160115697.
82. Laird, F. M., Cai, H., Savonenko, A. V., Farah, M. H., He, K., Melnikova, T., Wen, H., Chiang, H. C., Xu, G., Koliatsos, V. E., et al. (2005). BACE1, a major determinant of selective vulnerability of the brain to amyloid-beta amyloidogenesis, is essential for cognitive, emotional, and synaptic functions. *Journal of Neuroscience, 25*, 11693–11709. DOI: 10.1523/JNEUROSCI.2766-05.2005.
83. Ohno, M., Chang, L., Tseng, W., Oakley, H., Citron, M., Klein, W. L., Vassar, R., & Disterhoft, J. F. (2006). Temporal memory deficits in Alzheimer's mouse models: Rescue by genetic deletion of BACE1. *European Journal of Neuroscience, 23*, 251–260. DOI: 10.1111/j.1460–9568.2005.04551.x.
84. Ohno, M., Sametsky, E. A., Younkin, L. H., Oakley, H., Younkin, S. G., Citron, M., Vassar, R., & Disterhoft, J. F. (2004). BACE1 deficiency rescues memory deficits and cholinergic dysfunction in a mouse model of Alzheimer's disease. *Neuron, 41*, 27–33. DOI: 10.1016/S0896–6273(03).00810-9.
85. Araki, W. (2016). Post-translational regulation of the β-secretase BACE1. *Brain Research Bulletin, 126*, 170–177. DOI: 10.1016/j.brainresbull.2016.04.009.
86. Andrew, R. J., Fernandez, C. G., Stanley, M., Jiang, H., Nguyen, P., Rice, R. C., Buggia-Prévot, V., De Rossi, P., Vetrivel, K. S., Lamb, R., et al. (2017). Lack of BACE1 S-palmitoylation reduces amyloid burden and mitigates memory deficits in transgenic mouse models of Alzheimer's disease. *Proceedings of the National Academy of Sciences USA, 114*, E9665–E9674. DOI: 10.1073/pnas.1708568114.
87. Huse, J. T., Liu, K., Pijak, D. S., Carlin, D., Lee, V. M., & Doms, R. W. (2002). Beta-secretase processing in the trans-Golgi network preferentially generates truncated

amyloid species that accumulate in Alzheimer's disease brain. *Journal of Biological Chemistry, 277*, 16278–16284. DOI: 10.1074/jbc.M111141200.
88. Orobets, K. S., & Karamyshev, A. L. (2023). Amyloid Precursor Protein and Alzheimer's Disease. *International Journal of Molecular Sciences, 24*(19), 14794. https://doi.org/10.3390/ijms241914794
89. Nusse, R., & Varmus, H. (2012). Three decades of Wnts; A personal perspective on how a scientific field developed. *EMBO Journal, 31*, 2670–2682.
90. Clevers, H., & Nusse, R. (2012). Wnt/β-catenin signaling and disease. *Cell, 149*, 1192–1205.
91. Angers, S., & Moon, R. T. (2009). Proximal events in Wnt signal transduction. *Nature Reviews Molecular Cell Biology, 10*, 468–477.
92. Ciani, L., & Salinas, P. (2005). WNTs in the vertebrate nervous system nervous system: From patterning to neuronal connectivity. *Nature Reviews Neuroscience, 6*, 351–362.
93. Rosso, S. B., & Inestrosa, N. C. (2013). WNT signaling in neuronal maturation and synaptogenesis. *Frontiers in Cellular Neuroscience, 7*, 103.
94. Zhang, Z., Hartmann, H., Do, V. M., Abramowski, D., Sturchler-Pierrat, C., Staufenbiel, M., Sommer, B., van de Wetering, M., Clevers, H., Saftig, P., De Strooper, B., He, X., & Yankner, B. A. (1998). Destabilization of beta-catenin by mutations in presenilin-1 potentiates neuronal apoptosis. *Nature, 395*, 698–702.
95. De Ferrari, G. V., Chacon, M. A., Barria, M. I., Garrido, J. L., Godoy, J. A., Olivares, G., Reyes, A. E., Alvarez, A., Bronfman, M., & Inestrosa, N. C. (2003). Activation of Wnt signaling rescues neurodegeneration and behavioral impairments induced by beta-amyloid fibrils. *Molecular Psychiatry, 8*, 195–208.
96. Alvarez, A., Godoy, J. A., Mullendorff, K., Olivares, G. H., Bronfman, M., & Inestrosa, N. C. (2004). Wnt-3a overcomes beta-amyloid toxicity in rat hippocampal neurons. *Experimental Cell Research, 297*, 186–196.
97. Caricasole, A., Copani, A., Caruso, A., Caraci, F., Iacovelli, L., Sortino, M. A., Terstappen, G. C., & Nicoletti, F. (2003). The Wnt pathway, cell-cycle activation and beta-amyloid: Novel therapeutic strategies in Alzheimer's disease? *Trends in Pharmacological Sciences, 24*, 233–238.
98. Caricasole, A., Copani, A., Caraci, F., Aronica, E., Rozemuller, A. J., Caruso, A., Storto, M., Gaviraghi, G., Terstappen, G. C., & Nicoletti, F. (2004). Induction of Dickkopf-1, a negative modulator of the Wnt pathway, is associated with neuronal degeneration in Alzheimer's brain. *Journal of Neuroscience, 24*, 6021–6027.
99. Rosi, M. C., Luccarini, I., Grossi, C., Fiorentini, A., Spillantini, M. G., Prisco, A., Scali, C., Gianfriddo, M., Caricasole, A., Terstappen, G. C., & Casamenti, F. (2010). Increased Dickkopf-1 expression in transgenic mouse models of neurodegenerative disease. *Journal of Neurochemistry, 112*, 1539–1551.
100. Zenzmaier, C., Marksteiner, J., Kiefer, A., Berger, P., & Humpel, C. (2009). Dkk-3 is elevated in CSF and plasma of Alzheimer's disease patients. *Journal of Neurochemistry, 110*, 653–661.
101. Caruso, A., Motolese, M., Iacovelli, L., Caraci, F., Copani, A., Nicoletti, F., Terstappen, G. C., Gaviraghi, G., & Caricasole, A. (2006). Inhibition of the canonical Wnt signaling pathway by apolipoprotein E4 in PC12 cells. *Journal of Neurochemistry, 98*, 364–371.

102. De Ferrari, G. V., Papassotiropoulos, A., Biechele, T., Wavrant De-Vrieze, F., Avila, M. E., Major, M. B., Myers, A., Saez, K., Henriquez, J. P., Zhao, A., Wollmer, M. A., Nitsch, R. M., Hock, C., Morris, C. M., Hardy, J., & Moon, R. T. (2007). Common genetic variation within the low-density lipoprotein receptor-related protein 6 and late-onset Alzheimer's disease. *Proceedings of the National Academy of Sciences of the USA, 104*, 9434–9439.
103. Purro, S. A., Dickins, E. M., & Salinas, P. C. (2012). The secreted Wnt antagonist Dickkopf-1 is required for amyloid-beta-mediated synaptic loss. *Journal of Neuroscience, 32*, 3492–3498.
104. Purro, S. A., Galli, S., & Salinas, P. C. (2014). Dysfunction of Wnt signaling and synaptic disassembly in neurodegenerative diseases. *Journal of Molecular Cell Biology, 6*, 75–80.
105. Killick, R., Ribe, E., Ribe, E. M., Al-Shawi, R., Malik, B., Hooper, C., Fernandes, C., Dobson, R., Nolan, P. M., Lourdusamy, A., Furney, S., Lin, K., Breen, G., Wroe, R., To, A. W., Leroy, K., Causevic, M., Usardi, A., Robinson, M., Noble, W., Williamson, R., Lunnon, K., Kellie, S., Reynolds, C. H., Bazenet, C., Hodges, A., Brion, J. P., Stephenson, J., Simons, J. P., & Lovestone, S. (2014). Clusterin regulates β-amyloid toxicity via Dickkopf-1-driven induction of the Wnt-PCP-JNK pathway. *Molecular Psychiatry, 19*, 88–98.
106. Serrano-Pozo, A., Frosch, M. P., Masliah, E., & Hyman, B. T. (2011). Neuropathological alterations in Alzheimer disease. *Cold Spring Harbor Perspectives in Medicine, 1*, a006189.
107. Cerpa, W., Farias, G. G., Godoy, J. A., Fuenzalida, M., Bonansco, C., & Inestrosa, N. C. (2010). Wnt-5a occludes Abeta oligomer-induced depression of glutamatergic transmission in hippocampal neurons. *Molecular Neurodegeneration, 5*, 3.
108. Muñoz, F. J., Godoy, J. A., Cerpa, W., Poblete, I. M., Huidobro-Toro, J. P., & Inestrosa, N. C. (2014). Wnt-5a increases NO and modulates NMDA receptor in rat hippocampal neurons. *Biochemical and Biophysical Research Communications, 444*, 189–194.
109. Coutiño, L., Godoy, J. A., Couve, A., Fuenzalida, M., & Inestrosa, N. C. (2010). Wnt-5a modulates recycling of functional GABAA receptors on hippocampal neurons. *Journal of Neuroscience, 30*, 8411–8420.
110. Farias, G. G., Alfaro, I. E., Cerpa, W., Grabowski, C. P., Godoy, J. A., Bonansco, C., & Inestrosa, N. C. (2009). Wnt-5a/JNK signaling promotes the clustering of PSD-95 in hippocampal neurons. *Journal of Biological Chemistry, 284*, 15857–15866.
111. Varela-Nallar, L., Alfaro, I. E., Serrano, F. G., Parodi, J., & Inestrosa, N. C. (2010). Wingless-type family member 5A (Wnt-5a). stimulates synaptic differentiation and function of glutamatergic synapses. *Proceedings of the National Academy of Sciences USA, 107*, 21164–21169.
112. Silva-Alvarez, C., Arrazola, M. S., Godoy, J. A., Ordenes, D., & Inestrosa, N. C. (2013). Canonical Wnt signaling protects hippocampal neurons from Abeta oligomers: Role of non-canonical Wnt-5a/Ca(2+). in mitochondrial dynamics. *Frontiers in Cellular Neuroscience, 7*, 97.
113. Vargas, J., Fuenzalida, M., Inestrosa, N. C. (2014). In vivo activation of Wnt signaling pathway enhances cognitive function of adult mice and reverses cognitive deficits in an Alzheimer's disease model. *Journal of Neuroscience, 34*, 2191–2202.

114. Inestrosa, N. C., & Varela-Nallar, L. (2014). Wnt signaling in the nervous system and in Alzheimer's disease. *Journal of Molecular Cell Biology, 6*, 64–74.
115. Inestrosa, N. C., & Toledo, E. M. (2008). The role of Wnt signaling in neuronal dysfunction in Alzheimer's disease. *Molecular Neurodegeneration, 3*, 9.
116. Inestrosa, N. C., & Arenas, E. (2010). Emerging roles of Wnts in the adult nervous system. *Nature Reviews Neurosciences, 11*, 77–86.
117. Dumas, J. A., & Newhouse, P. A. (2011). The cholinergic hypothesis of cognitive aging revisited again: Cholinergic functional compensation. *Pharmacology Biochemistry & Behavior, 99*, 254–261.
118. Inestrosa, N. C., Alvarez, A., Perez, C. A., Moreno, R. D., Vicente, M., Linker, C., Casanueva, O. I., Soto, C., & Garrido, J. (1996). Acetylcholinesterase accelerates assembly of amyloid-beta-peptides into Alzheimer's fibrils: Possible role of the peripheral site of the enzyme. *Neuron, 16*, 881–891.
119. Munoz, F. J., & Inestrosa, N. C. (1999). Neurotoxicity of acetylcholinesterase amyloid beta-peptide aggregates is dependent on the type of Abeta peptide and the AChE concentration present in the complexes. *FEBS Letters, 450*, 205–209.
120. Cerpa, W., Hancke, J. L., Morazzoni, P., Bombardelli, E., Riva, A., Marin, P. P., & Inestrosa, N. C. (2010). The hyperforin derivative IDN5706 occludes spatial memory impairments and neuropathological changes in a double transgenic Alzheimer's mouse model. *Current Alzheimer Research, 7*, 126–133.
121. Dinamarca, M. C., Arrazola, M., Toledo, E., Cerpa, W. F., Hancke, J., & Inestrosa, N. C. (2008)) Release of acetylcholinesterase (AChE). from beta-amyloid plaques assemblies improves the spatial memory impairments in APP-transgenic mice. *Chemico-Biological Interactions, 175*, 142–149.
122. Carvajal, F. J., Zolezzi, J. M., Tapia-Rojas, C., Godoy, J. A., & Inestrosa, N. C. (2013). Tetrahydrohyperforin decreases cholinergic markers associated with amyloid-beta plaques, 4-hydroxynonenal formation, and caspase-3 activation in AbetaPP/PS1 mice. *Journal of Alzheimer's Disease, 36*, 99–118.
123. Jensen, M., Hoerndli, F. J., Brockie, P. J., Wang, R., Johnson, E., Maxfield, D., Francis, M. M., Madsen, D. M., & Maricq, A. V. (2012). Wnt signaling regulates acetylcholine receptor translocation and synaptic plasticity in the adult nervous system. *Cell, 149*, 173–187.
124. Jensen, M., Brockie, P. J., & Maricq, A. V. (2012). Wnt signaling regulates experiencedependent synaptic plasticity in the adult nervous system. *Cell Cycle, 11*, 2585–2586.
125. Lee, H. P., Zhu, X., Casadesus, G., Castellani, R. J., Nunomura, A., Smith, M. A., Lee, H. G., & Perry, G. (2010). Antioxidant approaches for the treatment of Alzheimer's disease. *Expert Review of Neurotherapeutics, 10*, 1201–1208.
126. Habib, L. K., Lee, M. T., & Yang, J. (2010). Inhibitors of catalase-amyloid interactions protect cells from beta-amyloid-induced oxidative stress and toxicity. *Journal of Biological Chemistry, 285*, 38933–38943.
127. Quintanilla, R. A., Munoz, F. J., Metcalfe, M. J., Hitschfeld, M., Olivares, G., Godoy, J. A., & Inestrosa, N. C. (2005). Trolox and 17beta-estradiol protect against amyloid beta-peptide neurotoxicity by a mechanism that involves modulation of the Wnt signaling pathway. *Journal of Biological Chemistry, 280*, 11615–11625.

128. Inestrosa, N. C., Carvajal, F. J., Zolezzi, J. M., Tapia-Rojas, C., Serrano, F., Karmelic, D., Toledo, E. M., Toro, A., Toro, J., & Santos, M. J. (2013). Peroxisome proliferators reduce spatial memory impairment, synaptic failure, and neurodegeneration in brains of a double transgenic mice model of Alzheimer's disease. *Journal of Alzheimer's Disease, 33*, 941–959.
129. Miyaoka, T., Seno, H., & Ishino, H. (1999). Increased expression of Wnt-1 in schizophrenic brains. *Schizophrenia Research, 38*, 1–6.
130. Wagner, U., Brownlees, J., Irving, N. G., Lucas, F. R., Salinas, P. C., & Miller, C. C. (1997). Overexpression of the mouse dishevelled-1 protein inhibits GSK-3betamediated phosphorylation of tau in transfected mammalian cells. *Febs Letters, 411*, 369–372.
131. Carling, D., Thornton, C., Woods, A., & Sanders, M. J. (2012). AMP-activated protein kinase: New regulation, new roles? *Biochemical Journal, 445*, 11–27.
132. Zhang, J., Xie, Z., Dong, Y., Wang, S., Liu, C., & Zou, M. H. (2008). Identification of nitric oxide as an endogenous activator of the AMP-activated protein kinase in vascular endothelial cells. *Journal of Biological Chemistry, 283*, 27452–27461.
133. Ruderman, N. B., Carling, D., Prentki, M., & Cacicedo, J. M. (2013). AMPK, insulin resistance, and the metabolic syndrome. *Journal of Clinical Investigation, 123*, 2764–2772.
134. Hardie, D. G., Ross, F. A., & Hawley, S. A. (2012). AMP-activated protein kinase: A target for drugs both ancient and modern. *Chemical Biology, 19*, 1222–1236.
135. Jager, S., Handschin, C., St-Pierre, J., & Spiegelman, B. M. (2007). AMP-activated protein kinase (AMPK). action in skeletal muscle via direct phosphorylation of PGC-1α. *Proceedings of the National Academy of Sciences USA, 104*, 12017–12022.
136. Greer, E. L., Oskoui, P. R., Banko, M. R., Maniar, J. M., Gygi, M. P., Gygi, S. P., & Brunet, A. (2007). The energy sensor AMP-activated protein kinase directly regulates the mammalian FOXO3 transcription factor. *Journal of Biological Chemistry, 282*, 30107–30119.
137. . Egan, D. F., Shackelford, D. B., Mihaylova, M. M., Gelino, S., Kohnz, R. A., Mair, W., Vasquez, D. S., Joshi, A., Gwinn, D. M., Taylor, R., Asara, J. M., Fitzpatrick, J., Dillin, A., Viollet, B., Kundu, M., Hansen, M., & Shaw, R. J. (2011). Phosphorylation of ULK1 (hATG1) by AMP-activated protein kinase connects energy sensing to mitophagy. *Science, 331*, 456–461.
138. Alers, S., Loffler, A. S., Wesselborg, S., & Stork, B. (2012). Role of AMPK-mTOR-Ulk1/2 in the regulation of autophagy: Cross talk, shortcuts, and feedbacks. *Molecular and Cellular Biology, 32*, 2–11.
139. Garelick, M. G., & Kennedy, B. K. (2011). TOR on the brain. *Experimental Gerontology, 46*, 155–163.
140. Caberlotto, L., Lauria, M., Nguyen, T. P., & Scotti, M. (2013). The central role of AMP-kinase and energy homeostasis impairment in Alzheimer's disease: A multifactor network analysis. *PloS One*, *8*, e78919.
141. Yates, S. C., Zafar, A., Hubbard, P., Nagy, S., Durant, S., Bicknell, R., Wilcock, G., Christie, S., Esiri, M. M., Smith, A. D., & Nagy, Z. (2013). Dysfunction of the mTOR pathway is a risk factor for Alzheimer's disease. *Acta Neuropathologica Communications, 1*, 3.

142. Greco, S. J., Sarkar, S., Johnston, J. M., & Tezapsidis, N. (2009). Leptin regulates tau phosphorylation and amyloid through AMPK in neuronal cells. *Biochemical and Biophysical Research Communications, 380*, 98–104.
143. Thornton, C., Bright, N. J., Sastre, M., Muckett, P. J., & Carling, D. (2011). AMP-activated protein kinase (AMPK) is a tau kinase, activated in response to amyloid beta-peptide exposure. *Biochemical Medicine, 434*, 503–512.
144. Vingtdeux, V., Davies, P., Dickson, D. W., & Marambaud, P. (2011). AMPK is abnormally activated in tangle- and pre-tangle-bearing neurons in Alzheimer's disease and other tauopathies. *Acta Neuropathologica, 121*, 337–349.
145. Vingtdeux, V., Giliberto, L., Zhao, H., Chandakkar, P., Wu, Q., Simon, J. E., Janle, E. M., Lobo, J., Ferruzzi, M. G., Davies, P., & Marambaud, P. (2010). AMP-activated protein kinase signaling activation by resveratrol modulates amyloid-beta peptide metabolism. *Journal of Biological Chemistry, 285*, 9100–9113.
146. Cai, Z., Yan, L. J., Li, K., Quazi, S. H., & Zhao, B. (2012). Roles of AMP-activated protein kinase in Alzheimer's disease. *Neuromolecular Medicine, 14*, 1–14.
147. Cai, Z., Li, B., Li, K., & Zhao, B. (2012). Down-regulation of amyloid-beta through AMPK activation by inhibitors of GSK-3beta in SH-SY5Y and SH-SY5Y-AbetaPP695 cells. *Journal of Alzheimer's Disease, 29*, 89–98.
148. Ghavami, S., Shojaei, S., Yeganeh, B., Ande, S. R., Jangamreddy, J. R., Mehrpour, M., Christoffersson, J., Chaabane, W., Moghadam, A. R., Kashani, H. H., Hashemi, M., Owji, A. A., & Los, M. J. (2014). Autophagy and apoptosis dysfunction in neurodegenerative disorders. *Progress in Neurobioloyg, 112*, 24–49.
149. Tan, C. C., Yu, J. T., Tan, M. S., Jiang, T., Zhu, X. C., Tan, L. (2013). Autophagy in aging and neurodegenerative diseases: Implications for pathogenesis and therapy. *Neurobiology Aging, 35*, 941–957.
150. Frisoni, G. B., Boccardi, M., Barkhof, F., et al. (2017). Strategic roadmap for an early diagnosis of Alzheimer's disease based on biomarkers. *Lancet Neurology, 16*, 661–676.
151. Villemagne, V. L., Doré, V., Burnham, S. C., Masters, C. L., & Rowe, C. C. (2018). Imaging tau and amyloid-β proteinopathies in Alzheimer disease and other conditions. *Nature Reviews Neurology, 14*, 225–236.
152. Clark, C. M., Pontecorvo, M. J., Beach, T. G., et al. (2012). Cerebral PET with florbetapir compared with neuropathology at autopsy for detection of neuritic amyloid-β plaques: A prospective cohort study. *Lancet Neurology, 11*, 669–678.
153. Wong, D. F., Rosenberg, P. B., Zhou, Y., et al. (2010). In vivo imaging of amyloid deposition in Alzheimer disease using the radioligand 18F-AV-45 (florbetapir F 18). *Journal of Nuclear Medicine, 51*, 913–920.
154. Duits, F. H., Martinez-Lage, P., Paquet, C., et al. (2016). Performance and complications of lumbar puncture in memory clinics: Results of the multicenter lumbar puncture feasibility study. *Alzheimer's & Dementia, 12*, 154–163.
155. Shaw, L. M., Arias, J., Blennow, K., et al. (2018). Appropriate use criteria for lumbar puncture and cerebrospinal fluid testing in the diagnosis of Alzheimer's disease. *Alzheimer's & Dementia, 14*, 1505–1521.
156. Hansson, O., Seibyl, J., Stomrud, E., et al. (2018). CSF biomarkers of Alzheimer's disease concord with amyloid-β PET and predict clinical progression: A study of

fully automated immunoassays in BioFINDER and ADNI cohorts. *Alzheimer's & Dementia, 14*, 1470–1481.

157. Blennow, K., Dubois, B., Fagan, A. M., Lewczuk, P., de Leon, M. J., & Hampel, H. (2015). Clinical utility of cerebrospinal fluid biomarkers in the diagnosis of early Alzheimer's disease. *Alzheimer's & Dementia, 11*, 58–69.
158. Lee, S. A., Sposato, L. A., Hachinski, V., & Cipriano, L. E. (2017). Cost-effectiveness of cerebrospinal biomarkers for the diagnosis of Alzheimer's disease. *Alzheimer's Research & Therapy, 9*, 18.
159. Dolgin, E. (2018). Alzheimer's disease is getting easier to spot. *Nature, 559*, S10–S12.
160. Morris, E., Chalkidou, A., Hammers, A., Peacock, J., Summers, J., & Keevil, S. (2016). Diagnostic accuracy of 18F amyloid PET tracers for the diagnosis of Alzheimer's disease: A systematic review and meta-analysis. *European Journal of Nuclear Medicine and Molecular Imaging, 43*, 374–385.
161. Engelborghs, S., Niemantsverdriet, E., Struyfs, H., et al. (2017). Consensus guidelines for lumbar puncture in patients with neurological diseases. *Alzheimer's & Dementia, 8*, 111–126.
162. Barthélemy, N. R., Bateman, R. J., Hirtz, C., et al. (2020). Cerebrospinal fluid phosphotau T217 outperforms T181 as a biomarker for the differential diagnosis of Alzheimer's disease and PET amyloid-positive patient identification. *Alzheimer's Research & Therapy, 12*, 26.
163. O'Bryant, S. E., Mielke, M. M., Rissman, R. A., et al. (2017). Blood-based biomarkers in Alzheimer disease: Current state of the science and a novel collaborative paradigm for advancing from discovery to clinic. *Alzheimer's & Dementia, 13*, 45–58.
164. Porsteinsson, A. P., Isaacson, R. S., Knox, S., Sabbagh, M. N., & Rubino, I. (2021). Diagnosis of early Alzheimer's disease: Clinical practice in 2021. *The Journal of Prevention of Alzheimer's Disease, 8*(3), 371–386. https://doi.org/10.14283/jpad.2021.23
165. Egan, M. F., et al. (2019). Randomized trial of verubecestat for mild-to-moderate Alzheimer's disease. *New England Journal of Medicine, 380*(15), 1444–1453.
166. Doody, R. S., et al. (2013). A phase 3 trial of semagacestat for treatment of Alzheimer's disease. *New England Journal of Medicine, 369*(4), 341–350.
167. Sevigny, J., et al. (2016). The antibody aducanumab reduces Aβ plaques in Alzheimer's disease. *Nature, 537*(7618), 50–56.
168. Wischik, C. M., et al. (2015). Tau aggregation inhibitor therapy: An exploratory phase 2 study in mild or moderate Alzheimer's disease. *Journal of Alzheimer's Disease, 44*(2), 705–720.
169. Congdon, E. E., & Sigurdsson, E. M. (2018). Tau-targeting therapies for Alzheimer disease. *Nature Reviews Neurology, 14*(7), 399–415.
170. Breitner, J. C., et al. (2011). Extended results of the Alzheimer's disease anti-inflammatory prevention trial. *Alzheimer's & Dementia, 7*(4), 402–411.
171. Ulland, T. K., & Colonna, M. (2018). TREM2—A key player in microglial biology and Alzheimer disease. *Nature Reviews Neurology, 14*(11), 667–675.
172. Dysken, M. W., et al. (2014). Effect of vitamin E and memantine on functional decline in Alzheimer disease: The TEAM-AD VA cooperative randomized trial. *JAMA, 311*(1), 33–44.

173. Galasko, D. R., et al. (2012). Antioxidants for Alzheimer disease: A randomized clinical trial with cerebrospinal fluid biomarker measures. *Archives of Neurology, 69*(7), 836–841.
174. Reisberg, B., et al. (2003). Memantine in moderate-to-severe Alzheimer's disease. *New England Journal of Medicine, 348*(14), 1333–1341.

CHAPTER 2

Advancement in Comprehending Cholinergic System and Emerging Tools to Novel Therapeutics

MOHAMMED NAZISH QUASMI and ASHOK JANGRA*

Department of Pharmaceutical Sciences, School of Interdisciplinary and Applied Sciences, Central University of Haryana, Mahendragarh, Haryana, India

*Corresponding author

ABSTRACT

The cholinergic system is an essential component of the autonomic nervous system and plays a significant role in learning and memory. Any disruption or dysregulation of this system can lead to different neurodegenerative disorders, like Parkinson's disease (PD) and Alzheimer's disease (AD). Acetylcholine is a major neurotransmitter in the cholinergic system and has different functions according to the cell and receptor where it binds. In the cholinergic system, there are two main types of receptors: nicotinic receptor and muscarinic receptor; they are ion gated channels and G-protein coupled receptor, respectively. Acetylcholine binds to these receptors and performs various functions, including cognitive functions. Disruption in the cholinergic system and low levels of acetylcholine can lead to the accumulation of the tau tangles and Aβ plaques, resulting in AD. The level of acetylcholine can be increased and cognitive impairments can be improved by blocking the activity of acetylcholinesterase (protein that breaks down acetylcholine in the postsynaptic cleft). This chapter will provide the detailed knowledge about the working of cholinergic system as well as how its dysregulation can cause cognitive deficits, leading to AD.

Cellular and Molecular Mechanisms in Alzheimer's Disease Therapeutics, Volume 2: Molecular Mechanisms. Prashant Tiwari, Sunil Kumar Kadiri, and Kunjbihari Sulakhiya (Eds)

DOI: 10.1201/9781779641816-2

Furthermore, we have discussed traditional acetylcholinesterase inhibitors used in AD treatment and their adverse effects, and why next generation AChE inhibitors are required. We have also discussed the novel therapeutic strategies targeting the cholinergic system including, nicotinic and muscarinic modulators which can be used in treating AD along with the advance techniques including advanced imaging methods, like PET, MRI, and EEG that can be useful in the early diagnosis and management of AD due to the cholinergic system defect.

2.1 INTRODUCTION

The first neurotransmitter which was identified in the past was acetylcholine (ACh). All cholinergic neurons in the peripheral nervous system (PNS) and in the central nervous system (CNS) use ACh as the neurotransmitter. ACh functions as a neurotransmitter in pre-ganglionic sympathetic neurons as well as in post-ganglionic parasympathetic neurons [1]. In the parasympathetic nervous system as well as in motor neurons, the cytoplasmic enzyme choline acetyltransferase uses the acetyl-coenzyme A that is generated by the mitochondria to esterify the choline. ACh is produced by the B cells, T cells, macrophages, as well as by the dendritic cells in the marginal and follicular zones of the spleen and by the vagus nerve. Immune cells produce ACh, which is subsequently delivered directly into the bloodstream. On the other hand, ACh is held by brain cells in the distinct neurosecretory vesicle and subsequently is released by the exocytosis at specific synaptic clefts [2]. Human blood contains 8.66 ± 1.02 pmol/mL of ACh, while plasma has 3.12 ± 0.36 pmol/mL of ACh. ACh acts and functions differently throughout the body due to its different ways that it binds and activates several types of receptors on the surfaces of different tissues and cells. Previous researchers have linked reduced levels of ACh to chronic inflammatory neurological disorders, including vascular dementia (VD), Alzheimer's disease (AD), and multiple sclerosis (MS) [3, 4].

ACh is a neurotransmitter which is vital in various cognitive functions like learning and memory. The ACh's concentration and release is controlled by the central cholinergic nervous system [5]. Basal forebrain cholinergic neurons (BFCNs) in the brain play an important role in learning, memory, and cognition. The hippocampal and cortical regions of the brain contain BFCNs, and the development and survival of these

BFCNs depends on the nerve growth factor (NGF) [6]. The primary cause of memory impairment in AD is the loss of the synapse between the target tissues of the hippocampus, cortex, and the basal forebrain.

Recent clinical data shows that AD patients have severe neurodegeneration, reduced cholinergic neurons, and ACh deficiency in their brains. Additionally there is decreased activity of the acetylcholine transferase, which indicates serious injury to the cholinergic system [7]. That's why current therapeutic treatment for AD includes the restoring of damaged cholinergic system. Acetylcholinesterase (AChE) stimulates presenilin-1 (PS-1), which is an enzyme responsible for the production of Aβ. When PS-1 is released it can cause high levels of Aβ, which can then impair cognitive functions [8]. Aberrant central cholinergic changes may also result in abnormal tau protein phosphorylation, inflammation of nerve cells, apoptosis, imbalance in the neurotransmitter, and other pathogenic problems [9]. The main objective of this chapter is to enhance the knowledge about the recent advancement in the cholinergic system in contrast to AD and demonstrating how the next generation AChE inhibitors and other cholinergic receptor modulators can be beneficial in the AD treatment. This chapter also discusses the advanced tools and techniques which can be helpful for diagnosing neurodegenerative disorders like AD caused by a defective cholinergic system in the brain efficiently.

2.2 ANATOMY AND PHYSIOLOGY OF CHOLINERGIC SYSTEM

The human brain is made up mostly of cholinergic neurons, which are involved in cognition. In memory and learning, ACh controls the transmission of cholinergic signals [10]. As shown in Figure 2.1, ACh is synthesized by choline acetyl transferase and substrates like acetyl-CoA and choline that are needed for the enzymatic activity of ACh. ACh signaling has a vital role in cognition as well as in the inflammation suppression. Proteins and non-coding short RNAs (CholinomiRs) control cholinergic signaling at several levels to maintain homeostasis. The enzyme AChE degrades ACh and prevents its transmission via postsynaptic signals [11]. Cholinergic system in the CNS is very important to regulate neuronal differentiation, neurogenesis, neuroprotection, and synaptic plasticity. The two types of Ach receptors in the human body are the muscarinic acetylcholine receptors (mAChR) and the nicotinic acetylcholine receptors (nAChR) and each serves a different function. Intracellular signaling is triggered by mAChRs and nAChRs via G-protein activity and ion influx, respectively [2].

nAChR is widely expressed in the hippocampus, particularly the α7 subtype. In the PFC, amygdala and hippocampus, BDNF, and p-CREB/CREB levels rise during memory formation [12]. Cognitive decline, cholinergic atrophy, and disruption of normal hippocampal function can result in the malfunction or degeneration of BFCNs which can be connected to the formation of the neuroinflammatory factors in disorders, such as Alzheimer's [13].

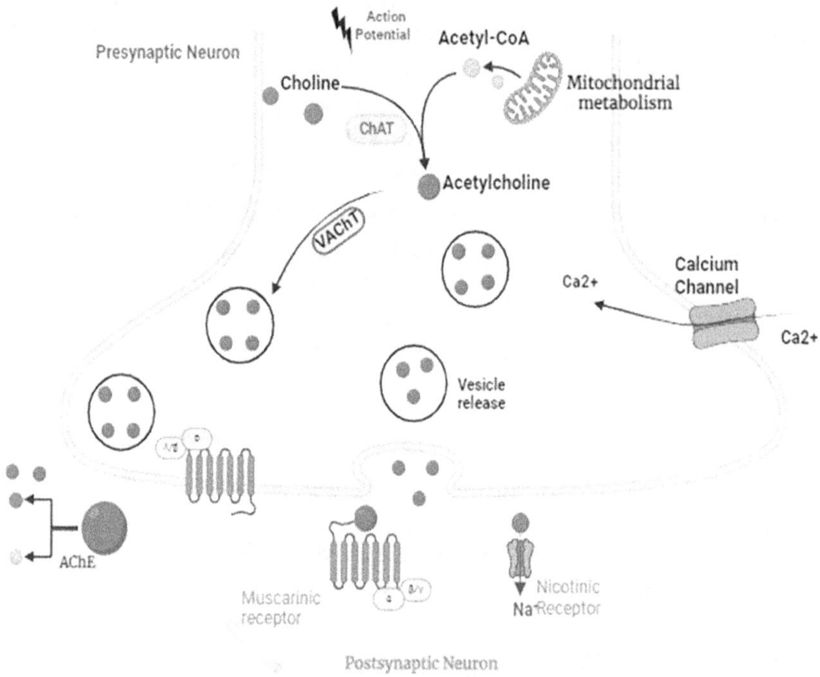

FIGURE 2.1 A schematic illustration of biological factors included in the ACh neurotransmission: The enzyme choline acetyltransferase (ChAT) changes choline and acetyl-coenzyme A (acetyl-CoA) into ACh in the cytosol of cholinergic presynaptic neurons. In synaptic vesicles, ACh is transferred via the vesicular acetylcholine transporter (VAChT). ACh is released from synaptic vesicles into the synaptic cleft by depolarization of the presynaptic neuron. There, it binds to nicotinic or muscarinic receptors to provide either stimulatory or inhibitory responses. In synaptic clefts, AChE quickly breaks down ACh to produce choline and acetate.

2.2.1 RECEPTORS OF ACETYLCHOLINE AND THEIR DISTRIBUTION

ACh is a neurotransmitter that acts on receptors to stimulate or inhibit the CNS depending on which kind of receptor it is and where the neuron is located. ACh receptors are classified as muscarinic or nicotinic based on their agonist selectivity and pharmacological characteristics [2].

2.2.2 MUSCARINIC ACETYLCHOLINE RECEPTORS

mAChR are the G protein-coupled receptors that control several ion channels. Five muscarinic receptor isoforms (M1–5) have currently been found to have distinct genes. While M4 and M2 are associated with Gαi, M5, M3, and M1are linked to Gαq [14]. When Gαq linked muscarinic receptors are stimulated, phospholipase C is activated, this results in the production of inositol phosphates and other second messengers. This encourages K^+ channels to close, which raises cell excitability [15]. Gαi related muscarinic receptor activation decreases cAMP levels and adenylyl cyclase, which in turn blocks voltage-gated Ca^{2+} channels and lessens cell excitability. Stimulating ACh muscarinic receptors can open or close Ca^{2+}, K^+ or Cl^- channels, leading to depolarization or hyperpolarization according to the type of the cell [16]. Presynaptic terminals are the primary locations of M1/M3 receptors [17]. M2/M4 receptors are situated presynaptically and operate as auto receptors, adversely controlling ACh release. They can also act as hetero receptors, altering synaptic transmission through Ca^{2+} channel activity [18].

Muscarinic receptor stimulation can cause either excitement or inhibition. However, the stimulation of muscarinic receptors typically results in excitement, particularly in the cortex [19]. Muscarinic mediated excitation occurs through many pathways. K^+ current suppression in postsynaptic areas is a common result of excitation through these receptors. By inhibiting the voltage-gated K^+ channels and inhibiting resting K^+ channels, muscarinic receptor activation efficiently blocks a variety of current [20]. Presynaptic muscarinic receptors have the ability to inhibit GABAergic neurotransmission as well as K^+ currents. NMDA currents are enhanced by intracellular second messengers produced when Gαq linked muscarinic receptors are stimulated. Excitation occurs when GABAergic neurons are inhibited and glutamatergic neurons are stimulated [21]. Muscarinic receptor activation rarely has inhibitory effects in the CNS. ACh has been shown to enhance inhibition in certain areas of the cortex. According to the suggested theory, muscarinic-mediated inhibition is thought to include a decrease in Ca^{2+} currents, a reduction in glutamatergic EPSPs, and an increase in K^+ current [1].

2.2.3 NICOTINIC ACETYLCHOLINE RECEPTORS

These receptors are ion-gated channels which only accept cations, such as Ca^{2+}, Na^+, K^+, and Na^+. Activation of the nicotinic receptor triggers a fast cellular response. There are currently 17 identified nicotinic receptors [22]. Nicotinic receptors consist of five subunits: 10 α (α1–10) and 4 β (β1-4), δ, γ or ε. nAChRs in neurons can be homo-pentamers (α9, α8, or α7) or hetero-pentamers (α2-α6 in combination with β2-β4 or α9 with α10 subunits). In contrast, hetero-pentameric muscle nAChRs consist of two α1 and include β1, ε, and δ subunits for adult muscle or γ subunits for fetal muscle. Because nicotinic receptors are composed of varied combinations of these subunits, there are possible variations in the kinds of nicotinic receptors. This demonstrates that the characteristics and roles of each kind of nicotinic receptor vary [23].

In the PNS, nicotinic receptor activation results in the fast synaptic transmission, where as in ACh receptors in the CNS, it modulates rather than excites or inhibits [24]. The long-term regulating effects of nAChRs in the CNS are primarily because of the high Ca^{2+} permeability. Most of nicotinic receptors in the PNS are post-synaptic, whereas others found in the CNS primarily regulate release of the neurotransmitters at the presynaptic neuronal membrane [25]. nAChRs activation in presynaptic neuronal membrane of the CNS causes an increase in presynaptic Ca^{2+} level and releases the neurotransmitters, like glutamate, GABA, ACh, dopamine, serotonin, and norepinephrine. ACh regulates neurotransmission across the CNS along with the synaptic strength and fidelity [26]. Nicotinic ACh receptor agonists increase the cognitive function, whereas antagonists decrease it [27]. According to research, ACh regulates glutamatergic neurotransmission that is involved in learning and memory [28]. Both neuroprotection and cognitive functions depend on the relationship between the cholinergic and the glutamatergic systems. nAChRs agonists have neuroprotective effect through Ca^{2+}dependent mechanism involving the glutamatergic system [29]. Ca^{2+} ions can impact nicotinic receptors by binding to its extracellular amino terminal domain and improving Ach dependent currents without changing any voltage [30]. Protein kinases A and C along with the tyrosine kinase can also influence the activity of nicotinic receptors [31].

2.3 CHOLINERGIC DYSFUNCTION AND ALZHEIMER'S DISEASE

Cholinergic system disruption is associated with a number of animal models of cognitive impairments and is essential to cognitive functioning [32, 33]. Cholinergic transmission deficits may impact cognition and behavior, including cortical and hippocampal information processing capabilities [34]. AD can lead to the alterations in cholinergic system, the development of Aβ, metal dyshomeostasis, tau hyperphosphorylation, and neuroinflammation [35]. AChE is an enzyme that degrades the ACh in the postsynaptic neuromuscular junction and can lead to memory and learning deficits by inhibiting the AChE's activity [36, 37]. Tau phosphorylation and the amyloid precursor protein metabolism alterations caused by dysfunctional cholinergic system can result in neuroinflammation, neurotoxicity, and neuronal death [38]. AD is associated with cognitive dysfunctions, including memory loss, which are produced through elevated deficiency in cholinergic and glutamate signaling in the brain [39]. The two essential proteins which were found in the lesions of AD are Aβ and α-synuclein [40]. Aβ was shown to bind to α7 nAChRs with significant affinity based on its location, especially in areas rich in cholinergic neurons [41]. Aβ has the ability to significantly impair memory and spatial working, as well as to increase cholinergic dysfunction along with the reduced ChAT and Ach activity, and elevated AChE activity due to which it dramatically increase the degree of apoptosis in the amygdala, prefrontal cortex, and the hippocampus [42]. The muscarinic receptor mediated disruption of cholinergic signals has been interlinked to various diseases, including AD. Potential targets of AD include M1/M4 muscarinic receptors [43]. Since the frontal brain and hippocampus express M1 mAChR at high levels, lowering these levels worsens AD like symptoms. Ablation of M1 mAChR was shown to boost amyloidogenic processing and tau hyperphosphorylation. These effects were linked to modifications in the GSK-3β and PKC activity, and increase in the microglial and astrocytic response linked to beta-amyloid plaques [44]. In APP/PS1 transgenic mice, elimination of M1 mAChR led to elevated brain Aβ levels and amyloid plaque deposition. In contrast, it has been discovered that M1 mAChR activation reverses cognitive impairments via regulating tau pathology and Aβ neurotoxicity through PKC activation and GSK-3β inactivation that is reliant on Gαq/11 [45]. Conditions related with cholinergic dysfunction like AD develop due to weakening of the forebrain cholinergic neurons that control the immune responses as well as the inflammation. Large amounts of cholinergic neurons in the brain can be found in the basal forebrain. BFCNs are liable for several cognitive functions, including memory, concentration, and learning. Loss of BFCNs can result in abnormal synapse activity, reduced cholinergic neurotransmission, and abnormalities in structural lipid metabolism [46]. The hippocampus and cerebral cortex are connected to the basal forebrain centers via cholinergic networks. A necessary condition for the survival, connection, and functioning of BFCNs is NGF. The kind of NGF found in the human brain is called proNGF. Because proNGF signaling and transport rely on retrograde nutritional support, BFCNs are especially vulnerable to AD. Memory loss can happen from the degradation of cholinergic neurons in the basal forebrain caused by a lack of NGF [47].

2.4 TRADITIONAL PHARMACOLOGICAL INTERVENTIONS TARGETING CHOLINERGIC SYSTEM IN AD

Tacrine, donepezil, rivastigmine, and galantamine are the four conventional acetylcholinesterase inhibitors (AChEI) that are USFDA approved. These drugs increase the amount of ACh in synapse by inhibiting the action of AChE in postsynaptic cleft. Various *in vitro* studies have shown the neuro-protective effects of donepezil, tacrine, rivastigmine, and galantamine which include reduced neuronal death, increased cell viability, and reduced inflammation [52]. In various animal AD models, donepezil showed improved cognitive and behavioral performance, reduced Aβ protein accumulation, and decreased neuronal death [53]. Previous studies state that these drugs can reduce the amount of Aβ protein deposited in the brain which can then enhance the memory and learning in Alzheimer's patients [52].

Neurofibrillary tangles (NFTs) can be formed in the brain as a result of tau hyperphosphorylation, which is mediated by the protein kinase GSK-3β. Research shows that donepezil provides neuroprotection by the inhibition of GSK-3β activity [54]. In a study, galantamine, rivastigmine, and donepezil all have shown neuroprotective effects against the okadaic acid induced toxicity [55]. Table 2.1 demonstrates the first (used for treating mild to severe symptoms of AD) and second generation (used for treating moderate to severe symptoms of AD) FDA-approved AChEI drugs for AD with their correlation to AD and their adverse effects. There are still a lot of challenges in the field of neuroprotection using cholinergic drugs, like AChEI, that have low tolerance, are less effective, and can only cure the symptomatic problems while having various adverse effects [56].

To overcome these problems, next generation AChEI drugs with less side effects or adverse effects are required.

TABLE 2.1 FDA Approved Acetylcholinesterase Inhibitor Drugs for AD.

Classification	Drug	Therapeutic Effect on Alzheimer's Disease	Side Effects	References
First generation	Tacrine	Suppresses AChE in tissues and plasma. Stimulates the M1 receptor to produce ACh.	Liver toxicity	[48]
Second generation	Donepezil	Reduces Aβ plaque that impacts NMDA (N-Methyl-D-aspartate) receptors. Reduced glutamate and increased ACh levels.	Nausea, vomiting, diarrhea, and fatigue	[49]
Second generation	Rivastigmine	Inhibition of *BACE1* and Aβ aggregation. Inhibition of GSK-3β.	Dizziness, vertigo, and upper respiratory tract infection	[50]
Second generation	Galantamine	Elevates ACh levels, particularly in the synaptic gap.	Salivation, bradycardia, dizziness, and abdominal pain	[51]

2.5 EMERGING THERAPEUTIC STRATEGIES TARGETING THE CHOLINERGIC SYSTEM

2.5.1 NEXT GENERATION ACETYLCHOLINESTERASE INHIBITORS

Many of the next generation AChEI's potential therapeutics are based on tacrine-1 to direct the development of anti-AD medications, even though all of the initial AChEI medications were used as scaffolds to suggest novel therapeutic molecules. Despite the fact that this compound's hepatotoxicity led to its discontinuation in 2013, it is still of interest currently [57]. Several new donepezil-tacrine like hybrids have recently been identified and shown to have inhibitory effect on AChE, such as hybrid tacrine compound 22d shows 80 times more inhibitory effect on AChE as compared to tacrine [58]. Another compound, H11 (a donepezil-tacrine compound) has 20 times more AChE inhibitory activity than tacrine alone, and has less side effects as compared to tacrine [59]. Compounds having the strongest AChE inhibition along with their IC50 values are listed in Table 2.2. AChE is specific to cholinergic synapses in the nervous system, hydrolyzing ACh predominantly. By inhibiting AChE, there will be elevated levels of ACh in the postsynaptic cleft [60]. Due to the several adverse effects of traditional AChEI, like hepatotoxicity, these medications have their own limitations. In order to overcome these limitations, new next generation AChEI are required with more efficacy and less side effects.

TABLE 2.2 Inhibitory Activity (IC50 nM) of AChE of Hybrid Tacrine and Donepezil Compounds.

Sr. No.	Hybrid Tacrine Derivatives	IC50 of AChE	References
1.	4.7. 2-(1-(9-amino-1,2,3,4-tetrahydroacridine-2-carbonyl) piperidin-4-yl)-N-benzylacetamide	23.66 ± 5.03 (μM)	[61]
2.	9- [2-(1-Benzylpiperidin-4-ylidene) hydrazinyl]-1,2,3,4-tetrahydroacridine trihydrochloride	2.550±0.530 (μM)	[59]
3.	9-{2- [1-(2-Methylbenzyl) piperidin-4-ylidene]hydrazinyl}-1,2,3,4-tetrahydroacridine trihydrochloride	2.970±0.090 (μM)	[59]
4.	9-{2- [1-(2-Fluorobenzyl) piperidin-4-ylidene]hydrazinyl}-1,2,3,4-tetrahydroacridine trihydrochloride	3.262±0.060 (μM)	[59]
5.	9-(2-{1- [2-(Trifluoromethyl)benzyl] piperidin-4-ylidene}hydrazinyl)-1,2,3,4-tetrahydroacridine trihydrochloride	6.101±0.860 (μM)	[59]
6.	2-((2-((7-Chloro-1,2,3,4-tetrahydroacridin-9-yl)amino)ethyl)amino)-1-methyl-[4,4′-bipyrimidin]-6(1H)-one	18.8 ± 0.18 (μM)	[58]
7.	9-{2- [1-(4-Bromobenzyl)piperidin-4-ylidene]hydrazinyl}-1,2,3,4-tetrahydroacridine trihydrochloride	3.384±0.610 (μM)	[59]
8.	3-Methyl-2-((2-((7-methyl-1,2,3,4-tetrahydroacridin-9-yl)amino)ethyl)amino)-6-(pyridin-4-yl)pyrimidin-4(3H)-one	20.0 ±2.40 (μM)	[58]
9.	9-{2- [1-(2,6-Dichlorobenzyl) piperidin-4-ylidene]hydrazinyl}-1,2,3,4-tetrahydroacridine trihydrochloride	2.703±0.400 (μM)	[59]
10.	2-((2-((6-Methoxy-1,2,3,4-tetrahydroacridin-9-yl)amino)ethyl)amino)-1-methyl-[4,4′-bipyrimidin]-6(1H)-one	23.0 ± 0.29 (μM)	[58]
11.	9-amino-6-(1-methyl1H-pyrazol-4-yl)-1,2,3,4-tetrahydroacridinium	40.89 ± 4.82 (μM)	[62]

2.5.2 NACHR AND MACHR MODULATORS IN AD

2.5.2.1 NACHR MODULATORS

Numerous studies on the cholinergic system have demonstrated a substantial correlation between nAChR signaling and the promotion of cognitive functioning, as well as neuroprotection [63]. These advantageous effects of nAChR signaling have been conclusively shown in recent years by genetic modification of subunits of the particular receptor in the animal models, as well as the use of positive allosteric modulators (PAMs), antagonists, and highly selective agonists. Table 2.3 summarizes some of the neuroprotective benefits of modulating nAChRs .ACh and nicotine, which are orthosteric cholinergic agonists, immediately trigger the nAChRs to allow the ligand-gated cation channel to open. This activates many intracellular signaling pathways primarily through calcium influx and causes membrane depolarization by sodium influx [64]. However, nAChRs, particularly α7 nAChRs, rapidly desensitize to orthosteric agonist activation. However, PAMs bind to certain allosteric sites to control AChR function. PAMs are classified into two categories: type I and type II. While type II PAMs improve ion permeability and lengthen the duration of channel opening, type I PAMs increase the agonist-induced current peak amplitude as well as the permeability of the channel ion [65]. PAMs, particularly the type II have significant influence on the roles of the nAChRs, which exhibit fast desensitization following agonist stimulation, particularly α7 nAChRs. With their high permeability to calcium, α7 nAChRs can cause membrane depolarization as well as the activation of several intracellular signaling pathways [66]. A research demonstrates that activating α7 and α4β2 nAChRs with the PNU-282987 and RJR-2403 oxalate can restore AMPAR cell surface expression in primary cultured mouse hippocampus neurons treated with Aβ1-42 oligomers. This is thought to restrict the amount of calcium that Aβ1-42 oligomers bring in, which would subsequently prevent the calcium-dependent AMPAR GluA1 dephosphorylation. It is unlikely that activating just one receptor subtype can restore expression back, though. Surprisingly, following Aβ1-42 treatment, AMPAR surface expression was not recovered when α3β4 nAChRs were stimulated with α7 nAChRs, α4β2 nAChRs, or both. This suggests that in order to have therapeutic benefits on AD, such as cognitive improvement, highly selective activation is required, and that distinct receptor subtypes may activate various and complex downstream pathways [67].

TABLE 2.3 Neuroprotective Effects of nAChR Associated Compounds on AD.

Sr. No.	Type	Compound	Therapeutic Effect	Animal Model	References
1	Selective α7 nicotinic ACh receptors agonist	PHA-543613	Restored the hippocampal-related memory deficit.	Conditional presenilin 1 and presenilin 2 double knockout mice	[68]
			Restored theta power and strength of theta-gamma PAC, which had been reduced.		
			Improved recognition memory in Aβ-treated animals.	Aβ25-35 AD mice model	[69]
2	Selective α7 nicotinic ACh receptors partial agonist	A582941	Elevated mRNA for Erk1/2 and the immediate-early gene response (including Arc), increased synaptic plasticity enhancing memory.	APdE9 AD mice model	[70]
3	Type I PAM for α7 nicotinic ACh receptors	JWX-A0108	Inhibition of NF-κB.	AD mice model (APP/PS1)	[71]
			Suppresses IL-1β, IL-6, and TNF-α. levels.		
4	Type II PAM for α7 nicotinic ACh receptors	PNU-120596	Restores the cognitive defects.	Scopolamine induced rat	[72]
5	Non selective nicotinic ACh receptors agonist	Ach	Activates Erk1/2.	AD model mice (3xTgAD)	[73]
			Inhibits the JNK and MAPKs.	Aβ1–42Mice model	[74]

2.5.2.2 MACHR MODULATORS

Alzheimer's patient's neocortex has damaged M1 AChR coupling to the G-proteins, and the degree of uncoupling is linked with the intensity of cognitive symptoms [75]. According to several studies, the M1 AChR agonists are a great treatment option for AD; M1 receptor signaling has a significant impact on AD symptoms, such as cholinergic deficiency, cognitive impairment, and tau and Aβ pathology. By inhibiting GSK-3β and triggering PKC, M1 receptor activation lowers tau hyperphosphorylation [76]. In an animal model of AD, AF267B, which is an M1 agonist, has been demonstrated to enhance cognitive performance and decrease Aβ42 and tau abnormalities [77]. A recent study has demonstrated that M1 mAChR PAM, VU0486846, can decrease Aβ-like pathologies and neurogliosis markers in female AD mice. Moreover, it modifies the progression of the disease by reducing neuroglial migration to amyloid beta plaques, potentially lowering neuroinflammation in the AD brain (Table 2.4) [78]. Another M1 mAChR PAM, SUVN-I6107, restored the delay-induced memory impairment in the object identification task. Additionally, it lessened the memory impairments induced on by scopolamine in the tasks of contextual fear conditioning and object recognition. SUVN-I6107 improved cerebral blood flow, enhanced donepezil's effects, and increased non-amyloidogenic amyloid precursor protein processing in rats. SUVN-I6107 does not cause any negative effects linked with cholinergic system activation. SUVN-I6107 was shown to be safe in both safety pharmacology and phase 1 animal toxicology trials [79]. In a placebo-controlled clinical research, xanomeline which is a muscarinic agonist that targets M1/M4 receptors specifically was found to be an effective therapy for the psychiatric and cognitive symptoms of AD, including mood disorder, memory loss, hallucinations, and agitation [80]. M4 selective allosteric agonists, such as VU0152100 and VU0010010 have been demonstrated to mitigate amphetamine induced hyperlocomotion in rats and lessen excitatory synaptic transmission in the hippocampus. This shows that M4 PAMs might also function as antipsychotic modulators [81]. M1/M4 PAMs provide promising treatment alternatives for CNS disorders like AD [82].

nAChR and mAChR modulators are the ones which are specific to a particular target receptor subtype in the cholinergic system and have a great potential to improve the cognitive function by providing neuroprotection against conditions like AD.

TABLE 2.4 M1-mAChR Allosteric Modulators and Their Importance in AD.

Sr. No.	Compound Name	Property of Ligand	Role in Alzheimer's Disease	References
1	VU0486846	PAM	Restores the cognitive dysfunction and lowers the amount of Aβ.	[78]
2	PF-06764427	PAM-agonist	Decreases the hyperlocomotion in rats.	[83]
3	BQCA	PAM-agonist	Recovers rodents with cognitive impairments.	[84]
4	MK-7622	PAM-agonist	Restores cognitive deficiencies in animals.	[85]
5	PQCA	PAM-agonist	Reverses recognition memory deficiencies in the AD mouse model.	[86]
6	PF-06767832	PAM-agonist	Restores rodents' impaired cognitive function caused by scopolamine.	[87]
7	PF-06827443	PAM-agonist	(a) Strong neural penetration; (b) Induces procognitive effects in rodents.	[88]

2.6 ADVANCED TOOLS TO STUDY THE CHOLINERGIC SYSTEM IN AD

2.6.1 POSITRON EMISSION TOMOGRAPHY (PET) IMAGING TECHNIQUE

PET imaging is an important technique for studying various neurodegenerative disorders like AD as it provides in depth *in vivo* information about the cholinergic transmission functions [89]. A major achievement in PET imaging of human brain VAChT binding was the effective clinical translation of the FFEOBV. When comparing AD patients to healthy controls, F-FEOBV PET showed decreased VAChT binding in the cholinergic terminals in the cingulate cortices and front temporoparietal, suggesting caudorostral dysfunction of nucleus basalis of Meynert [90]. VAChT in AD was linked to overall cognitive impairment. In AD patients, 18F-FEOBV PET detected AD-related alterations more accurately than F-FDG PET or b-amyloid PET [91]. Most of the α7 nAChR PET tracers that are investigated are the derivatives of diazabicyclononane. C-CHIBA-1001 was the first α7 nAChR PET tracers for human use [92]. However, more research on this tracer revealed minimal brain dispersion, low selectivity, and low binding affinity. Diazabicyclononane molecules with high brain absorption were developed in a series (F-NS14490, F-NS10743, and C-NS14492). Nevertheless, there was little evidence of particular brain binding in preclinical studies. Dibenzothiophene sulfone derivatives like F-ASEM and its paraisomer, FDBT10, are other alternatives. Both tracers produced the same brain dynamics in the primates other than humans and demonstrated strong specific binding inside the human brain [93]. Azabicyclooctylester, C-MeQAA is an alternative α7 nAChR PET tracer. This tracer showed selective *in vitro* binding, but its *in vivo* selectivity was reduced because it binds to 5-hydroxytryptamine-3 receptors off-target which is a similar receptor to α7 nAChR [94]. The synthesis of certain mAChR PET ligands is crucial yet difficult. Recently C-GSK1034702 is an M1-selective mAChR PET ligand that has been developed for therapeutic use by using PET. The suggested tracer advancement and the clinical PET application of the CGSK1034702 aim to monitor B.B.B permeability and reduce the risk of the medication development for GSK1034702. CGSK1034702 was not found to be a viable PET ligand because of the restricted specific binding and poor affinity to the target [95]. In previous research, C-MK-6884, which is a new C-labeled positive allosteric modulator of M4 mAChR was synthesized and tested preclinically [96]. The binding of C-MK-6884 to activate M4 mAChR makes it as an effective positive allosteric modulator of M4. Recently C-MK-6884 was developed for clinical PET applications [97]. For the first time, non-selective mAChR PET ligand (4-F-fluorodexetimide) was administered to humans and positive outcomes were seen, including high image quality, specific binding in the cerebellum, high brain uptake, low interindividual variability in healthy subjects, and irreversible tracer distribution in mAChR-rich brain areas [98]. The level of AChE activity in the brain can be measured by using the PET tracers. Two types of AChE substrates and inhibitors are currently available: 5-C-methoxy-donepezil and C-physostigmine. PET radioligands, N-C-methylpiperidinyl4-acetate (C-MP4A), and C-methylpiperidin-4-yl propionate (C-PMP) are examples of the ACh analogues. As compared to C-PMP, C-MP4A is more selective for the AChE, but its flow is limited and has a greater relative hydrolysis rate, that is, (k3). C-PMP is more effective than C-MP4A at differentiating between different AChE brain regions. While C-MP4A only measures cortical AChE, C-PMP can detect both cortical and thalamic AChE. In an experimental model of AD, both tracers have been shown to be reliable as well as sensitive enough to identify a 30–50% decline in AChE activity [89].

With the help of PET imaging technique, changes in cholinergic markers inside the brain can be monitored and can provide valuable insights into many kinds of neurodegenerative diseases including AD. However, number of limitations still exist for PET scans, including tracer selectivity and off-target binds which can interfere with its accuracy. To develop better PET ligands that can accurately identify neurotransmitter activity in neurodegenerative diseases, more research is required.

2.6.2 ELECTROENCEPHALOGRAPHY (EEG)

The electroencephalogram (EEG) is a non-invasive tool which measures the activity of the brain at different frequencies by showing the rate of deterioration. In order to differentiate between different neurodegenerative diseases, visual analysis of EEG recordings may be helpful. Due to its extensive use, non-invasiveness, and low cost, EEG is also a very promising diagnostic tool with superior capabilities than other tools [99]. AD results in irregular brain wave rhythms. Alzheimer's patient's EEG activity is less complex and has a slower mean frequency and poor cortical area coherence [100]. Studies have shown that there is change in power spectrum in the brain of AD patients which moves from high-frequency waves (alpha, beta, and gamma) to low-frequency waves (theta and delta) [101]. Various research studies have revealed that individuals with AD have decreased activity in the beta and alpha bands, and there is an increased activity of the theta and delta bands. In the early phase of AD, theta activity rises while alpha activity falls. Studies have revealed that cholinesterase inhibitors can considerably modify the EEG frequency bands [102]. AChEI temporarily reverse EEG slowness in AD patients and increase cholinergic neurotransmission. After one to two weeks of treatment with AChEI (rivastigmine or donepezil), there was a drop in theta power and an enhancement in the AD cognitive assessment scale. In posterior area of the AD patients, rivastigmine enhances the spectral strength of alpha waves [103]. Additionally, in AD patients, galantamine decreases absolute frontal EEG values in theta, alpha, and beta frequencies [102]. Furthermore, donepezil medication has been linked to large reductions in mean alpha and delta frequencies in AD patients [104].

Through monitoring of brain activity, EEG helps to identify a number of neurodegenerative disorders like AD. However, interpreting data can be difficult and may not always distinguish Alzheimer's from other diseases. Alzheimer's patients have abnormal frequency of waves in their brain, with fewer high-frequency waves and more low-frequency waves. Drugs like cholinesterase inhibitors may temporarily change EEG patterns and improve cognitive performance.

2.6.3 MAGNETIC RESONANCE IMAGING (MRI)

An MRI generates detailed images of the brain and other organs of the body using radio waves and a large magnet, and these images are formed and visualized using a computer. Magnetic resonance imaging (MRI) and the diffusion tensor imaging (DTI) when used together can offer an evaluation of the cholinergic system's integrity *in vivo*. They supplement PET with additional information on the cholinergic system, but they are more widely available, less expensive, and non-invasive [105]. With structural MRI, the extensively examined region is the cBF nuclei (Ch1-4) [106]. Voxel-based morphometry (VBM) was a method used in early *in vivo* cBF imaging to identify brain shrinkage. The methodology relied on quantifying an extremely large region called substantia innominata. Morphometric analysis of the cBF at a sub-regional scale is now feasible due to the availability of stereotactic cytoarchitectonic maps of the human cBF [107]. Subregionally precise cBF maps reveal the moderate cognitive impairment (MCI) stage of AD. The posterior part of the Ch4 is where degradation of neurons is the most noticeable and in later stages of AD, it progresses toward more anteromedial cBF areas [108]. Researchers found that when comparing AD patients to those with late MCI, simple linear estimate of the Ch4 thickness demonstrated excellent diagnostic accuracy [109]. Moreover, a substantial decrease in cBF atrophy was observed to be associated with AChE-treated patients' improved cognitive function in prodromal AD. This finding shows the possibility that structural MRI data might be a reliable indicator of cholinergic therapeutic responses [110].

A non-invasive method to assess the health of the brain's cholinergic system, MRI provides precise and reliable images which help in the diagnosis of a number of neurological conditions. It is more affordable than other techniques like PET imaging and is generally available. Sometimes small changes in a specific brain area cannot be detected by the MRI. Despite this, MRI is still a highly helpful tool for studying a variety of neurological diseases, and recent developments in the field might make it even more beneficial.

2.6.4 MULTIMODAL IMAGING (MMI) TECHNIQUES

There are various similarities in the molecular imaging and in MRI cholinergic imaging. Multimodal imaging methods can provide more precise information about the anatomical and functional characteristics of the subcortical and cortical regions, and the cholinergic system. The combined MRI and PET imaging techniques showed that the hypometabolism in a specific cortical area produced cognitive dysfunction. In MCI, cBF shrinkage was revealed to be substantially linked to these conditions [111]. Reduced structural and functional connectivity between the Ch3/4 related frontal, peri-insular, and medial/pulvinar thalamic and occipital regions has been linked to reduced overall cognition and fronto-executive functions. Conversely, memory impairments have been linked to a decreased degree of functional connectivity between the medial temporal lobe and Ch1/2 [112].

Combining molecular imaging as well as MRI cholinergic imaging helps to understand the brain function as compared to PET or MRI when used alone. Multimodal imaging improves the ability to detect even small changes in the brain which allow the earlier diagnosis and treatment of the neurological condition. So, using MMI technique can improve our understanding related to the various neurological conditions.

2.6.5 CHEMOGENETICS AND OPTOGENETICS

Designer receptors that are triggered by the designer drugs also known as DREADDs are used in the chemogenetic methods to influence the cholinergic signaling in a cell type specific manner. By genetically engineering cholinergic neurons to express DREADDs, they can be used to selectively modulate cholinergic activity. In AD, optogenetic methods can be used to selectively activate or inhibit cholinergic neurons through the use of light-sensitive proteins [113, 114]. The brain's cholinergic system is a significant neuromodulator that covers numerous nuclei. Up to 90% of these cholinergic projections, which regulate episodic memory, end up in the hippocampus. The cholinergic neurons come from two primary sources. In the basal forebrain, there are two kinds of nuclei: the vertical limb of the diagonal band nucleus (VDB) and the medial septal nucleus (MS) [115]. Results from studies reveal that ACh from the basal forebrain binds to the muscarinic acetylcholine receptor in hippocampal pyramidal neurons. This causes theta waves to fluctuate, which helps to improve memory and learning [116].

Another study suggested that stimulation by optogenetics in the medial septal nucleus region of the brain improves cognitive function in a mouse model of AD via enhancing the hippocampal Schaeffer collateral pathway's synaptic plasticity and stimulating cholinergic input in the hippocampus [117]. These results show that cholinergic circuit which connects the hippocampus to the VDB/MS of the basal forebrain projections may have a positive effect on the processing episodic memory. In the lateral phase of AD, cholinergic networks get damaged by the NFTs and amyloid beta. AD related changes associated with the impaired neural structure and cholinergic function can be visualized by the magnetic resonance spectroscopy (MRS) [118]. Patients with AD show decreased levels of the neurochemical metabolite, N-acetyl aspartate (NAA), which is an important chemical marker of the neuronal integrity and can be assessed by using the MRS. Low levels of choline (Cho), which is a precursor of Ach, were found in the AD patient's hippocampi and basal forebrains [119]. APP/PS1 mice model of AD shows that when chemogenetic, "rAAV-ChAT-cre-2a-eGFP-WPRE-pA" (1×1012 particles/mL) combined with an adeno-associated virus in 1:3 ratios was injected to the MS/VDB area of mice brain, the cholinergic neuronal stimulation from the MS/VDB nuclei enhanced the memory and the learning and in the mice [120]. Furthermore, in another research, the introduction of chemical genetic therapy results in a substantial reduction in the formation of Aβ plaque in 5XFAD mice and increased cognitive function by triggering the locus coeruleu's noradrenergic neurons [121].

Chemogenetic and optogenetic techniques allow the precise regulation of the cholinergic signaling, which can lead to improved cognitive functions in AD. In a number of animal models of AD, chemogenetic medications that target the cholinergic circuits in the MS/VDB nucleus enhance cognitive performance. These methods offer promising pathways for the development of novel therapeutics for cholinergic dysfunction in AD and can be used for the management and treatment of the neurodegenerative disease.

2.7 FUTURE DIRECTIONS AND CHALLENGES

While developing cholinergic drugs for AD presents opportunities, various challenges still remain. This is because AD is a multifaceted cascade involving neuroinflammation, neuronal death, and tau and amyloid-beta proteins accumulation in the brain. Synaptic and cholinergic system dysfunction is only one aspect of AD. So, addressing this complex pathogenic process with only the cholinergic treatment is difficult. Although, cholinergic system's role in cognitive function still makes it a target of interest for developing new therapeutic drugs for the treatment of AD.

There are various drawbacks of the existing cholinergic drugs like the traditional or existing AChEI that have various adverse effects, like nausea, vomiting, dizziness, and so forth. Moreover, these drugs only provide symptomatic relief for short period of time. Developing new drugs with high tolerance and minimal adverse effects is a must for the treatment of AD. Recent advancements in the neuroimaging, genetics, and in biomarker research have deepened our knowledge regarding the working of the cholinergic system, and these findings point toward the possible therapeutic targets and biomarkers for AD treatment by modifying the cholinergic system. One approach to treat AD in the future can be through personalized medications as AD is a multifaceted cascade which includes various complex pathogenic pathways, and targeting these pathways is difficult using a single standard drug. But for personalized treatment to be effective, it is very important to identify various biomarkers and genetic variables which can predict the response of these medications in the disease progression.

Various research studies have shown that innovative techniques like gene editing hold great promise for repairing the damaged cholinergic neurons in AD animal models by improving the cognitive functions. Additionally, advanced neuroimaging using PET scan, MRI, EEG, and MMI helps to understand the cholinergic system in different areas of the brain and how small changes can lead to various neurological disorders. These tools and techniques together offer potential opportunities for the novel cholinergic therapies to treat AD.

2.8 CONCLUSION

In conclusion, the cholinergic system is an important neuronal communication network that plays a vital role in memory and learning, and loss of cholinergic neurons or cholinergic dysfunction can lead to AD. Currently, there are only three AChEI drugs that are FDA approved and are used to treat AD, but these drugs only provide temporary relief along and have various adverse effects, like nausea, vomiting, and potential hepatotoxicity.

Due to these reasons, developing new drugs with less adverse effects and having disease modifying properties is required. Furthermore, the selective modulators also known as (PAMs) of ACh receptors have the potential to increase the effectiveness of the medication by modulating the receptors accordingly. Advance imaging techniques, like PET scan, EEG, MMI, and MRI are very useful and important tools in the early diagnosis of the neurodegenerative disorders like AD and PD. These imaging techniques can detect even small changes in the brain and in the cholinergic system. By overcoming these challenges and using new tools and techniques, we can better understand the cholinergic system and can develop more effective medications and therapies for AD.

KEYWORDS

- **acetylcholine**
- **acetylcholinesterase inhibitors**
- **Alzheimer's disease**
- **cholinergic system**
- **muscarinic receptor**
- **nicotinic receptor**
- **neuroimaging techniques**

REFERENCES

1. Ferreira-Vieira, T. H., Guimaraes, I. M., Silva, F. R., & Ribeiro, F. M. (2016). Send orders for reprints to reprints@benthamscience.ae Alzheimer's disease: Targeting the cholinergic system. *Current Neuropharmacology, 14*, 101–115.
2. Halder, N., & Lal, G. (2021). Cholinergic system and its therapeutic importance in inflammation and autoimmunity. *Frontiers in Immunology, 12*, 660342. DOI: 10.3389/FIMMU.2021.660342/BIBTEX.
3. Lombardo, S., & Maskos, U. (2015). Role of the nicotinic acetylcholine receptor in Alzheimer's disease pathology and treatment. *Neuropharmacology, 96*, 255–262. DOI: 10.1016/J.NEUROPHARM.2014.11.018.
4. Reale, M. et al. (2012). Relation between pro-inflammatory cytokines and acetylcholine levels in relapsing-remitting multiple sclerosis patients. *International Journal of Molecular Sciences, 13*(10), 12656–12664. DOI: 10.3390/IJMS131012656.

5. Bekdash, R. A. (2021). The cholinergic system, the adrenergic system and the neuropathology of Alzheimer's disease. *International Journal of Molecular Sciences, 22*(3), 1–18. DOI: 10.3390/IJMS22031273.
6. Latina, V. et al. (2018). NGF-dependent changes in ubiquitin homeostasis trigger early cholinergic degeneration in cellular and animal AD-model. *Frontiers in Cellular Neuroscience, 12*. DOI: 10.3389/FNCEL.2018.00487/FULL.
7. Chen, Z. R., Huang, J. B., Yang, S. L., & Hong, F. F. (2022). Role of cholinergic signaling in Alzheimer's disease. *Molecules, 27*(6). DOI: 10.3390/MOLECULES27061816.
8. Campanari, M. L., García-Ayllón, M. S., Belbin, O., Galcerán, J., Lleó, A., & Sáez-Valero, J. (2014). Acetylcholinesterase modulates presenilin-1 levels and γ-secretase activity. *Journal of Alzheimer's Disease, 41*(3), 911–924. DOI: 10.3233/JAD-140426.
9. Ramos-Rodriguez, J. J. et al. (2013). Rapid β-amyloid deposition and cognitive impairment after cholinergic denervation in APP/PS1 mice. *Journal of Neuropathology & Experimental Neurology, 72*(4), 272. DOI: 10.1097/NEN.0B013E318288A8DD.
10. Arora, M. K. et al. (2022). Plumbagin alleviates intracerebroventricular-quinolinic acid induced depression-like behavior and memory deficits in wistar rats. *Molecules, 27*(6). DOI: 10.3390/MOLECULES27061834.
11. Simchovitz, A., Heneka, M. T., & Soreq, H. (2017). Personalized genetics of the cholinergic blockade of neuroinflammation. *Journal of Neurochemistry, 142*(Suppl 2), 178. DOI: 10.1111/JNC.13928.
12. Eivani, M., Alijanpour, S., Arefian, E., & Rezayof, A. (2019). Corticolimbic analysis of microRNAs and protein expressions in scopolamine-induced memory loss under stress. *Neurobiology of Learning and Memory, 164*. DOI: 10.1016/J.NLM.2019.107065.
13. Yegla, B., & Parikh, V. (2017). Developmental suppression of forebrain trkA receptors and attentional capacities in aging rats: A longitudinal study. *Behavioural Brain Research, 335*, 111. DOI: 10.1016/J.BBR.2017.08.017.
14. Block, M., & Blanchard, D. L. (2023). Aminoglycosides. *Encyclopedia of Toxicology—3rd Ed.*, 191–193. DOI: 10.1016/B978-0-12-386454-3.00234-7.
15. Yohn, S. E., Weiden, J., Felder, C. C., & Stahl, S. M. (2022). Muscarinic acetylcholine receptors for psychotic disorders: Bench-side to clinic. *Trends in Pharmacological Sciences, 43*, 1098–1112. DOI: 10.1016/j.tips.2022.09.006.
16. De Angelis, F., & Maria Tata, A. (2016). Analgesic effects mediated by muscarinic receptors: Mechanisms and pharmacological approaches. *Central Nervous System Agents in Medicinal Chemistry, 16*(3), 218–226.
17. Carlson, A. B., & Kraus, G. (2024). Physiology, cholinergic receptors. *StatPearls*. https://www.ncbi.nlm.nih.gov/books/NBK526134/. Accessed: Mar. 30, 2024.
18. Pepeu, G. (2004). From messengers to molecules. Acetylcholine: I Muscarinic receptors. ndl.ethernet.edu.et. http://ndl.ethernet.edu.et/bitstream/123456789/45306/1/283.pdf#page=113. Accessed: Mar. 30, 2024.
19. Andrade, R. (1991). Cell excitation enhances muscarinic cholinergic responses in rat association cortex. *Brain Research, 548*(1–2), 81–93. DOI: 10.1016/0006-8993(91)91109-E.
20. Dannenberg, H., Young, K., & Hasselmo, M. (2017). Modulation of hippocampal circuits by muscarinic and nicotinic receptors. *Frontiers in Neural Circuits, 11*. DOI: 10.3389/FNCIR.2017.00102.

21. Li, D., Chen, S. R., Pan, Y. Z., Levey, A. I., & Pan, H. L. (2002). Role of presynaptic muscarinic and GABAB receptors in spinal glutamate release and cholinergic analgesia in rats. *Journal of Physiology, 543*, 807. DOI: 10.1113/JPHYSIOL.2002.020644.
22. Le Gratiet, K. L. (2021). Differential distribution of co-transmitted cholinergic and GABAergic synaptic inputs onto substantia nigra dopaminergic neurons. http://hdl.handle.net/1828/12887. Accessed: Mar. 30, 2024.
23. Ho, T. N. T., Abraham, N., & Lewis, R. J. (2020). Structure-function of neuronal nicotinic acetylcholine receptor inhibitors derived from natural toxins. *Frontiers in Neuroscience, 14*. DOI: 10.3389/FNINS.2020.609005.
24. Chung, B. Y. T., Bignell, W., Jacklin, D. L., Winters, B. D., & Bailey, C. D. C. (2016). Postsynaptic nicotinic acetylcholine receptors facilitate excitation of developing CA1 pyramidal neurons. *Journal of Neurophysiology, 116*(5), 2043. DOI: 10.1152/JN.00370.2016.
25. Dineley, K. T., Pandya, A. A., & Yakel, J. L. (2015). Nicotinic ACh receptors as therapeutic targets in CNS disorders. *Trends in Pharmacological Science, 36*(2), 96. DOI: 10.1016/J.TIPS.2014.12.002.
26. Jiang, L., López-Hernández, G. Y., Lederman, J., Talmage, D. A., & Role, L. W. (2014). Optogenetic studies of nicotinic contributions to cholinergic signaling in the central nervous system. *Reviews in the Neuroscience, 25*(6), 755–771. DOI: 10.1515/REVNEURO-2014-0032/XML.
27. Upright, N. A., & Baxter, M. G. (2021). Effects of nicotinic antagonists on working memory performance in young rhesus monkeys. *Neurobiology of Learning and Memory, 184*, 107505. DOI: 10.1016/J.NLM.2021.107505.
28. Huang, Q., Liao, C., Ge, F., Ao, J., & Liu, T. (2022). Acetylcholine bidirectionally regulates learning and memory. *Journal of Neurorestoratology, 10*(2), 100002. DOI: 10.1016/J.JNRT.2022.100002.
29. Cheng, Y. J., Lin, C. H., & Lane, H. Y. (2021). Involvement of cholinergic, adrenergic, and glutamatergic network modulation with cognitive dysfunction in Alzheimer's disease. *International Journal of Molecular Sciences*, *22*(5), 1–15. DOI: 10.3390/IJMS22052283.
30. Uteshev, V. V. (2012). α7 nicotinic ACh receptors as a ligand-gated source of Ca2+ ions: The search for a Ca2+ optimum. *Advances in Experimental Medicine and Biology, 740*, 603. DOI: 10.1007/978-94-007-2888-2_27.
31. Chrestia, J. F., Turani, O., Araujo, N. R., Hernando, G., Esandi, M. del C., & Bouzat, C. (2023). Regulation of nicotinic acetylcholine receptors by post-translational modifications. *Pharmacology Research, 190*, 106712. DOI: 10.1016/J.PHRS.2023.106712.
32. Arora, M. K., Kisku, A., & Jangra, A. (2020). Mangiferin ameliorates intracerebroventricular-quinolinic acid-induced cognitive deficits, oxidative stress, and neuroinflammation in Wistar rats. *Indian Journal of Pharmacology, 52*(4), 296–305. DOI: 10.4103/IJP.IJP_699_19.
33. Singh, T., Kwatra, M., Kushwah, P., Pant, R., Bezbaruah, B. K., & Jangra, A. (2022). Binge alcohol consumption exacerbates high-fat diet-induced neurobehavioral anomalies: Possible underlying mechanisms. *Chemico-Biological Interactions, 364*, 110039. DOI: 10.1016/J.CBI.2022.110039.

34. Newman, E. L., Gupta, K., Climer, J. R., Monaghan, C. K., & Hasselmo, M. E. (2012). Cholinergic modulation of cognitive processing: insights drawn from computational models. *Frontiers in Behavioral Neuroscience, 6*. DOI: 10.3389/FNBEH.2012.00024.
35. Huat, T. J., Camats-Perna, J., Newcombe, E. A., Valmas, N., Kitazawa, M., & Medeiros, R. (2019). Metal toxicity links to Alzheimer's disease and neuroinflammation. *Journal of Molecular Biology, 431*(9), 1843–1868. DOI: 10.1016/J.JMB.2019.01.018.
36. Jangra, A., Chadha, V., Kumar, D., Kumar, W., & Arora, M. K. (2021). Neuroprotective and acetylcholinesterase inhibitory activity of plumbagin in ICV-LPS induced behavioral deficits in rats. *Current Research in Behavioral Sciences*, *2*, 100060. DOI: 10.1016/J.CRBEHA.2021.100060.
37. Jangra, A. et al. (2016). Edaravone alleviates cisplatin-induced neurobehavioral deficits via modulation of oxidative stress and inflammatory mediators in the rat hippocampus. *European Journal of Pharmacology*, *791*, 51–61. DOI: 10.1016/J.EJPHAR.2016.08.003.
38. Majdi, A. et al. (2020). Amyloid-β, tau, and the cholinergic system in Alzheimer's disease: Seeking direction in a tangle of clues. *Reviews in the Neuroscience, 31*(4), 391–413. DOI: 10.1515/REVNEURO-2019-0089.
39. Gasiorowska, A. et al. (2021). The biology and pathobiology of glutamatergic, cholinergic, and dopaminergic signaling in the aging brain. *Frontiers in Aging Neurosciences, 13*, 654931. DOI: 10.3389/FNAGI.2021.654931/BIBTEX.
40. Shim, K. H., Kang, M. J., Youn, Y. C., An, S. S. A., & Kim, S. Y. (2022). Alpha-synuclein: a pathological factor with Aβ and tau and biomarker in Alzheimer's disease. *Alzheimer's Research Therapy, 14*(1). DOI: 10.1186/S13195-022-01150-0.
41. Lee, C. H., & Hung, S. Y. (2022). Physiologic functions and therapeutic applications of α7 nicotinic acetylcholine receptor in brain disorders. *Pharmaceutics, 15*(1), 31. DOI: 10.3390/PHARMACEUTICS15010031.
42. Zong, B. et al. (2022). Understanding how physical exercise improves alzheimer's disease: Cholinergic and monoaminergic systems. *Frontiers in Aging Neurosciences, 14*. DOI: 10.3389/FNAGI.2022.869507.
43. Montani, C. et al. (2021). The M1/M4 preferring muscarinic agonist xanomeline modulates functional connectivity and NMDAR antagonist-induced changes in the mouse brain. *Neuropsychopharmacology*, *46*(6), 1194. DOI: 10.1038/S41386-020-00916-0.
44. Shiozaki, K., Iseki, E., Hino, H., & Kosaka, K. (2001). Distribution of m1 muscarinic acetylcholine receptors in the hippocampus of patients with Alzheimer's disease and dementia with Lewy bodies-an immunohistochemical study. *Journal of the Neurological Sciences, 193*(1), 23–28. DOI: 10.1016/S0022-510X(01)00638-4.
45. Medeiros, R. et al. (2011). Loss of muscarinic M1 receptor exacerbates Alzheimer's disease–like pathology and cognitive decline. *American Journal of Pathology, 179*(2), 980. DOI: 10.1016/J.AJPATH.2011.04.041.
46. Llorente-Ovejero, A. et al. (2021). Specific phospholipid modulation by muscarinic signaling in a rat lesion model of Alzheimer's disease. *ACS Chemical Neurosciences, 12*(12), 2167–2181. DOI: 10.1021/ACSCHEMNEURO.1C00169/SUPPL_FILE/CN1C00169_SI_001.PDF.

47. Claudio Cuello, A., Pentz, R., & Hall, H. (2019). The brain NGF metabolic pathway in health and in Alzheimer's pathology. *Frontiers in Neurosciences, 13*, 441218. DOI: 10.3389/FNINS.2019.00062/BIBTEX.
48. Mitra, S. et al. (2022). Tacrine derivatives in neurological disorders: Focus on molecular mechanisms and neurotherapeutic potential. *Oxidative Medicine and Cellular Longevity*. DOI: 10.1155/2022/7252882.
49. Saeedi, M., & Mehranfar, F. (2022). Challenges and approaches of drugs such as memantine, donepezil, rivastigmine, and aducanumab in the treatment, control and management of Alzheimer's disease. *Recent Patents on Biotechnology, 16*(2), 102–121. DOI: 10.2174/1872208316666220302115901.
50. Bortolami, M. et al. (2021). Acetylcholinesterase inhibitors for the treatment of Alzheimer's disease—A patent review (2016–present). *Expert Opinion on Therapeutic Patents, 31*(5), 399–420. DOI: 10.1080/13543776.2021.1874344.
51. Müller et al. (2021). Structural MRI of the basal forebrain as predictor of cognitive response to galantamine in healthy older adults—A randomized controlled double-blinded crossover study. *Alzheimer's & Dementia: Translational Research & Clinical Interventions, 7*(1), e12153. DOI: 10.1002/TRC2.12153.
52. Moreira, N. C. D. S., Lima, J. E. B. de F., Marchiori, M. F., Carvalho, I., & Sakamoto-Hojo, E.T. (2022). Neuroprotective effects of cholinesterase inhibitors: Current scenario in therapies for Alzheimer's disease and future perspectives. *Journal of Alzheimer's Disease Reports, 6*(1), 177–193. DOI: 10.3233/ADR-210061.
53. Çinar, E. et al. (2022). Donepezil-loaded PLGA-b-PEG nanoparticles enhance the learning and memory function of beta-amyloid rat model of Alzheimer's disease. *Archieves of Neuropsychiatry, 59*(4), 281. DOI: 10.29399/NPA.28275.
54. Zheng, H., Niu, S., Zhao, H., Li, S., & Jiao, J. (2018). Donepezil improves the cognitive impairment in a tree shrew model of Alzheimer's disease induced by amyloid-β1-40 via activating the BDNF/TrkB signal pathway. *Metabolic Brain Disease, 33*(6), 1961–1974. DOI: 10.1007/S11011-018-0303-6.
55. Talib, L. L., Hototian, S. R., Joaquim, H. G., Forlenza, O. V., & Gattaz, W. F. (2015). Increased iPLA2 activity and levels of phosphorylated GSK3B in platelets are associated with donepezil treatment in Alzheimer's disease patients. *European Archives of Psychiatry and Clinical Neuroscience, 265*(8), 701–706. DOI: 10.1007/S00406-015-0600-6.
56. Benfante, R., Di Lascio, S., Cardani, S., & Fornasari, D. (2021). Acetylcholinesterase inhibitors targeting the cholinergic anti-inflammatory pathway: A new therapeutic perspective in aging-related disorders. *Aging Clinical and Experimental Research, 33*(4), 823–834. DOI: 10.1007/S40520-019-01359-4.
57. Sharma, K. (2019). Cholinesterase inhibitors as Alzheimer's therapeutics (Review). *Molecular Medicine Reports, 20*(2), 1479–1487. DOI: 10.3892/MMR.2019.10374.
58. Yao, H. et al. (2021). Discovery of novel tacrine-pyrimidone hybrids as potent dual AChE/GSK-3 inhibitors for the treatment of Alzheimer's disease. *Journal of Medicinal Chemistry, 64*(11), 7483–7506. DOI: 10.1021/ACS.JMEDCHEM.1C00160/SUPPL_FILE/JM1C00160_SI_002.CSV.
59. Bayraktar, G. et al. (2024). Novel multifunctional tacrine–donepezil hybrids against Alzheimer's disease: Design synthesis and bioactivity studies. *Arch. Pharm. (Weinheim)*.. DOI: 10.1002/ARDP.202300575.

60. Mishra, A. K., & Panda, G. (2019). Anti-cholinesterase hybrids as multi-target-directed ligands against Alzheimer's disease (1998–2018). *Bioorganic & Medicinal Chemistry, 27*(6), 895–930. DOI: 10.1016/J.BMC.2019.01.025.
61. Dogga, B. et al. (2022). Design, synthesis and SAR studies of novel tacrine derivatives as potent cholinesterase inhibitors. *European Journal of Medicinal Chemistry Reports, 6*, 100094. DOI: 10.1016/J.EJMCR.2022.100094.
62. Remya, C. et al. (2021). Neuroprotective derivatives of tacrine that target NMDA receptor and acetyl cholinesterase-design, synthesis and biological evaluation-NC-ND license (http://creative-commons.org/licenses/by-nc-nd/4.0/). *Computational and Structural Biotechnology Journal, 19*, 4517–4537. DOI: 10.1016/j.csbj.2021.07.041.
63. Shimohama, S. (2018). Roles of nicotinic acetylcholine receptors in the pathology and treatment of Alzheimer's and Parkinson's diseases. DOI: 10.1007/978-981-10-8488-1.
64. Buckingham, S. D., Jones, A. K., Brown, L. A., & Sattelle, D. B.. (2009). Nicotinic acetylcholine receptor signalling: Roles in alzheimer's disease and amyloid neuroprotection. *Pharmacological Reviews, 61*(1), 39–61. DOI: 10.1124/PR.108.000562.
65. Papke, R. L., & Horenstein, N. A.. (2021). Therapeutic targeting of α7 nicotinic acetylcholine receptors. *Pharmacological Reviews, 73*(3), 1118–1149. DOI: 10.1124/PHARMREV.120.000097.
66. Takata, K. et al. (2022). Nicotinic acetylcholine receptors and microglia as therapeutic and imaging targets in alzheimer's disease. *Molecules, 27*(9), 2780. DOI: 10.3390/MOLECULES27092780.
67. Roberts, J. P., Stokoe, S. A., Sathler, M. F., Nichols, R. A., & Kim, S. (2021). Selective coactivation of α7-and α4β2-nicotinic acetylcholine receptors reverses beta-amyloid-induced synaptic dysfunction. *Journal of Biological Chemistry, 296*, 100402. DOI: 10.1016/J.JBC.2021.100402/ATTACHMENT/55760FCB-931F-4205-895D-CDB298C00023/MMC1.DOCX.
68. Lv, J. et al. (2023). Alpha7 nicotinic acetylcholine receptor agonist PHA-543613 improves memory deficits in presenilin 1 and presenilin 2 conditional double knockout mice. *Experimental Neurology, 359*, 114271. DOI: 10.1016/J.EXPNEUROL.2022.114271.
69. Sadigh-Eteghad, S., Mahmoudi, J., Babri, S., & Talebi, M. (2015). Effect of alpha-7 nicotinic acetylcholine receptor activation on beta-amyloid induced recognition memory impairment. Possible role of neurovascular function. *Acta Cirurgica Brasileira, 30*(11), 736–742. DOI: 10.1590/S0102-865020150110000003.
70. Potasiewicz, A., Faron-Gorecka, A., Popik, P., & Nikiforuk, A. (2021). Repeated treatment with alpha 7 nicotinic acetylcholine receptor ligands enhances cognitive processes and stimulates Erk1/2 and arc genes in rats. *Behavioural Brain Research, 409*, 113338. DOI: 10.1016/J.BBR.2021.113338.
71. Li, H. et al. (2021). JWX-A0108, a positive allosteric modulator of α7 nAChR, attenuates cognitive deficits in APP/PS1 mice by suppressing NF-κB-mediated inflammation. *International Immunopharmacology, 96*, 107726. DOI: 10.1016/J.INTIMP.2021.107726.
72. Potasiewicz, A., Krawczyk, M., Gzielo, K., Popik, P., & Nikiforuk, A. (2020). Positive allosteric modulators of alpha 7 nicotinic acetylcholine receptors enhance

procognitive effects of conventional anti-Alzheimer drugs in scopolamine-treated rats. *Behavioural Brain Research, 385*, 112547. DOI: 10.1016/J.BBR.2020.112547.

73. Fonar, G. et al. (2021). Modified snake α-neurotoxin averts β-amyloid binding to α7 nicotinic acetylcholine receptor and reverses cognitive deficits in Alzheimer's disease mice. *Molecular Neurobiology, 58*(5), 2322–2341. DOI: 10.1007/S12035-020-02270-0/FIGURES/10.
74. Chang, K. W. et al. (2018). Activation of α7 nicotinic acetylcholine receptor alleviates Aβ1-42-induced neurotoxicity via downregulation of p38 and JNK MAPK signaling pathways. *Neurochemistry Internatiional, 120*, 238–250. DOI: 10.1016/J.NEUINT.2018.09.005.
75. Tsang, S. W. Y. et al. (2006). Impaired coupling of muscarinic M1 receptors to G-proteins in the neocortex is associated with severity of dementia in Alzheimer's disease. *Neurobiol. Aging, 27*, no. 9, 1216–1223. DOI: 10.1016/J.NEUROBIOLAGING.2005.07.010.
76. Fisher, A. (2007). M1 muscarinic agonists target major hallmarks of Alzheimer's disease—An update. *Current Alzheimer's Research, 4*(5), 577–580. DOI: 10.2174/156720507783018163.
77. Caccamo, A. et al. (2006). M1 receptors play a central role in modulating AD-like pathology in transgenic mice. *Neuron, 49*(5), 671–682. DOI: 10.1016/J.NEURON.2006.01.020.
78. Abd-Elrahman, K. S., Colson, T. L. L., Sarasija, S., & Ferguson, S. S. G. (2024). A M1 muscarinic acetylcholine receptor-specific positive allosteric modulator VU0486846 reduces neurogliosis in female Alzheimer's mice. *Biomedicine & Pharmacotherapy, 173*, 116388. DOI: 10.1016/J.BIOPHA.2024.116388.
79. Palacharla, V. R. C. et al. (2023). SUVN-I6107: A true positive allosteric modulator (PAM) at muscarinic M1 receptors for the treatment of cognitive disorders. *Alzheimer's & Dementia, 19*(S21), e074091. DOI: 10.1002/ALZ.074091.
80. Russell, J. K., Ingram, S. M., Teal, L. B., Lindsley, C. W., & Jones, C. K. (2023). M1/M4-preferring muscarinic cholinergic receptor agonist xanomeline reverses wake and arousal deficits in nonpathologically aged mice. *ACS Chemical Neurosciences, 14*(3), 435–457. DOI: 10.1021/ACSCHEMNEURO.2C00592/SUPPL_FILE/CN2C00592_SI_004.PDF.
81. Johnson, C. R., Kangas, B. D., Jutkiewicz, E. M., Bergman, J., & Coo, A. (2022). Drug design targeting the muscarinic receptors and the implications in central nervous system disorders. *Biomedicnes, 10*(2), 398. DOI: 10.3390/BIOMEDICINES10020398.
82. Yohn, S. E., & Conn, J. (2018). Positive allosteric modulation of M1 and M4 muscarinic receptors as potential therapeutic treatments for schizophrenia. *Neuropharmacology, 136*, 438. DOI: 10.1016/J.NEUROPHARM.2017.09.012.
83. Nguyen, H. T. M. et al. (2022). Opportunities and challenges for the development of M1 muscarinic receptor positive allosteric modulators in the treatment for neurocognitive deficits. *British Journal of Pharmacology*. DOI: 10.1111/BPH.15982.
84. Dwomoh, L. et al. (2021). Activation of M1 muscarinic receptors reduce pathology and slow progression of neurodegenerative disease. *bioRxiv*. DOI: 10.1101/2021.07.30.454298.

85. Moran, S. et al. (2019). Biased M1 receptor–positive allosteric modulators reveal role of phospholipase D in M1-dependent rodent cortical plasticity. *Science Signaling, 12*(610). DOI: 10.1126/SCISIGNAL.AAX2057.
86. Puri, V., Wang, X., Vardigan, J. D., Kuduk, S. D., & Uslaner, J. M. (2015). The selective positive allosteric M1 muscarinic receptor modulator PQCA attenuates learning and memory deficits in the Tg2576 Alzheimer's disease mouse model. *Behavioural Brain Research, 287*, 96–99. DOI: 10.1016/J.BBR.2015.03.029.
87. Davoren, J. E. et al. (2017). Design and synthesis of γ- and δ-lactam M1 positive allosteric modulators (PAMs): Convulsion and cholinergic toxicity of an M1-selective PAM with weak agonist activity. *Journal of Medicinal Chemistry, 60*(15), 6649–6663. DOI: 10.1021/ACS.JMEDCHEM.7B00597.
88. Rook, J. M. et al. (2017). Diverse effects on M1 signaling and adverse effect liability within a series of M1 Ago-PAMs. *ACS Chemical Neurosciences, 8*(4), 866. DOI: 10.1021/ACSCHEMNEURO.6B00429.
89. Tiepolt, S. et al. (2022). PET imaging of cholinergic neurotransmission in neurodegenerative disorders. DOI: 10.2967/jnumed.121.263198.
90. Xia, Y. et al. (2022). Reduced cortical cholinergic innervation measured using [18F]-FEOBV PET imaging correlates with cognitive decline in mild cognitive impairment. *Neuroimage, 34*. DOI: 10.1016/J.NICL.2022.102992.
91. Aghourian, M. et al. (2017). Quantification of brain cholinergic denervation in Alzheimer's disease using PET imaging with [18F]-FEOBV. *Molecular Psychiatry, 22*(11), 1531–1538. DOI: 10.1038/MP.2017.183.
92. Toyohara, J. et al. (2009). Preclinical and the first clinical studies on [11C]CHIBA-1001 for mapping alpha7 nicotinic receptors by positron emission tomography. *Annals of Nuclear Medicine, 23*(3), 301–309. DOI: 10.1007/S12149-009-0240-X.
93. van Waarde, A., Marcolini, S., de Deyn, P. P., & Dierckx, R. A. J. O. (2021). PET agents in dementia: An overview. *Seminars in Nuclear Medicine, 51*(3), 196–229. DOI: 10.1053/J.SEMNUCLMED.2020.12.008.
94. Ogawa, M. et al. (2010). Synthesis and evaluation of new imaging agent for central nicotinic acetylcholine receptor α7 subtype. *Nuclear Medicine and Biology, 37*(3), 347–355. DOI: 10.1016/J.NUCMEDBIO.2009.11.007.
95. Ozenil, M. et al. (2021). Update on PET tracer development for muscarinic acetylcholine receptors. *Pharmaceuticals, 14*(6). DOI: 10.3390/PH14060530.
96. Tong, L. et al. (2020). Discovery of [11C]MK-6884: A positron emission tomography (PET) imaging agent for the study of m4muscarinic receptor positive allosteric modulators (PAMs) in neurodegenerative diseases. *Journal of Medicinal Chemistry, 63*(5), 2411–2425. DOI: 10.1021/ACS.JMEDCHEM.9B01406/SUPPL_FILE/JM9B01406_SI_002.CSV.
97. Li, W. et al. (2022). The PET tracer [11C]MK-6884 quantifies M4 muscarinic receptor in rhesus monkeys and patients with Alzheimer's disease. *Science Translational Medicine, 14*(627). DOI: 10.1126/SCITRANSLMED.ABG3684.
98. Rowe, C. C. et al. (2021). PET Imaging of brain muscarinic receptors with 18F-fluorobenzyl-dexetimide: A first in human study. *Psychiatry Research Neuroimaging, 316*. DOI: 10.1016/J.PSCYCHRESNS.2021.111354.
99. Arjmandi-Rad, S., Vestergaard Nieland, J. D., Goozee, K. G., & Vaseghi, S. (2023). The effects of different acetylcholinesterase inhibitors on EEG patterns in patients

with Alzheimer's disease: A systematic review. *Neurological Sciences, 45*(2), 417–430. DOI: 10.1007/S10072-023-07114-Y.

100. Jeong, J. (2004). EEG dynamics in patients with Alzheimer's disease. *Clinical Neurophysiology, 115*(7), 1490–1505. DOI: 10.1016/J.CLINPH.2004.01.001.
101. Cassani, R., Estarellas, M., San-Martin, R., Fraga, F. J., & Falk, T. H. (2018). Systematic review on resting-state EEG for Alzheimer's disease diagnosis and progression assessment. *Disease Markers*. DOI: 10.1155/2018/5174815.
102. Baakman, A. C. et al. (2022). Acute response to cholinergic challenge predicts long-term response to galantamine treatment in patients with Alzheimer's disease. *British Journal of Clinical Pharmacology, 88*(6), 2814–2829. DOI: 10.1111/BCP.15206.
103. Moretti, D. V., Frisoni, G. B., Giuliano, B., & Zanetti, O. (2014). Comparison of the effects of transdermal and oral rivastigmine on cognitive function and EEG markers in patients with Alzheimer's disease. *Frontiers in Aging Neurosciences, 6*, 86723. DOI: 10.3389/FNAGI.2014.00179/BIBTEX.
104. Reeves, R. R., Struve, F. A., & Patrick, G. (2002). The effects of donepezil on quantitative EEG in patients with Alzheimer's disease. *Clinical Electroencephalography, 33*(2), 93–96. DOI: 10.1177/155005940203300209.
105. Chouliaras, L., & O'Brien, J. T. (2023). The use of neuroimaging techniques in the early and differential diagnosis of dementia. *Molecular Psychiatry, 28*(10), 4084–4097. DOI: 10.1038/s41380-023-02215-8.
106. Bohnen, N. I., Grothe, M. J., Ray, N. J., Müller, M. L. T. M., & Teipel, S. J. (2018). Recent advances in cholinergic imaging and cognitive decline—Revisiting the cholinergic hypothesis of dementia. *Current Geriatrics Reports, 7*(1), 1. DOI: 10.1007/S13670-018-0234-4.
107. Kilimann, I. et al. (2014). Subregional basal forebrain atrophy in Alzheimer's disease: A multicenter study. *Journal of Alzheimer's Disease, 40*(3), 687. DOI: 10.3233/JAD-132345.
108. Craig, C. E., Ray, N. J., Müller, M. L. T. M., & Bohnen, N. I. (2020). New developments in cholinergic imaging in Alzheimer and lewy body disorders. *Current Behavioral Neuroscience Reports, 7*(4), 278. DOI: 10.1007/S40473-020-00221-6.
109. Jethwa, K. D., Dhillon, P., Meng, D., & Auer, D. P. (2019). Are linear measurements of the nucleus basalis of meynert suitable as a diagnostic biomarker in mild cognitive impairment and Alzheimer disease?. *Americal Journal of Neuroradiology, 40*(12), 2039. DOI: 10.3174/AJNR.A6313.
110. Cavedo, E. et al. (2017). Reduced basal forebrain atrophy progression in a randomized Donepezil trial in prodromal Alzheimer's disease. *Scientific Reports, 7*(1). DOI: 10.1038/S41598-017-09780-3.
111. Grothe, M. J., Heinsen, H., Amaro, E., Grinberg, L. T., & Teipel, S. J. (2016). Cognitive correlates of basal forebrain atrophy and associated cortical hypometabolism in mild cognitive impairment. *Cerebral Cortex, 26*(6), 2411. DOI: 10.1093/CERCOR/BHV062.
112. Gargouri, F. et al. (2019). Multimodal magnetic resonance imaging investigation of basal forebrain damage and cognitive deficits in Parkinson's disease. *Movement Disorders, 34*(4), 516. DOI: 10.1002/MDS.27561.
113. Atasoy, D., & Sternson, S. M. (2018). Chemogenetic tools for causal cellular and neuronal biology. *Physiological Review, 98*(1), 391. DOI: 10.1152/PHYSREV.00009.2017.

114. Mirzayi, P., Shobeiri, P., Kalantari, A., Perry, G., & Rezaei, N. (2022). Optogenetics: Implications for Alzheimer's disease research and therapy. *Molecular Brain, 15*(1). DOI: 10.1186/S13041-022-00905-Y.
115. Li, X. et al. (2017). Generation of a whole-brain atlas for the cholinergic system and mesoscopic projectome analysis of basal forebrain cholinergic neurons. *Proceedings of the. National Academy Science USA, 115*(2), 415–420. DOI: 10.1073/PNAS.1703601115/-/DCSUPPLEMENTAL.
116. Gu, Z., Alexander, G. M., Dudek, S. M., & Yakel, J. L. (2017). Hippocampus and entorhinal cortex recruit cholinergic and NMDA receptors separately to generate hippocampal theta oscillations. *Cell Reports, 21*(12), 3585. DOI: 10.1016/J.CELREP.2017.11.080.
117. Gu, Z., & Yakel, J. L. (2011). Timing-dependent septal cholinergic induction of dynamic hippocampal synaptic plasticity. *Neuron, 71*(1), 155. DOI: 10.1016/J.NEURON.2011.04.026.
118. Ballinger, E. C., Ananth, M., Talmage, D. A., & Role, L. W. (2016). Basal forebrain cholinergic circuits and signaling in cognition and cognitive decline. *Neuron, 91*(6), 1199. DOI: 10.1016/J.NEURON.2016.09.006.
119. Su, L., Blamire, A. M., Watson, R., He, J., Hayes, L., & O'Brien, J. T. (2016). Whole-brain patterns of 1H-magnetic resonance spectroscopy imaging in Alzheimer's disease and dementia with Lewy bodies. *Translational Psychiatry, 6*(8), e877. DOI: 10.1038/TP.2016.140.
120. Liu, W. et al. (2022). Chemical genetic activation of the cholinergic basal forebrain hippocampal circuit rescues memory loss in Alzheimer's disease. *Alzheimer's Research Therapy, 14*(1). DOI: 10.1186/S13195-022-00994-W.
121. Yuan, P., & Grutzendler, J. (2016). Attenuation of β-amyloid deposition and neurotoxicity by chemogenetic modulation of neural activity. *Journal of Neuroscience, 36*(2), 632. DOI: 10.1523/JNEUROSCI.2531-15.2016.

CHAPTER 3

Current Progress in Monoamine Oxidase Inhibitors as Anti-Alzheimer's: An Understanding into Therapeutic Prospectives of Inhibitors

REENA GUPTA[1,*], JITENDRA GUPTA[1], and ITTISHREE[1,2]

[1]*Institute of Pharmaceutical Research, GLA University, Mathura, Uttar Pradesh, India*

[2]*SRM Modinagar College of Pharmacy, SRMIST Delhi-NCR Campus, Modinagar, Ghaziabad, Uttar Pradesh, India*

*Corresponding author

ABSTRACT

Monoamine oxidase inhibitors (MAOIs) have been recognized for their potential in treating neurological ailments, including Alzheimer's disease (AD). MAOIs, which inhibit monoamine oxidase enzymes that break down neurotransmitters like serotonin, dopamine, and norepinephrine, have recently emerged as a potential avenue for alleviating cognitive decline and behavioral symptoms associated with AD. Recent research has focused on understanding the therapeutic potential of MAOIs, particularly their ability to combat neuroinflammation, oxidative stress, and the accumulation of amyloid-beta, all of which are hallmarks of AD pathology. Additionally, MAOIs have been found to have neuroprotective effects, enhance synaptic plasticity, and play a crucial role in maintaining cognitive function. Nevertheless, there are challenges in optimizing MAOIs' efficacy and safety profile, selectivity for specific MAO isoforms, and potential adverse effects, such as hypertensive crisis. Nonetheless, ongoing preclinical and

Cellular and Molecular Mechanisms in Alzheimer's Disease Therapeutics, Volume 2: Molecular Mechanisms. Prashant Tiwari, Sunil Kumar Kadiri, and Kunjbihari Sulakhiya (Eds)

DOI: 10.1201/9781779641816-3

clinical investigations hold promise for developing novel MAOIs with improved therapeutic outcomes for AD patients. This chapter emphasizes the importance of further research into MAOIs as a viable therapeutic avenue for AD, highlighting their multifaceted pharmacological actions and potential benefits in mitigating the complex pathophysiology of the disease.

3.1 INTRODUCTION

Alzheimer's disease (AD) is a brain disorder that progressively affects memory, cognition, and behavior, and is a part of the aging process. It is the predominant cause of dementia, accounting for 60–80% of all dementia cases. Although a definitive cure remains elusive, ongoing research sheds light on the disease's complex biology, offering hope for future therapeutic interventions. Additionally, women have a higher susceptibility to developing Alzheimer's compared to men. While the longer lifespan of women may contribute to this difference, researchers are investigating how genetic variations might influence the threat of Alzheimer's differently in men and women [1].

3.2 BRIEF OVERVIEW OF ALZHEIMER'S DISEASE

AD is a progressive neurological disorder that leads to an incremental decrease in cognitive function and the ability to carry out daily activities. Symptoms typically start to appear in individuals with late-onset Alzheimer's in their mid-60s, indicating the beginning of cognitive drop. On the other hand, early-onset Alzheimer's, although less common, can manifest between a person's 30s and mid-60s, presenting unique challenges due to its occurrence at a younger age.

As Alzheimer's progresses, individuals may experience memory loss, confusion, disorientation, difficulties in problem-solving, and changes in behavior and personality. These symptoms profoundly impact daily life, often leading to dependency on others for primary care [2]. It's essential to recognize that Alzheimer's is not simply a part of aging. This debilitating condition primarily affects quality of life and independence. However, age is not the only risk factor. Encouragingly, ongoing research is steadily uncovering various lifestyle, hereditary, and environmental factors contributing to the disease's development and progression.

In the complexity of Alzheimer's, a comprehensive approach is necessary for diagnosis, management, and potential treatment. Early markers and interventions are essential for improving results and enhancing the well-being of both patients and their caregivers [3, 4].

3.3 IMPORTANCE OF MONOAMINE OXIDASE INHIBITORS (MAOIS) IN ALZHEIMER'S TREATMENT

Monoamine oxidase (MAO) is a crucial enzyme primarily found within the mitochondria and is notably abundant in neuronal and gastrointestinal tissues. This enzyme consists of two isoforms, MAO-A and MAO-B, which have consequential series homology but vary in their substrate-binding and inhibitor recognition areas and tissue dispersal patterns. MAO-A and MAO-B are essential in promoting the oxidative deamination of a diverse array of monoamines, thereby regulating neurotransmitter levels in the nervous system and contributing to various physiological processes. Monoamine oxidase inhibitors (MAOIs) have emerged as a promising approach in the treatment of AD, offering several potential benefits (Figure 3.1) [5].

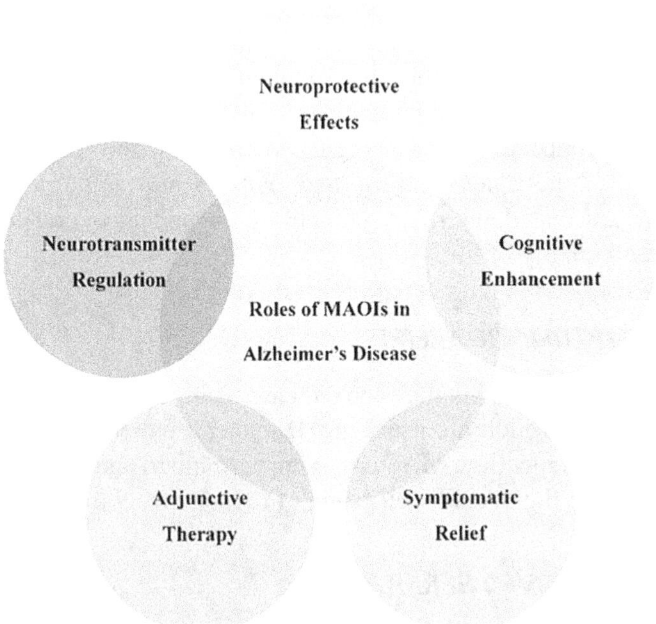

FIGURE 3.1 Roles of monoamine oxidase inhibitors in the treatment of AD.

3.3.1 NEUROTRANSMITTER REGULATION

MAOIs inhibit the activity of MAO and are accountable for damaging down neurotransmitters, like serotonin, dopamine, and norepinephrine. By blocking MAO activity, these remedies improve the availability of these neurotransmitters in the brain. This modulation can potentially enhance synaptic transmission and neuronal function, which may help alleviate cognitive deficits associated with AD.

3.3.2 NEUROPROTECTIVE EFFECTS

Research suggests that MAOIs may have neuroprotective properties by reducing oxidative stress and inflammation in the brain. AD is characterized by the expansion of oxidative damage and neuroinflammation, contributing to neuronal degeneration. By inhibiting MAO activity, MAOIs may help mitigate these pathological processes, thus potentially protecting against neuronal loss and disease progression.

3.3.3 COGNITIVE ENHANCEMENT

Increasing neurotransmitter levels in the brain through MAOI therapy may lead to cognitive enhancement in Alzheimer's patients. Serotonin, dopamine, and norepinephrine are involved in various cognitive functions, like recollection, concentration, and organizational function. By modulating the activity of these neurotransmitters, MAOIs may improve cognitive symptoms associated with AD, potentially enhancing overall cognitive function and quality of life.

3.3.4 SYMPTOMATIC RELIEF

MAOIs don't cure AD, but they can reduce specific symptoms like depression and apathy, which are usual in Alzheimer's patients. By enhancing neurotransmitter function, MAOIs can improve mood and motivation, and in turn, improve the overall well-being of individuals with AD.

3.3.5 ADJUNCTIVE THERAPY

MAOIs are often used as adjunctive therapy alongside other medications for AD, like cholinesterase inhibitors or NMDA receptor antagonists.

Combining different classes of drugs may offer synergistic effects and provide comprehensive symptom management for Alzheimer's patients, addressing multiple aspects of the disease simultaneously.

MAOIs have shown promise in treating AD. However, their use should be carefully observed due to the possibility of side effects and interactions with different medicines. Further research is necessary to better understand MAOIs' effectiveness and safety in treating AD. Nonetheless, MAOIs are a valuable therapeutic option for addressing the complex pathology of this condition [4–7].

3.4 UNDERSTANDING MONOAMINE OXIDASE INHIBITORS (MAOIS)

MAO is an essential enzyme that breaks down monoamine neurotransmitters, like norepinephrine, serotonin, and dopamine. These neurotransmitters regulate behavior, cognition, and mood. MAO helps in the oxidative deamination of monoamines, converting them into their respective metabolites.

MAO enzymes have MAO-A and MAO-B isoforms, have distinct preferences for substrates, and are found in various tissues. MAO-A has a greater affinity for norepinephrine and serotonin, primarily found in the gastrointestinal tract and specific brain regions. MAO-B prefers to metabolize phenylethylamine and benzylamine and is more abundant in the brain, especially in areas with many dopaminergic neurons. Dysregulation of MAO activity has been linked to various neuropsychiatric disorders. For example, increased MAO-A activity has been found in people with depression, while decreased MAO-B activity is associated with Parkinson's disease (PD). In AD, changes in MAO activity may contribute to the upsurge of amyloid-beta plaques and neurofibrillary knots, two vital pathological features of the disorder.

Pharmacological inhibitors of MAO have been invented as therapeutic agents for the therapy of depression, PD, and other psychiatric conditions. By inhibiting MAO activity, these drugs boost the grades of monoamine neurotransmitters in the brain, alleviating manifestation associated with neurotransmitter imbalances. However, the use of MAO inhibitors requires careful monitoring because of potential side effects and drug interactions. Certain foods and medications can interact with MAO inhibitors, leading to hypertensive crises or serotonin syndrome. Therefore, patients prescribed with MAO inhibitors must adhere to dietary and medication restrictions to minimize these risks. Despite these challenges, MAO inhibitors remain valuable therapeutic options for managing neuropsychiatric disorders characterized by monoamine dysregulation [8–10].

3.5 ROLE OF MAOIS IN ALZHEIMER'S DISEASE

MAO is an enzyme mainly found in mitochondria and is present in high levels in neuronal and gastrointestinal tissues. MAO-A and MAO-B are identical in structure while differing in their allocation within the body and their recognition areas for substrates and inhibitors. These enzymes are accountable for the oxidative deamination of different monoamines, which is essential for the metabolism of neurotransmitters. Changes in neurotransmitter levels in the brain resulting from irregular MAO activity have been linked to several neurodegenerative diseases, like PD, depression, and AD. Flavin-binding MAO helps in the oxidative deamination of monoamine analogues and neurotransmitters, such as adrenaline, noradrenaline, serotonin, dopamine, and others. This process generates neurotoxic molecules, like hydrogen peroxide, aldehydes, substituted amines, or ammonia [11, 12].

MAOs are a type of flavoenzyme oxidases that use molecular oxygen (O_2) to build hydrogen peroxide (H_2O_2) in the existence of a catalyst. The specific properties of the substrate molecule determine the substrate specificity of both the isoforms (MAO-A and MAO-B). Although these isoforms have a 73% succession resemblance, they are significantly distinct in their substrate-inhibitor particularity and tissue dispersal. MAO-A is more inclined toward significant amines, such as epinephrine, serotonin, and norepinephrine, although MAO-B primarily acts on more diminutive amines, like benzylamine and β-phenylethylamine. Dopamine and tyramine are typical substrates for both isoforms [13–15].

3.6 THERAPEUTIC PROSPECTS OF MAOIS IN ALZHEIMER'S DISEASE

MAOIs have tremendous therapeutic potential for treating AD, and researchers are currently studying them. MAOIs are generally used for treating depression and PD because they help increase neurotransmitter levels, like dopamine, norepinephrine, and serotonin in the brain achieved by inhibiting the enzyme MAO. However, their potential in AD stems from their broader neuroprotective effects and the ability to modulate neurotransmitter systems implicated in the disease's pathophysiology. Both MAO isoforms metabolize the substrates dopamine, tyramine, and tryptamine, and medications, such as clorgiline and selegiline, which act as MAO inhibitors, find applications in treating various neurodegenerative and neurological disorders. Specifically, MAO-A inhibitors are commonly employed in depression therapy, while MAO-B inhibitors are utilized in managing PD. There's ongoing research exploring the potential efficacy of MAO-B inhibitors in treating AD and therapeutic approaches, as shown in Figure 3.2 (A–F) [16].

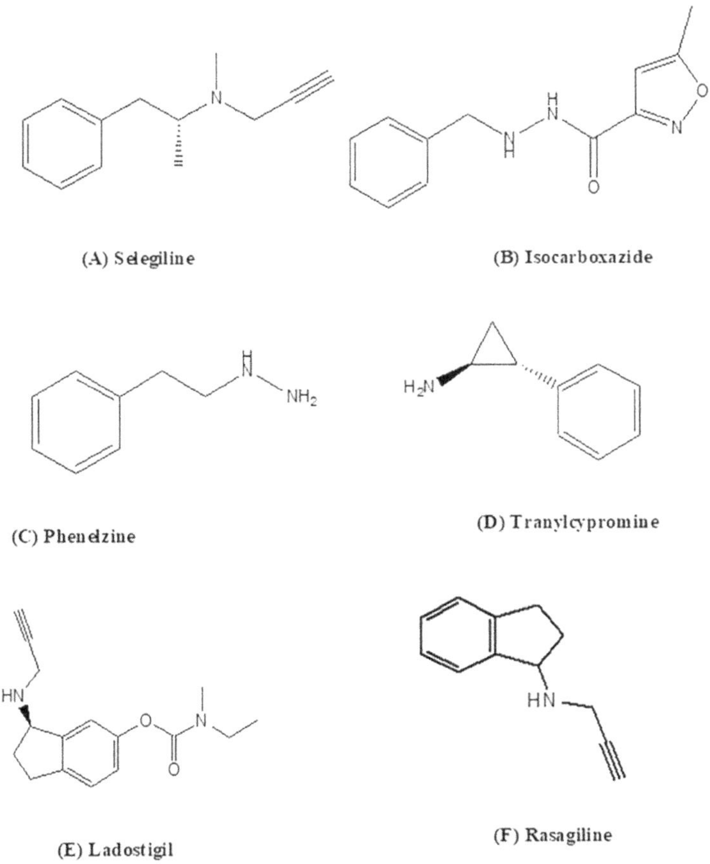

FIGURE 3.2 Monoamine oxidase inhibitors along with their structure.

As life expectancy rises, the incidence of neurodegenerative conditions like Parkinson's and Alzheimer's is also increasing, leading to a growing elderly population. Consequently, these diseases present significant societal health challenges, as affected individuals often require caregiving assistance due to declining cognitive and motor function. Addressing these conditions has become a pressing concern in healthcare systems worldwide, prompting continued research and innovation in treatment approaches [17–19].

3.7 ROLE OF HERBAL DRUGS AND NUTRACEUTICALS IN AD

Nutraceuticals and pharmacological interventions like MAOIs are becoming a potential supplementary therapy for AD (Figure 3.3) [16]. Nutraceuticals are functional nutrition or dietary supplements offering health benefits beyond essential nutrition. According to research, some of these supplements have shown promising results in modulating AD pathology and improving cognitive function [20]. Here's how specific nutraceuticals could complement the therapeutic prospects of MAOIs in AD:

3.7.1 CURCUMIN

Curcumin is a compound found in turmeric and is known for its potent anti-inflammatory and antioxidant properties. Research studies indicate that curcumin may help reduce the formation of amyloid-beta plaques and tau tangles, two characteristic pathological features of AD. Combining curcumin with MAOIs is believed to enhance neuroprotection and reduce neuroinflammation in individuals with AD [21].

3.7.2 OMEGA-3 FATTY ACIDS

Eicosapentaenoic acid (EPA) and docosahexaenoic acid (DHA) are the essential omega-3 fatty acids for good brain health. A report showed that these fatty acids can increase mental function and decrease the risk of cognitive decline in AD. When combined with MAOIs, omega-3 fatty acids may provide even better results by targeting multiple pathways involved in AD development.

3.7.3 RESVERATROL

Resveratrol, a polyphenol found in red grapes and berries, exhibits antioxidant and anti-inflammatory properties. Resveratrol may help protect against neuronal damage and cognitive decline in AD. Combining resveratrol with MAOIs could provide additional neuroprotective benefits and enhance therapeutic efficacy [22, 23].

3.7.4 GREEN TEA EXTRACT

Catechins are potent antioxidants of green tea that have been shown to protect against neurodegeneration and improve cognitive function in AD animal models. Combining green tea extract with MAOIs may help enhance neuroprotection and mitigate AD-related cognitive decline [23].

3.7.5 VITAMIN E

Vitamin E aids in defending against oxidative stress and neurodegeneration due to its powerful antioxidant action. Vitamin E supplements may help slow AD progression and improve cognitive function. Combining Vitamin E with MAOIs could enhance antioxidant defenses and promote neuroprotection in AD patients [20].

3.7.6 GINKGO BILOBA

Ginkgo biloba extract has traditionally been employed to enhance cognition and memory. Investigations suggest that *Ginkgo biloba* may help enhance cerebral blood flow, reduce oxidative stress, and improve neurotransmitter function in AD. Combining *Ginkgo biloba* with MAOIs could improve cognitive function and ameliorate behavioral manifestation interrelated with AD [24].

3.8 CHALLENGES AND LIMITATIONS

Cognitive and operational assessments, genotyping, positron emission tomography (PET), cerebrospinal fluid (CSF) imaging biomarkers, and magnetic resonance imaging (MRI) are valuable tools for stratifying patients based on their pathology and clinical stage of AD.

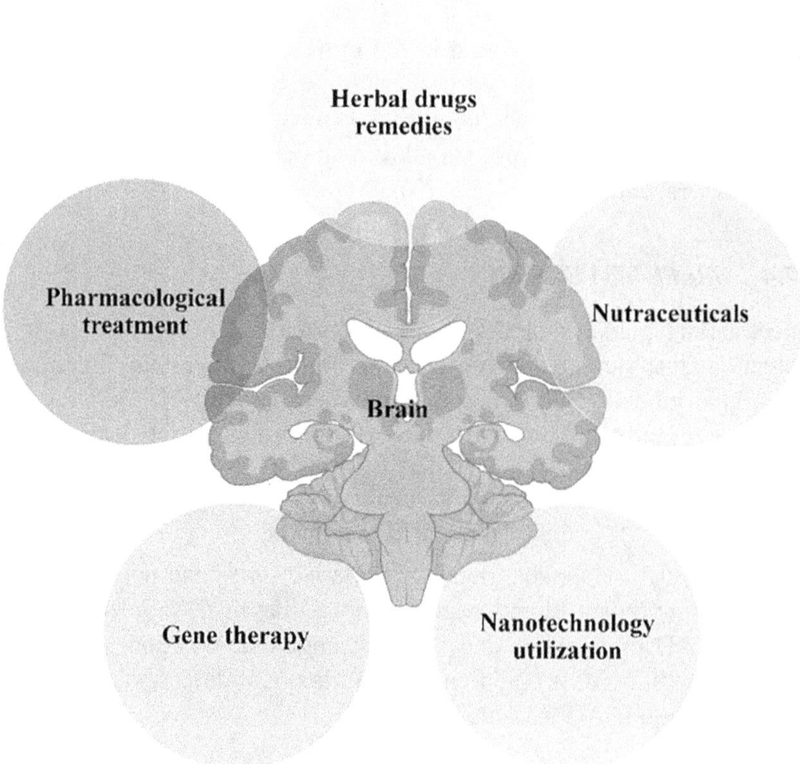

FIGURE 3.3 Treatment strategies of Alzheimer's disease.

3.8.1 *COGNITIVE AND FUNCTIONAL ASSESSMENTS*

These assessments evaluate many aspects of cognitive function, such as language, executive, memory function, and visuospatial skills, as well as functional abilities like activities of daily living. They provide insights into the severity of cognitive impairment and its impact on daily functioning.

3.8.2 *GENOTYPING*

Genetic testing can identify specific gene variants associated with AD, such as the apolipoprotein E (APOE) ε4 allele, an understood threat element for late-onset AD. Genotyping allows for threat review and personalized management strategies based on genetic susceptibility.

3.8.3 CSF AND PET IMAGING BIOMARKERS

CSF biomarkers like total tau (t-Tau), amyloid-beta 42 (Aβ42), and phosphorylated tau (p-Tau) provide information about the underlying pathology of AD, including amyloid plaques and tau tangles. PET imaging with radiotracers targeting the amyloid or tau entireties permits *in vivo* visualization and quantification of pathological changes in the brain.

3.8.4 MRI

Structural MRI delivers detailed photographs of the brain's anatomy, allowing for the assessment of hippocampal volume, cortical thickness, and white matter integrity, which are often altered in AD. Functional MRI (FMRI) can measure brain activity and connectivity patterns, providing insights into neural network dysfunction associated with AD.

These patient stratification methods can yield "dynamic" or "static" results. Dynamic measures, such as cognitive assessments and functional tests, may change over time as AD progresses, reflecting the evolving nature of the disease and its impact on cognition and behavior. In contrast, static measures, such as genotyping or specific imaging biomarkers, may remain relatively stable throughout the disease.

When interpreting these results and using them for patient stratification, it's essential to consider the dynamic nature of AD progression. Longitudinal assessments and integrating multiple biomarkers and imaging modalities can furnish a more exhaustive understanding of illness progression and help develop personalized treatment strategies tailored to individual patient needs [25–27].

AD is an advanced neurodegenerative ailment with a gradual onset of cognitive impairments that worsen over time. Current therapeutic approaches aim to slow disease progression rather than cure it. Given the insidious nature of AD, prevention strategies are crucial, especially considering that initial brain lesions develop during the preclinical stage and progress over many years.

Transporting drugs to the central nervous system (CNS) is challenging because of the blood-brain barrier (BBB), which limits the path of numerous implications from the bloodstream into the brain. Enhancing medicine delivery to the brain is essential for maximizing therapeutic benefits. In recent research, novel delivery systems, such as liposomes and exosomes have shown promise in effectively delivering drugs to the CNS. Among these delivery systems, intranasal (IN) administration is a promising method for AD treatment. IN administration offers several advantages, including rapid medicine delivery to the brain because of the proximity of nasal cavity to the CNS. Liposomes administered via the IN route can bypass the BBB, potentially improving the bioavailability and efficacy of the drug while protecting the drug from degradation in the periphery [27–29].

By improving drug delivery to the brain, particularly in AD, researchers aim to optimize treatment outcomes and potentially slow disease progression more effectively. Choosing the most appropriate drug delivery pathway, such as IN administration of liposomes, is critical for maximizing therapeutic efficiency and targeting specific brain regions affected by AD pathology. This approach holds promise for advancing the development of effective treatments for AD and other CNS disorders [30, 31].

3.9 FUTURE DIRECTIONS AND RESEARCH OPPORTUNITIES

In advanced stages of AD, behavioral symptoms such as aggression and agitation can become more pronounced, posing challenges for both patients and caregivers. While atypical antipsychotic medications may be considered in some cases, their use should be approached cautiously due to studies linking them to increased mortality in elderly dementia patients. Similarly, benzodiazepines should be used sparingly due to their potential to exacerbate agitation, as well as the risk of dependence and withdrawal. Non-pharmacological approaches play a crucial role in managing behavioral symptoms in AD. These may include maintaining a consistent and calming environment, avoiding triggers that may lead to agitation, and simplifying daily routines to minimize stressors. Additionally, addressing underlying conditions, such as depression and anxiety is essential, as these are common in AD patients and can significantly impact their quality of life.

Healthcare providers and caregivers must remain vigilant for signs of depression and anxiety in AD patients and provide appropriate treatment and support. Organizations like the Alzheimer's Association offer valuable resources, including informational sessions, support groups, and referral services, to assist patients and caregivers in managing the challenges associated with AD. Such resources can provide valuable support and guidance for individuals affected by the disease [32–35].

The emergence of new biomarkers has revolutionized our ability to diagnose AD at very early stages, and in some cases, even before symptoms manifest. However, despite advancements in early detection, therapeutic interventions aimed at treating symptomatic AD have thus far been unsuccessful in halting or reversing disease progression.

As a result, there is a growing consensus within the scientific community that remedy strategies for AD must be instigated as early as possible in the disease span. The rationale behind this approach is to intervene during the early stages of neurodegeneration before irreversible damage occurs in the brain. Early intervention offers several potential benefits as discussed in the following sections.

3.9.1 PRESERVATION OF BRAIN FUNCTION

By targeting AD pathology at its earliest stages, there is a greater likelihood of preserving cognitive function and delaying the onset of debilitating symptoms.

3.9.2 MAXIMIZING THERAPEUTIC EFFICACY

Treatment interventions may be more effective when initiated before extensive neuronal damage, potentially increasing the likelihood of therapeutic success.

3.9.3 PREVENTING DISEASE PROGRESSION

Early intervention may help prevent or slow the passage of AD pathology, thereby reducing the overall burden of the disease on individuals and society.

3.9.4 IMPROVING QUALITY OF LIFE

Delaying the onset of cognitive decline and functional impairment can significantly enhance the quality of vitality for individuals affected by AD and their guardians.

Ongoing research is needed to identify effective therapeutic targets and develop interventions that can modify the underlying disease processes of AD to achieve successful early intervention that involves targeting key pathological hallmarks, such as amyloid-beta and tau protein accumulation and addressing other factors contributing to neurodegeneration, such as neuroinflammation and oxidative stress.

Aside from making lifestyle changes, there are other ways to intervene early in AD, such as using medication and non-pharmacological methods like cognitive stimulation and physical exercise. Early intervention is a promising approach to fighting AD, as it can lead to more effective treatments and better outcomes for those who are affected by this devastating condition [36–40]. Research is vital to clarifying the causal links between potential novel hazard elements and AD, comprehending underlying mechanisms, and exploring their association with AD neuropathogenesis. Clarifying causal directions, identifying underlying mechanisms, and assessing clinical implications are key research priorities [41].

Numerous research opportunities in AD can advance our understanding and treatment of the condition and explore new targets beyond amyloid and tau, such as neuroinflammation and synaptic dysfunction, for drug development, developing innovative drug delivery systems to improve drug penetration across the BBB, and enhancing therapeutic efficacy. Leveraging machine learning and AI algorithms to analyze large-scale datasets can improve diagnosis, prognosis, and treatment prediction [42, 43].

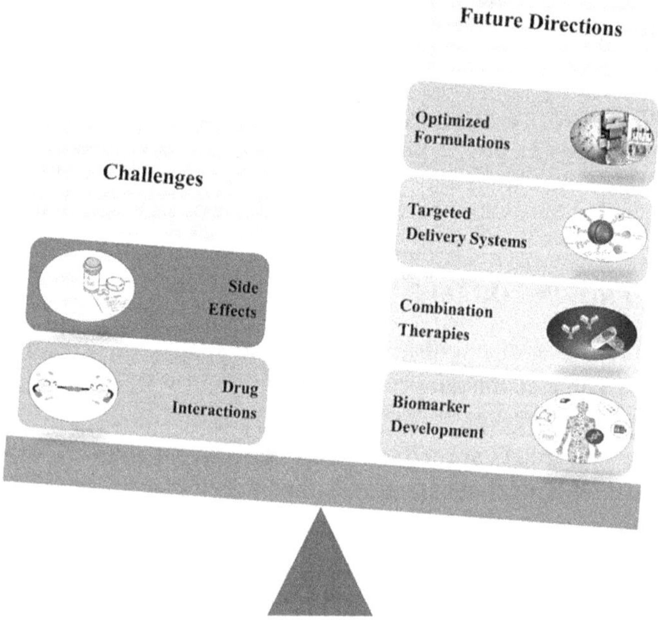

FIGURE 3.4 Challenges and future directions associated with MAOIs.

Investigating the synergistic effects of combining pharmacological and non-pharmacological interventions for comprehensive AD management is essential. Conducting well-designed clinical trials and promoting collaboration between academia, government agencies, and industries is essential to encourage translation into clinical practice. Addressing ethical, legal, and societal considerations related to AD research, including patient autonomy, informed consent, and access to treatment is critical. Exploring these research opportunities can significantly advance our ability to diagnose, treat, and ultimately prevent AD [5, 39, 40, 44–50].

3.10 CONCLUSION

The exhaustive investigation spotlights the prospect of MAOIs as a restoring alternative in AD treatment. MAOIs offer promise in addressing the neurodegenerative processes associated with AD, potentially enhancing cognitive function and slowing disease progression. However, the use of MAOIs in AD treatment also presents challenges and limitations, including side effects and drug interactions, which must be carefully managed (Figure 3.4). Despite these challenges, the therapeutic prospects of MAOIs in AD are encouraging, and further research is warranted to optimize their efficacy and safety. Future directions and research opportunities include exploring novel drug delivery systems, combination therapies, and precision medicine approaches to enhance the therapeutic potential of MAOIs in AD treatment. Finally, the study highlights the importance of continued research and innovation in seeking effective therapies for AD.

KEYWORDS

- **Alzheimer's disease**
- **dopamine**
- **monoamine oxidase inhibitors**
- **neurotransmitters**
- **norepinephrine**
- **neurodegeneration**
- **serotonin**

REFERENCES

1. Alzheimer's Association (Internet). (2024). Alzheimer's disease. https://www.alz.org/media/documents/alzheimers-dementia-about-alzheimers-disease-ts.pdf (accessed March 5, 2024).
2. National Institute on Aging (NIA) (Internet). (2021). Alzheimer's & related dementias?. https://www.nia.nih.gov/health/alzheimers-and-dementia/what-alzheimers-disease (accessed March 5, 2024).
3. National Institute on Aging (NIA) (Internet). (2024). Alzheimer's disease. https://www.alzheimers.gov/alzheimers-dementias/alzheimers-disease (accessed March 5, 2024).
4. Manzoor, S., & Hoda N. (2020). A comprehensive review of monoamine oxidase inhibitors as Anti-Alzheimer's disease agents: A review. *European Journal of Medicinal Chemistry, 206*, 112787. doi:10.1016/j.ejmech.2020.112787
5. National Institute on Aging (NIA) (Internet). (2024). Research on Alzheimer's disease and related dementias. https://www.alzheimers.gov/taking-action/research-activities (accessed May 2, 2024).
6. Nedic, G., Pivac, N., Hercigonja, D. K., Jovancevic, M., Curkovic, K. D., & Muck-Seler, D. (2010). Platelet monoamine oxidase activity in children with attention-deficit/hyperactivity disorder. *Psychiatry Research, 175*(3), 252–255. doi:10.1016/j.psychres.2009.08.013
7. Marziniak, M., Mössner, R., Benninghoff, J., Syagailo, Y. V., Lesch, K. P., & Sommer, C. (2004). Association analysis of the functional monoamine oxidase A gene promotor polymorphism in migraine. *Journal of Neural Transmission (Vienna), 111*(5), 603–609. doi:10.1007/s00702-004-0108-0
8. Youdim, M. B., Edmondson, D., & Tipton, K. F. (2006). The therapeutic potential of monoamine oxidase inhibitors. *Nature Review Neuroscience, 7*(4), 295–309. doi:10.1038/nrn1883
9. Tripathi, A. C., Upadhyay, S., Paliwal, S., & Saraf, S. K. (2018). Privileged scaffolds as MAO inhibitors: Retrospect and prospects. *European Journal of Medicinal Chemistry, 145*, 445–497. doi:10.1016/j.ejmech.2018.01.003
10. Ramsay, R. R. (2012). Monoamine oxidases: the biochemistry of the proteins as targets in medicinal chemistry and drug discovery. *Current Topics in Medicinal Chemistry, 12*(20), 2189–2209. doi:10.2174/156802612805219978
11. Livingston, M. G. (1995). Interactions with selective MAOIs. *Lancet, 345*(8949), 533–534. doi:10.1016/s0140-6736(95)90460-3
12. Hesselink, J. M. (1991). Safer use of MAOIs with nifedipine to counteract potential hypertensive crisis. *The American Journal of Psychiatry, 148*(11), 1616. doi:10.1176/ajp.148.11.1616a
13. Fox, H. H. (1952). The chemical approach to the control of tuberculosis. *Science, 116*(3006), 129–134. doi:10.1126/science.116.3006.129
14. Jo, S., Yarishkin, O., Hwang, Y. J., et al. (2014). GABA from reactive astrocytes impairs memory in mouse models of Alzheimer's disease. *Nature Medicine, 20*(8), 886–896. doi:10.1038/nm.3639
15. Mieo, W., Wang, H., Jia, W., et al. (2021). Monoamine oxidase inhibitors contribute to the treatment of Alzheimer's disease. Journal of Comprehensive Molecular Science

and Genetics, *2*(1), 1–11. https://mbgm.journals.publicknowledgeproject.org/index.php/mbgm/article/view/1398
16. Hassan, N. A., Alshamari, A. K., Hassan, A. A., et al. (2022). Advances on therapeutic strategies for alzheimer's disease: From medicinal plant to nanotechnology. *Molecules, 27*(15), 4839. doi:10.3390/molecules27154839
17. Calfio, C., Gonzalez, A., Singh, S. K., Rojo, L. E., & Maccioni, R. B. (2020). The emerging role of nutraceuticals and phytochemicals in the prevention and treatment of alzheimer's disease. *Journal of Alzheimers Disease, 77*(1), 33–51. doi:10.3233/JAD-200443
18. Shytle, R. D., Bickford, P. C., Rezai-zadeh, K., et al. (2009). Optimized turmeric extracts have potent anti-amyloidogenic effects. *Current Alzheimer Research, 6*(6), 564–571. doi:10.2174/156720509790147115
19. Özdemir, Z., Alagöz, M. A., Bahçecioğlu, Ö. F., & Gök, S. (2021). Monoamine oxidase-B (MAO-B) inhibitors in the treatment of alzheimer's and parkinson's disease. *Current Medicinal Chemistry, 28*(29), 6045–6065. doi:10.2174/0929867328666210203204710
20. Singh, S. K., Barreto, G. E., Aliev, G., & Echeverria, V. (2017). Ginkgo biloba as an alternative medicine in the treatment of anxiety in dementia and other psychiatric disorders. *Current Drug Metabolism, 18*(2), 112–119. doi:10.2174/1389200217666161201112206
21. Schweiger, S., Matthes, F., Posey, K., et al. (2017). Resveratrol induces dephosphorylation of Tau by interfering with the MID1-PP2A complex. *Scientific Reports, 7*(1), 13753. doi:10.1038/s41598-017-12974-4
22. Shati, A. A., & Alfaifi, M. Y. (2019). Trans-resveratrol inhibits Tau phosphorylation in the brains of control and cadmium chloride-treated rats by activating PP2A and PI3K/Akt induced-inhibition of GSK3β. *Neurochemistry Research, 44*(2), 357–373. doi:10.1007/s11064-018-2683-8
23. Dhouafli, Z., Cuanalo-Contreras, K., Hayouni, E. A., Mays, C. E., Soto, C., & Moreno-Gonzalez, I. (2018). Inhibition of protein misfolding and aggregation by natural phenolic compounds. *Cellular and Molecular Life Science, 75*(19), 3521–3538. doi:10.1007/s00018-018-2872-2
24. Dubois, B., Villain, N., Frisoni, G. B., et al. (2021). Clinical diagnosis of Alzheimer's disease: recommendations of the International Working Group. *Lancet Neurology, 20*(6), 484–496. doi:10.1016/S1474-4422(21)00066-1
25. Frisoni, G. B., Fox, N. C., Jack, C. R. Jr., Scheltens, P., & Thompson, P. M. (2010). The clinical use of structural MRI in Alzheimer disease. *Nature Reviews Neurology, 6*(2), 67–77. doi:10.1038/nrneurol.2009.215
26 Niemantsverdriet, E., Valckx, S., Bjerke, M., & Engelborghs, S. (2017). Alzheimer's disease CSF biomarkers: Clinical indications and rational use. *Acta Neurologic Belgica, 117*(3), 591–602. doi:10.1007/s13760-017-0816-5
27. Bandiwadekar, A., Jose, J., Khayatkashani, M., Habtemariam, S., Khayat Kashani, H. R., & Nabavi, S. M. (2022). Emerging novel approaches for the enhanced delivery of natural products for the management of neurodegenerative diseases. *Journal of Molecular Neuroscience, 72*(3), 653–676. doi:10.1007/s12031-021-01922-7
28. Yang, A., Kantor, B., & Chiba-Falek, O. (2021). *APOE*: The new frontier in the development of a therapeutic target towards precision medicine in late-onset

alzheimer's. *International Journal of Molecular Science, 22*(3), 1244. doi:10.3390/ijms22031244

29. Song, C., Shi, J., Zhang, P., et al. (2022). Immunotherapy for Alzheimer's disease: Targeting β-amyloid and beyond. *Translational Neurodegeneration, 11*(1), 18. doi:10.1186/s40035-022-00292-3
30. Wickner, R. B., Bezsonov, E. E., Son, M., Ducatez, M., DeWilde, M., & Edskes, H. K. (2028). Anti-prion systems in yeast and inositol polyphosphates. *Biochemistry, 57*(8), 1285–1292. doi:10.1021/acs.biochem.7b01285
31. Perrin, R. J., Fagan, A. M., & Holtzman, D. M. (2009). Multimodal techniques for diagnosis and prognosis of Alzheimer's disease. *Nature*, *461*(7266), 916–922. doi:10.1038/nature08538
32. Verghese, P. B., Castellano, J. M., & Holtzman, D. M. Apolipoprotein E in alzheimer's disease and other neurological disorders. *Lancet Neurology, 10*(3), 241–252. doi:10.1016/S1474-4422(10)70325-2
33. Mayeux, R. (2010). Clinical practice. Early Alzheimer's disease. The *New England Journal of Medicine, 362*(23), 2194–2201. doi:10.1056/NEJMcp0910236
34. McKhann, G. M., Knopman, D. S., Chertkow, H., & et al. (2011). The diagnosis of dementia due to Alzheimer's disease: Recommendations from the national institute on aging-alzheimer's association workgroups on diagnostic guidelines for Alzheimer's disease. *Alzheimers and Dementia, 7*(3), 263–269. doi:10.1016/j.jalz.2011.03.005
35. Haydar, S. N., & Dunlop, J. (2010). Neuronal nicotinic acetylcholine receptors - targets for the development of drugs to treat cognitive impairment associated with schizophrenia and Alzheimer's disease. *Current Topics in Medicinal Chemistry, 10*(2), 144–152. doi:10.2174/156802610790410983
36. Musiek, E. S., & Schindler, S. E. Alzheimer disease: Current concepts & future directions. *Missouri Medicine, 110*(5), 395–400.
37. Yaffe, K., Laffan, A. M., Harrison, S. L., et al. (2011). Sleep-disordered breathing, hypoxia, and risk of mild cognitive impairment and dementia in older women. *JAMA, 306*(6), 613–619. doi:10.1001/jama.2011.1115
38. Morris, J. C., Aisen, P. S., Bateman, R. J., et al. (2012). Developing an international network for alzheimer research: The dominantly inherited alzheimer network. *Clinics in Investigation (London)*, *2*(10), 975–984. doi:10.4155/cli.12.93
39. Arafah, A., Khatoon, S., Rasool, I., et al. (2023). The future of precision medicine in the cure of Alzheimer's disease. *Biomedicines*, *11*(2), 335. doi:10.3390/biomedicines11020335
40. Vrahatis, A. G., Skolariki, K., Krokidis, M. G., Lazaros, K., Exarchos, T. P., & Vlamos, P. (2023). Revolutionizing the early detection of alzheimer's disease through non-invasive biomarkers: The role of artificial intelligence and deep learning. *Sensors (Basel)*, *23*(9), 4184. doi:10.3390/s23094184
41. Alzheimer's Association (Internet). (2024). https://www.alz.org/ (accessed March 18, 2024).
42. Gupta, J., & Km, R. (2020). Nanotechnology approach intargeting and treatment of neurological disorders by overcome blood brain barrier. *International Journal of Advancesd Science and Technology, 29*(8), 2852–2870.
43. Bastianetto, S., Ramassamy, C., Doré, S., Christen, Y., Poirier, J., & Quirion R. The Ginkgo biloba extract (EGb 761) protects hippocampal neurons against cell death

induced by beta-amyloid. *European Journal of Neuroscience, 12*(6), 1882–1890. doi:10.1046/j.1460-9568.2000.00069.x

44. Dubois, B., von Arnim, C. A. F., Burnie, N., Bozeat, S., & Cummings, J. (2023). Biomarkers in Alzheimer's disease: role in early and differential diagnosis and recognition of atypical variants. *Alzheimers Research Therapy, 15*(1), 175. doi:10.1186/s13195-023-01314-6
45. Gupta, J., Gupta R, & Sharma P. Alzheimer's Disease: Role of amyloid-β peptide in the pathogenesis of neurodisorder. *European Journal of Molecular and Clinical Medicine, 9*(7), 7282–7290
46. Altomare, D., Ferrari, C., Caroli, A., et al. (2019). Prognostic value of alzheimer's biomarkers in mild cognitive impairment: The effect of age at onset. *Journal of Neurology, 266*(10), 2535–2545. doi:10.1007/s00415-019-09441-7
47. Bergeron, D., Sellami, L., Poulin, S., Verret, L., Bouchard, R. W., & Laforce, R. Jr. (2020). The behavioral/dysexecutive variant of Alzheimer's disease: A case series with clinical, neuropsychological, and FDG-PET characterization. *Dementia and Geriatric Cognitive Disorders, 49*(5), 518–525. doi:10.1159/000511210
48. O'Bryant, S. E., Lista, S., Rissman, R. A., et al. (2015). Comparing biological markers of alzheimer's disease across blood fraction and platforms: Comparing apples to oranges. *Alzheimers Dement (Amst), 3*, 27–34. doi:10.1016/j.dadm.2015.12.003
49. Ong, S. S., Doraiswamy, P. M., & Lad, E. M. (2018). Controversies and future directions of ocular biomarkers in alzheimer disease. *JAMA Neurology, 75*(6), 650–651. doi:10.1001/jamaneurol.2018.0602
50. Brenowitz, W. D., Xiang, Y., McEvoy, C. T., et al. (2021). Current Alzheimer disease research highlights: evidence for novel risk factors. *Chinese Medical Journal, (English), 134*(18), 2150–2159. doi:10.1097/CM9.0000000000001706

CHAPTER 4

Assimilation of Transgenic Modeling Methods of Alzheimer's Disease: Mechanism of Neurodegeneration

JUNAID TANTRAY[1], AKHILESH PATEL[1], SANJAY NAGDEV[2,*], DEVENDRA SINGH LODHI[3], PRADEEP GOLANI[3], and SHASHI ALOK[4]

[1]*NIMS Institute of Pharmacy, NIMS University, Jaipur, Rajasthan, India*

[2]*Shri Prakashchand Jain College of Pharmacy and Research, Palaskhede (Bk), Jamner, Maharashtra, India*

[3]*Gyan Ganga Institute of Technology and Sciences, Jabalpur, Madhya Pradesh, India*

[4]*Institute of Pharmacy, Bundelkhand University, Jhansi, Uttar Pradesh, India*

*Corresponding author

ABSTRACT

To identify the causes of neurodegeneration, this chapter explores the incorporation of transgenic modeling techniques for Alzheimer's disease (AD). Neuronal loss and increasing cognitive deterioration are hallmarks of AD, a complicated neurodegenerative illness. Our understanding of AD pathophysiology and the development of possible treatment strategies have greatly benefited from using transgenic animal models, especially mice. This study focuses on the several transgenic modeling techniques used to simulate important pathological features of AD, including tau and amyloid-beta (Aβ) protein buildup, neuroinflammation, synaptic dysfunction, and cognitive impairments. It also talks about the benefits and drawbacks of

Cellular and Molecular Mechanisms in Alzheimer's Disease Therapeutics, Volume 2: Molecular Mechanisms.
Prashant Tiwari, Sunil Kumar Kadiri, and Kunjbihari Sulakhiya (Eds)

DOI: 10.1201/9781779641816-4

these models and covers current developments in the field. The application of transgenic modeling techniques lays the groundwork for the creation of innovative treatment approaches by giving insightful knowledge about the molecular and cellular pathways underlying AD pathogenesis.

4.1 INTRODUCTION

4.1.1 OVERVIEW OF ALZHEIMER'S DISEASE

Incurable conditions associated with amyloid fibril production are numerous [1]. The first case of an amyloid disease was reported by Nicolaus Fontanus in 1639 [2]. In 1838, Matthias Scheiden incorrectly linked them with starch and referred to them as "amyloids" because of their blue staining with iodine [2]. Despite being chemically distinct from starch, the name "amyloid" persisted for these fibrils since their discovery in 1859. Among the most well-known amyloid diseases, Alzheimer's disease (AD) was initially discovered in 1906 by Alois Alzheimer. Accordingly, references are provided [3].

According to different studies, AD impacts more than 12 million people globally and stands as the primary cause of dementia among those aged over 60 [4]. Despite the potential to alleviate symptoms of amyloid diseases, effective cures remain elusive. Extensive research potential is dedicated to developing therapeutics to address amyloid diseases. However, understanding amyloid fibrils' various structural and physiochemical properties is essential to designing treatments for this particular category of diseases.

Amyloid fibrils represent self-assembled, (Figure 4.1) fibrillar structures composed of proteins adopting an alternative, β-rich conformation. Typically, proteins maintain their native state under physiological conditions, which allows them to refold into this state if unfolded [5]. However, exposure to denaturing conditions can lead proteins to adopt alternative structural conformations conducive to self-assembly into amyloid fibrils [6, 7]. This phenomenon is believed to be a common property among proteins, as diverse proteins like Aβ1–40 [8], lysozyme, and insulin [9] have demonstrated being able to make amyloid fibrils under appropriate incubation conditions, both *in vitro* and in nature. These fibrils can form plaques in organs, causing a variety of illnesses [10]. Amyloid disorders, including Alzheimer's, are often caused by a single protein or peptide that forms amyloid fibrils, such as amyloid beta (Aβ) [11].

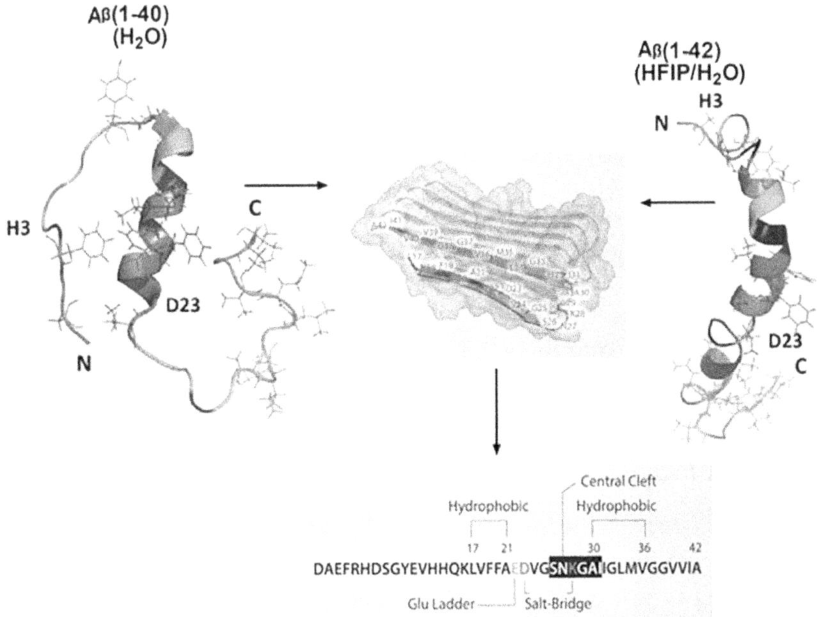

FIGURE 4.1 Aβ 1–40's structure according to solid-state NMR.
Source: Adapted from Ref. [11]

a. Amyloid fibril features a unique cross-β shape with parallel β-sheets stacked in a double-layered manner. The orange arrows indicate the β-sheet shape made up of residues 12–24, whereas the blue arrows show it from residues 30–40.
b. An Aβ molecule in its most energy-efficient form. The charges of red, blue, green, and magenta residues are negative, positive, hydrophobic, and polar, respectively.

The hallmark symptoms of AD include memory impairment and cognitive decline, significantly impacting the individual's daily functioning. Traditionally, diagnosis relied on recognizing these symptoms [12]. However, differentiating AD from other kinds of dementia can be difficult because of the complexity of dementia types related to aging. In recent times, more precise diagnostic methods have been employed, such as positron emission tomography (PET) and magnetic resonance imaging (MRI) [13]. These techniques can identify hypometabolism, hypoperfusion, and medial temporal lobe atrophy, all of which are symptoms of AD.

Moreover, the diagnostic method of biomarker analysis of cerebrospinal fluid is being explored for AD, which is indicated by elevated levels of tau protein and phosphorylated tau protein and lower levels of Aβ1-42. However, these developments have not eliminated the need for brain biopsies as the gold standard for AD diagnosis.

4.1.2 TRANSGENIC MODELING OF AD

Preclinical studies continue to confront the issue of creating animal models that replicate the complicated pathophysiology and clinical signs of AD. Although no one model perfectly represents AD, several models could be useful in finding a cure by facilitating the screening of possible treatment drugs.

4.1.2.1 AGING ANIMALS (RODENTS AND MONKEYS)

The fact that animals' learning and memory capacities decline with age is in complete contrast to the observed behavioral changes seen in the elderly. Because AD is a neurodegenerative condition that worsens with age, it is usual practice to study medications that may affect the disease in animals that are quite old. There are a few benefits to using this model, such as (a) the fact that older rodents and monkeys exhibit cognitive deterioration and (b) the fact that they are easy to procure and very inexpensive. The final stage before clinical trials is to employ old monkeys as models. Additionally, (c) rhesus monkeys that are between the ages of 23–31 have been shown to exhibit amyloid beta plaques, which are conventional pathological markers. Therefore, testing in non-human primates is an important model [14, 15].

4.1.2.2 TRANSGENIC MICE

The identification of four genes on chromosomes 21, 19, 14, and 1 that are associated with AD has allowed for the rational design of transgenic animal models and the testing of some of the proposed mechanisms of amyloidogenesis. Amyloidosis develops in the brain when these genes mutate and alter amyloid precursor protein (APP) processing [16].

Several genetically modified Alzheimer's mouse models exist. The Minnesota mouse and exemplar/Athena mouse are transgenic mice with familial AD mutations in all APP segments. Minnesota mice at 7–12 months and Athena mice at 4–8 months accumulate and deposit amyloid.

Data from these two models' scenes show that certain ADR traits are established, but neither model changes the number of neurons in amyloid-deposited brain regions [17, 18].

4.1.2.3 TRANSGENIC MODELING WITH PRESENILIN MUTATION

Given the capability to introduce genetic modifications, researchers have extensively pursued transgenic modeling in mice based on the amyloid hypothesis. Before familial Alzheimer's disease (FAD) mutations, pronuclear injection was used to overexpress wild-type APP in transgenic mice. However, these efforts, which included a variety of promoters, failed to produce amyloid plaques or any other hallmark of AD [19].

After the discovery of FAD mutations in APP, several research groups shifted their focus to creating AD models by overexpressing transgenes containing these mutations. Using a platelet-derived growth factor-β promoter, Games et al. effectively expressed a human APP transgenic with a V717F mutation related to FAD. The PDGF promoter was chosen because it strongly expresses foreign transgenes in neurons and the CNS. In comparison to endogenous mice APP, the PDAPP transgenic line showed a substantial rise in human APP protein and APP RNA levels, as well as corresponding increases in human amyloid beta (A) levels [20].

Like human AD, PDAPP mice produced compact, dense-cored thioflavin-S-positive plaques. Near plaques appeared dystrophic neurites, reactive astrocytes, and activated microglia. At 6 months of age, there was hardly any plaque buildup, but it was very noticeable at 9 months and increased considerably between 12 and 15 months. Later research confirmed that PDAPP mice exhibited synapse loss and age-related cognitive disabilities [21]. Overexpressing a human APP transgene carrying the Swedish FAD mutation (K670N/M671L) was also used by Hsiao et al., to induce AD-like disease in transgenic mice [22, 23].

4.1.3 TAUOPATHY MODELS

Normal, fully developed neurons have three microtubule-associated proteins (MAPs): tau, MAP1, and MAP2 [24]. Splicing alternatives of pre-mRNA via a single gene on the 17th chromosome creates all six-brain tau molecular isoforms [25]. Tau facilitates tubulin aggregation into microtubules and stabilizes their structural integrity as a phosphoprotein [26].

In AD and tauopathies, tau protein hyperphosphorylation forms filamentous bundles [27]. Tau pathology causes intraneuronal neurofibrillary tangles of PHF and straight filaments. Hyperphosphorylated filaments are found in dystrophic neurites, β-amyloid plaque cores, and neuropil threads [28]. The clinical presentation of AD and similar tauopathies is connected to neurofibrillary degeneration and hyperphosphorylated tau aggregates [29]. Thus, understanding the genesis of this key lesion in AD and tauopathies is essential for developing effective CNS treatments. Many researchers, including ourselves, have reviewed tau's function in neurodegeneration and its therapeutic targets [30].

4.1.3.1 NORMAL BRAIN TAU STRUCTURE AND FUNCTION

Six different molecular variants of tau protein are produced in the brain of a human as a consequence of alternate splicing of tau pre-mRNA [24]. There are a few key differences between these types. One is the number of microtubule-binding repetitions (R). Some have three (3R taus), while others have four (4R taus). Another difference is in amino acid count in the carboxy-terminal half, which can range from 31 to 32. Each version can also contain an amino-terminal insert of one, two, or no amino acids at all, with a total of 29 amino acids. R2 of 4R taus is where you'll find the extra repetition. This splicing produces three 3R taus and three 4R taus denoted as 0N3R, 1N3R, and 2N3R, and 0N4R, 1N4R, and 2N4R. The 441-amino-acid 2N4R tau isoform is most frequent in humans. Only human fatal brain tissue includes the tiniest tau isoform (0N3R; tau352), without amino-terminal inserts.

Tau stabilizes microtubules by interacting with tubulin and making it easier for them to assemble. Phosphorylation levels regulate tau's biological function, as they do with MAP1 and MAP2 [31]. In a normal tau brain state, the tau protein contains two to three moles of phosphate per mole of protein, promoting the assembly of tubulin and microtubules. Alternative splicing affects tau's phosphorylation and activity. The 4R tau isoforms' amino-terminal insertion (N1 and N2) and repeat (Repeat 2) improve tau-tubulin binding [32]. Consequently, 2N4R tau (tau441) exhibits the highest effectiveness in promoting microtubule assembly, while 0N3R tau (tau352), the fetal tau isoform, is comparatively less effective. All six tau isoforms are highly hydrophilic, rendering them stable for heat and solubility [33].

In a neuron, the abundance of tubulin exceeds that of tau by over tenfold. The concentration of tau within neurons is approximately 2 μM, and it binds to microtubules with an affinity of around 100 nM, indicating that most tau molecules are likely bound to microtubules within the cell [34]. When tau is overexpressed in cultured cells, it can lead to the bundling of microtubules. However, there have been no reports of such microtubule bundling occurring in AD or any related tauopathies [35].

4.1.3.2 PATHOLOGICAL TAU STATES AND FORMS

4.1.3.2.1 Abnormal Hyperphosphorylation and Tau Oligomerization

Some of the conditions that fall under this category are as follows: Pick's disease, corticobasal degeneration, dementia pugilistica, and FTDP-17-linked Parkinsonism. Tauopathies cause neurofibrillary abnormalities due to abnormal tau protein phosphorylation. These changes in the neocortex occur alongside dementia. FTDP-17-tau is linked to certain tau gene missense changes. The mutations G272V, P301L, V337M, and R406W have been thoroughly researched and demonstrated to increase tau's vulnerability to aberrant hyperphosphorylation under experimental conditions [36]. Nevertheless, neurodegenerative diseases in humans are characterized by neurofibrillary degeneration similar to AD. So far, animals who are elderly or have cognitive impairments have infrequently shown signs of neurofibrillary degeneration associated with abnormally hyperphosphorylated tau [33].

The hallmark tau pathology in all recognized human tauopathies, including AD, is the existence of aberrantly hyperphosphorylated tau protein [37]. For instance, in AD, paired helical filaments (PHF) are formed when all six tau isoforms are hyperphosphorylated and aggregated. Though tau hyperphosphorylation causes several structural changes and tau truncation in AD and relatives of tauopathies, aberrant tau hyperphosphorylation is still the predominant and best-established cause of tau dysfunction in these diseases [38].

In normal brain tissue, the majority of tau is soluble and can be found in the cytosol even after centrifugation at 200,000 × g [39, 40]. Nevertheless, there are primarily three forms of tau recovered in AD brain tissue: soluble, oligomeric, and fibrillated. Though the amount of normal cytosolic tau in the AD brain is kind of similar to that in the normal aged brain, the total tau level is four to eight times higher in AD, mostly because of abnormally hyperphosphorylated protein. Centrifugation at 200,000 × g can separate oligomeric tau from AD brain, which accounts for a significant amount (up to 40%). Using phosphocellulose chromatography, one may separate the hyperphosphorylated tau oligomers from the non-hyperphosphorylated tau oligomers [41]. The hyperphosphorylated tau purified from these oligomers is found to be three- to four fold more hyperphosphorylated compared to the non-hyperphosphorylated or normal tau [42].

4.1.3.2.2 Neurotoxic State of Tau

Tau serves two fundamental purposes: facilitating the formation of microtubules and preserving their architecture. Tau polymerizes into neurofibrillary tangles without binding tubulin or assembling microtubules. In AD, 40% of hyperphosphorylated tau protein develops neurofibrillary tangles and 60% remains in the cytoplasm. Tubulin does not interact with cytosolic hyperphosphorylated tau (AD P-tau) or induce microtubule formation. Microtubules break and formation is inhibited. AD P-tau traps normal tau and other neuronal microtubule proteins, such as MAP1 A/B and MAP2. Adverse P-tau behavior is mostly caused by aberrant hyperphosphorylation. Dephosphorylating ill tau creates a protein with normal properties [43, 44].

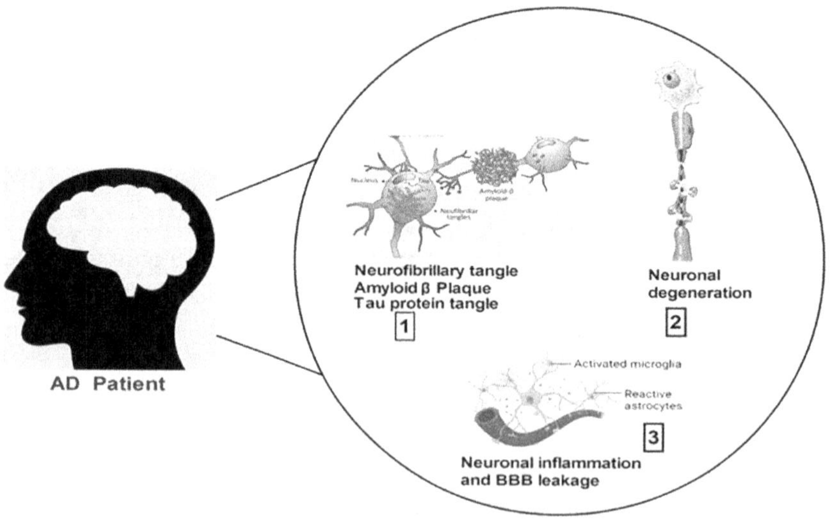

FIGURE 4.2 Several pathological conditions of tau protein, starting from its normal state in the brain, leading to the loss of normal functions and the development of potentially dangerous characteristics.

The usual phosphate stoichiometry of brain tau is two to three moles phosphate per mole of protein. It enhances tubulin assembly and stabilizes microtubule structures. Tau may be transiently hyperphosphorylated during development, anaesthesia, or hypothermia (e.g., hibernation). The large level of brain tubulin makes this hyperphosphorylation less important throughout development, but it decreases the microtubule network and its activities after anaesthesia and hypothermia. An imbalance in phosphorylation/dephosphorylation, likely caused by diminished protein phosphatase-2A activity, causes tau hyperphosphorylation in AD brains. AD P-tau has many negative effects. Normal MAPs are sequestered, inhibiting and disrupting microtubule formations. AD P-tau interacts with MAP1 or MAP2 to produce disordered clusters and with normal tau to generate oligomers. AD P-tau oligomers have the unique ability to be sedimented under strong centrifugal forces and spontaneously form PHF or SF, leading to the development of neurofibrillary tangles [45].

4.1.4 COMBINED MODELS

4.1.4.1 BIGENIC MICE

Only APP transgenic or APP/ps1 bigenic mice are used for future treatment approaches. Models with amyloid buildup and progressive memory impairment are unique. The paradigm is less effective in-app or app ps1 mice due to the lack of neurofibrillary tangles. By breeding APP with tau transgenic mice, amyloid plaques and neurofibrillary tangles can be produced. Paralysis makes these mice unfit to perform learning or memory tasks [46].

4.1.4.2 TRIPLE TRANSGENIC MOUSE MODEL

The 3xTg-AD mouse is a triple transgenic mouse model of AD mouse, which shows pathology similar to the progression of the disease in humans. An age-dependent increase in insoluble Aβ neuronal accumulation and extracellular deposition is recorded by the age of 7 months. The relative abundance is seen in the following orders subiculum>amygdale>CA1 region of hippocampus. Concomitantly, changes in behavioral paradigms are recorded from 3 months onward [47].

4.1.4.3 USING FEMTOSECOND LASER ABLATION TO INDUCE VASCULAR DEMENTIA IN TRANSGENIC MICE

Vascular dementia (VaD) constitutes around 15–20% of all dementia cases, positioning it as the most prevalent type of dementia after AD. Common pathological symptoms of VaD include big artery infarctions, small artery subcortical infarctions, cerebral amyloid angiopathy, hippocampal sclerosis, and chronic subcortical ischemia resulting in specific loss of neurons, glial cells, and endothelial cells. Depression, apathy, and mood changes lead to personality changes; these changes accompany this disease and are important behavioral components of VaD. The risk factors for VaD are stroke, microbleeds, hypertension, type 2 diabetes, and atherosclerosis.

Transgenic mice with fluorescent neuron proteins can be injured by femtosecond laser ablation to cause vascular endothelial damage and clotting. Follow red blood cell mobility in the 2-photon image to determine blood flow change. Penetrating arteriole occlusions reduces downstream capillary flow and triggers neuronal death. Real-time vascular occlusion shows dendrite deterioration within a minute. Stepwise degeneration leads to cellular death in the ischemic zone caused by tiny artery occlusions and then demonstrable behavioral and functional impairment. Leukocyte invasion and microglia activation accompany these lesions [48].

4.1.5 MECHANISMS OF NEURODEGENERATION IN AD

As the global population ages, AD prevalence is expected to triple by 2050, posing a major socioeconomic issue [49]. Current therapy only addresses symptoms and does not slow or stop the disease progression, emphasizing the urgent need for disease-modifying therapies [50].

Research has mainly concentrated on comprehending the harmful functions of microtubule-associated protein tau (MAPT) and Aβ in AD pathogenesis [51]. Aβ accumulation leads to synaptic damage and cognitive deficits [52], while tau hyperphosphorylation and aggregation contribute to neurotoxic effects. Consequently, therapeutic efforts have concentrated on targeting these proteins to prevent aggregation, reduce production, or enhance clearance [53].

Clinical trials were stopped due to severe consequences, such as liver damage, encephalitis, and worsening of cognitive deterioration, despite initial positive preclinical results. The role of these "pathological" proteins in AD is becoming more apparent and complex. Increasing evidence indicates that these proteins may play crucial roles in the body's normal processes, which could be negatively affected by AD or treatments targeting Aβ or tau [54].

4.1.5.1 NEUROINFLAMMATION

4.1.5.1.1 Chemokines

Chemokines have a crucial impact on AD by controlling the movement of microglia to areas of neuroinflammation, which worsens the inflammation in specific regions [55]. Higher levels of CCL2, CCR3, and CCR5 in activated microglia and CCL4 in activated astrocytes around Aβ plaques are observed in AD. In lab trials, Aβ exposure leads to the production of CXCL8 (IL-8), CCL2, CCL3, and CCL4 in human macrophages and astrocytes. In AD autopsy-derived cells, microglia exposed to Aβ express higher levels of CXCL8, CCL2, and CCL3. The CX3CR1/CX3CL1 system may affect neuronal survival in AD mice [56], plaque load, and cognition. Furthermore, receptors like CCR5 and CCR2 influence the progression of AD by affecting microglial positioning and function [57].

4.1.5.1.2 Caspases

Caspases carefully regulate apoptosis and inflammation. Caspase-1's catalytic function is controlled by signal-induced auto-activation in "inflammasomes," which cleave IL-1β and IL-18 precursors into active cytokines [58]. Lysosomal degradation of amyloid-beta (Aβ) fibrils in mouse microglia activates the NLRP3 inflammasome [59]. AD patients and APP/PS1 mice have brain caspase-1 elevations. NLRP3 and caspase-1-deficient APP/PS1 mice had better spatial memory, hippocampal synaptic plasticity, and behavioral abnormalities. In mice lacking NLRP3 or caspase-1, microglial polarization shifts from pro-inflammatory "M1-like" to neuroprotective "M2-like" [60]. Microglia activate caspase-8 and 3/7 following inflammation. Caspase-3 activation alters nuclear factor-κB activation, this leads to neurotoxic pro-inflammatory mediators, such as IL-1β, TNF-α, and nitric oxide. Neurotoxicity and microglia activation are reduced by caspase inhibition. The caspases are activated by microglia within the AD. Neuroprotective medications are beneficial to AD mice [61].

4.1.5.1.3 Prostanoids and Neuroprotectin D1

Prostanoids are principally synthesized by microglia from arachidonic acid with the help of cyclooxygenase-1 and cyclooxygenase-2. In AD, blocking cyclooxygenase-1 has been demonstrated to prevent glial activation, lower amyloid deposition, and reduce the presence of inflammatory markers, causing a change in microglia to a different phenotype in AD mouse models. In people with probable AD, prostaglandin PGE2, which binds to EP1-4 receptors, is increased. EP1-3 receptors are found in microglia and other brain cells, including neurons. Research indicates that microglial EP2 receptors reduce Aβ phagocytic clearance and increase neurotoxicity in experimental settings. In AD animal models, depletion of prostaglandin EP2 or EP3 receptors reduces OS, neuroinflammation, Aβ build-up, and BACE expression [62].

Research suggests that using an EP4 receptor agonist on microglia may decrease inflammation and improve synthetic Aβ absorption. In mice with AD, deleting the EP4 receptor enhanced plaque development and released proinflammatory markers, such as IL-1β and CCL3. Mild cognitive impairment (MCI) and AD brains had lower EP4 receptor expression, suggesting an inflammatory involvement. The role of PGE2 in neurodegeneration is likely complex because it affects multiple cell types, including neurons [63].

4.1.6 OXIDATIVE STRESS

4.1.6.1 ROLE OF OXIDATIVE STRESS (OS) IN AD

Brains in AD suffer from oxidative damage due to Aβ accumulation and neurofibrillary tangles. A recent study suggests that biometals play a crucial role in Aβ aggregation and neurodegeneration. Copper and zinc bind to specific Aβ and precursor sites, causing OS in AD brains. Copper, which produces extremely reactive hydroxyl radicals, is abundant in amyloid plaques. Aβ (1–42) fragments, especially toxic, contribute to the formation of reactive oxygen species (ROS) such as H_2O_2 [64]. Increased zinc concentrations, particularly in brain areas associated with memory and cognitive functions, such as the neocortex and hippocampus, stimulate the development of harmful Aβ clusters. Disruption of zinc homeostasis by Aβ plaques leads to uncontrolled release of cerebral zinc, exacerbating OS.

The vulnerability of the brain to oxidative damage is exacerbated by its composition of polyunsaturated fatty acids, making it susceptible to lipid peroxidation, a hallmark of AD pathology. Proteins are also targets of oxidative modification, affecting critical enzymes like glutamine synthetase and creatine kinase, essential for neuronal function and energy metabolism. OS leads to protein aggregation and excessive phosphorylation of tau proteins, which contributes to the formation of neurofibrillary tangles. Oxidation can harm DNA, causing different types of DNA damage [65].

An excessive amount of ROS is produced and OS plays significant roles in cellular damage and disease progression, including AD. However, antioxidant treatments have shown promise in mitigating OS, highlighting the complex nature of AD pathogenesis [66].

4.1.7 MITOCHONDRIAL DYNAMIC CHANGES SEEN IN AD

Mitofusin 1 and 2 generally merge the outer mitochondrial membranes during mitochondrial fusion, while Optic Atrophy 1 (OPA1) merges the inner membranes. Four receptors located on the external surface of the mitochondrial membrane—Fis1, Mff, MiD49/MiD51—recruit *Drp1* during mitochondrial fission. After that, *Drp1* contracts and divides the mitochondrion with an oligomeric ring. Post-translational changes, especially phosphorylation, affect Drp1's function. Specifically, *Drp1* translocation depends on its phosphorylation status; dephosphorylation at serine 637 and phosphorylation at serine 616 is necessary for its action [67].

Various research studies have produced conflicting results on the quantity of fusion proteins including OPA1 and Mfn2. For example, Calkins et al. [67] found that primary neurons from a Tg2576 mouse model had lower levels of Mfn1 and Mfn2, but APPsw/PS1dE9 animal model hippocampus tissue had higher levels of OPA1 and Mfn2 at 12 months. On the other hand, Trushina et al. [68] did not observe any alteration in these proteins in mice homozygotes for *APP*, *PSEN1*, or *APP/PSEN1*. In addition, 3xTg mice showed changes in Mfn2 levels; however, only males showed a decrease in levels in their cerebral mitochondria at 2 months, while females showed no change at 6 or 14 months [69].

Different transgenic and streptozotocin-induced AD mouse models have higher *Drp1* levels, varying by sex. Female 3xTg mice had elevated *Drp1* levels in cortical and hippocampal mitochondria at 2 and 6 months, while male mice had decreased levels at 6 months. No hippocampal mitochondrial differences were found. Lower levels of *Drp1* have been observed in neuroblastoma cell lines that have an increased expression of normal APP and in post-mortem brain tissue [70]. Dynamin-related protein 1 levels vary in sporadic and familial AD fibroblasts. Notably, *Drp1* levels were unaffected in sporadic AD-induced pluripotent stem cell (iPSC)-derived neurons. Conversely, astrocytes presenting apolipoprotein E4 (ApoE4), a risk factor for sporadic AD, had lower *Drp1* levels [71]. Moreover, *Drp1* localized in the mitochondria was observed to decrease in sporadic AD, potentially suggesting compromised recruitment to the mitochondria. [72–74].

4.1.8 ADVANTAGES AND LIMITATIONS OF TRANSGENIC MODELS

Transgenic models, especially genetically engineered mice, have been instrumental in advancing our understanding of AD. Here are some of the key advantages and limitations associated with using transgenic models in AD research.

4.1.8.1 ADVANTAGES

1. Relevance to human disease: Transgenic models are designed to express genes associated with AD, such as presenilin genes, which closely mimic the genetic basis of the disease in humans [75].
2. Pathological and behavioral phenotypes: These models generate amyloid-beta plaques, neurofibrillary tangles, neuronal death, and cognitive abnormalities, allowing researchers to investigate disease development and test potential treatments.
3. Temporal control: Researchers can manipulate transgenic expression timing to examine illness genesis and progression, which is crucial for understanding the temporal sequence of pathological events in AD.
4. Manipulability: Transgenic models offer the opportunity to manipulate specific genes or pathways implicated in AD, enabling researchers to investigate the effects of genetic modifications on disease pathology and phenotype.
5. Preclinical drug testing: These models are widely used for preclinical drug testing, which allows the efficacy and safety evaluation of potential therapeutics before advancing to human clinical trials.

6. High-throughput screening: Transgenic models enable high-throughput screening of potential drug candidates, facilitating the rapid identification of promising compounds for further development.

4.1.8.2 LIMITATIONS

1. Incompleteness of disease phenotype: Transgenic models typically do not fully recapitulate the complex spectrum of pathological and clinical features observed in human AD, limiting their translational relevance to certain aspects of the disease [76].
2. Species differences: Although transgenic mice share genetic similarities with humans, there are significant differences in brain anatomy, physiology, and immune response between mice and humans, which may affect the interpretation and generalization of findings from these models to human disease.
3. Limited predictive validity: Although useful for examining disease causes and evaluating treatment strategies, the predictive validity of transgenic models for human clinical outcomes remains a challenge, with many promising treatments failing to translate from preclinical studies to clinical trials.
4. High cost and resource intensiveness: Generating and maintaining transgenic mouse colonies can be costly and labor-intensive, requiring specialized expertise, infrastructure, and resources, which may limit their accessibility to certain research laboratories.
5. Ethical considerations: The creation and use of transgenic animals raise ethical concerns related to animal welfare, particularly regarding the suffering and welfare of genetically modified animals used in research.
6. Limited applicability to late-onset AD: Transgenic models focus on family AD with early-onset genetic alterations, which account for a minority of cases, limiting its applicability to late-onset sporadic AD.

Despite these limitations, transgenic models remain valuable tools for investigating specific aspects of AD pathogenesis, identifying novel therapeutic targets, and testing potential treatments, albeit within the context of their inherent constraints and considerations. It is crucial to combine results from transgenic models with various experimental methods, like cellular models, human brain imaging investigations, and clinical research to enhance our comprehension of AD and provide successful treatments for patients.

4.1.9 RECENT ADVANCEMENTS IN TRANSGENIC MODELING

4.1.9.1 NOVEL TRANSGENIC APPROACHES

4.1.9.1.1 Aβ Immunotherapy

Aβ immunotherapy seeks to decrease the accumulation of Aβ in AD by either actively using vaccination or passively employing monoclonal antibodies. The first active anti-Aβ vaccination, AN1792, was stopped because 6% of AD patients who received it developed meningoencephalitis. ACI-24, CAD106, UB-311, and ABVac40, which are subsequent active vaccines, did not demonstrate any therapeutic advantages throughout phase II trials. Passive immunotherapy targeted Aβ oligomers using three humanized monoclonal antibodies: BAN2401, gantenerumab, and aducanumab. BAN2401 showed slight cognitive enhancements in individuals with APOE4 carrying MCI to mild AD. Gantenerumab did not demonstrate any clinical effectiveness in those with prodromal to moderate AD. The phase III studies of aducanumab were stopped initially because they were deemed ineffective. However, after further data analysis, the results were found to be promising. In June 2021, the FDA approved its usage in treating AD, which has since sparked ongoing discussions. Upcoming oral small-molecule drugs, such as CT1812, PQ912, and ALZ-801, which focus on Aβ oligomers, are currently being developed [77].

4.1.9.1.2 Anti-tau Therapy

Tau protein is essential for axon and dendritic microtubule structure. Hyper-phosphorylation causes tau proteins to aggregate and form intracellular neurofibrillary tangles (NFTs) [78]. These NFTs increase neurotoxicity and inflammation. In AD, tau pathology strongly correlates with cognitive loss, unlike Aβ pathology. Various strategies for anti-tau therapy have been proposed, including modifiers of phosphatase, inhibitors of kinases, compounds that inhibit tau aggregation, microtubule-stabilizing agents, and immunotherapy targeting tau [79].

4.1.9.1.3 Phosphates Modifiers

Protein phosphatase 2A is activated by phosphatase modifiers to lower phosphorylation. Brain function depends on the PP2A activator sodium selenate. Cognitive impairment and oxidative damage are associated with sodium selenite insufficiency. Sodium selenate did not improve cognitive performance in a phase II study of mild to moderate AD patients. High sodium selenate doses increased CNS selenium levels, but AD treatment was restricted [80].

4.1.10 INNOVATIVE IMAGING TECHNIQUES

Numerous neuroimaging studies have identified AD-related brain structural and functional changes in recent decades. In suspected AD patients, Table 4.1 summarizes the practical use of several neuroimaging techniques.

TABLE 4.1 Neuroimaging Modalities in Individuals Suspected of AD.

Neuroimaging Modality	Information Provided	Clinical Utility in Suspected Alzheimer's Disease	Current Availability
Structural MRI	Detailed visualization of brain structures and abnormalities	Useful for ruling out other causes of cognitive impairment	Widely available
fMRI (functional MRI)	Mapping of brain activity and connectivity	Helps identify regions affected by AD, assess functional changes	Widely available
PET	A determination of the presence of tau tangles and Aβ plaques	Aids in early diagnosis and tracking disease progression	Available in specialized centers
SPECT (single photon emission computed tomography)	Cerebral blood flow assessment	Useful for differential diagnosis and assessing the severity of AD	Available, but less common than PET
CT (computed tomography)	Imaging of brain structures and detection of atrophy	Provides structural information, less sensitive to early changes	Widely available
DTI (diffusion tensor imaging)	Visualizing white matter connection and integrity	Helps understand white matter alterations in AD	Available in research settings

These studies, which compare groups of dementia patients and older volunteers who are not afflicted, have done thorough evaluations of the human brain *in vivo*, made possible by significant advancements in image capture, processing, and analysis. Table 4.2 presents a very swift review of clinically important neuroimaging results in samples of AD patients.

TABLE 4.2 Neuroimaging Results in AD Patients that are Clinically Meaningful.

Neuroimaging Finding	Description	Clinical Relevance
Hippocampal atrophy	Reduction in volume of the hippocampus, often seen in the early stages of AD	Correlates with cognitive decline and serves as a biomarker for AD progression
Cortical thinning	Decrease in cortical thickness, particularly in regions associated with memory and cognition	Reflects neuronal loss and is indicative of AD pathology
Increased amyloid deposition	Presence of amyloid-beta plaques, detectable using PET imaging	Early AD diagnosis, progression tracking, treatment guidance
Tau accumulation	PET-detected NFTs with hyperphosphorylated tau proteins	Helps explain AD pathogenesis by predicting cognitive loss and disease severity
Disrupted brain connectivity	Alterations in functional connectivity patterns between brain regions, observed using Fmri	Reflects neuronal dysfunction and provides insights into cognitive impairment in AD
White matter hyperintensities	Hyperintense lesions in white matter regions, typically seen on MRI	Linked to vascular disease and AD cognitive impairment

4.1.11 THERAPEUTIC POTENTIALS

Various therapeutic approaches are being explored for AD, including targeting NMDA receptors, adenosine receptors, microtubule stabilization, tau protein oligomerization inhibition, antioxidant treatment, calcium homeostasis modulation, anti-inflammatory strategies, statin therapy, and more. Promising ongoing clinical trials involve antibodies targeting amyloid-beta, such as solanezumab, gantenerumab, and crenezumab, while early-stage trials are investigating vaccines targeting the tau protein like AADvac1 and ACI-35. Small molecule inhibitors of *BACE-1* (MK8931, E2609) and combination therapies, such as idalopirdine with donepezil, have also shown interesting results. It is difficult to identify the most promising agent among the tested drugs [80]. Currently undergoing clinical studies for specific anti-Alzheimer's medications are represented in Table 4.3.

Unfortunately, despite positive outcomes in phase II trials, several prospective treatments often do not get to phase III due to adverse reactions or lack of effectiveness. Creating novel medications for AD is exceptionally difficult, with a success rate of less than 5%. No new medications have been approved for the treatment despite significant research and multiple compounds advancing to phase II trials. Past medicines focusing on beta-amyloid protein have not shown clear advantages, indicating the necessity for other techniques. Researchers are investigating selective ligands for multiple targets as a potential approach to target many biological processes involved in AD development. Early intervention in the disease process, including prevention strategies, probably won't let subsequent treatments fail [81].

TABLE 4.3 Currently Undergoing Clinical Studies for Specific Anti-Alzheimer's Medications.

Drug Name	Mechanism of Action	Clinical Trial Phase	Current Status	References
Aducanumab	Anti-amyloid beta	Phase III	Approved by FDA (June 2021) for AD treatment	FDA press release
BAN2401	Anti-amyloid beta	Phase III	Ongoing clinical trials	ClinicalTrials.gov [82]
Gantenerumab	Anti-amyloid beta	Phase III	Ongoing clinical trials	ClinicalTrials.gov [82]
Solanezumab	Anti-amyloid beta	Phase III	Discontinued after failing to meet primary endpoints	Salloway et al. [79]
Verubecestat	Beta-secretase inhibitor	Phase III	Stopped due to ineffectiveness and safety concerns	Egan et al. [80]
AAB-003 (Crenezumab)	Anti-amyloid beta	Phase III	Ongoing clinical trials	ClinicalTrials.gov [82]
LMTX (LMTM)	Tau aggregation inhibitor	Phase III	Ended due to ineffectiveness	Gauthier et al. [83]

4.1.12 CONCLUSION AND FUTURE DIRECTIONS

Transgenic modeling has significantly contributed to our understanding of AD by replicating the genetic abnormalities linked to familial forms of the disease. Through these models, researchers have gained crucial insights into the progression of AD, including the development of hallmark pathological features, such as amyloid-beta plaques, NFTs, neuronal loss, and cognitive deficits. Additionally, transgenic models have served as indispensable tools in preclinical drug testing, aiding in the evaluation of potential therapeutics and informing clinical trial design. Moving forward, future directions in transgenic modeling will likely focus on enhancing translational relevance, refining disease phenotypes, incorporating multi-omics approaches, exploring non-amyloid targets, and utilizing induced pluripotent stem cells (iPSCs). The goal and effort of these activities is to overcome existing restrictions, expand our knowledge of AD development, and discover new treatment options to enhance patient outcomes.

KEYWORDS

- **Alzheimer's disease**
- **amyloid-beta**
- **neurodegeneration**
- **synaptic dysfunction**
- **tauopathy**
- **transgenic models**

REFERENCES

1. Fändrich, M. (2007). On the structural definition of amyloid fibrils and other polypeptide aggregates. *Cellular and Molecular Life Sciences, 64*, 2066–2078.
2. Kyle, R. A. (2001). Amyloidosis: A convoluted story. *British Journal of Haematology, 114*, 529–538.
3. Möller, H. J., & Graeber, M. B. (1998). The case described by Alois Alzheimer in 1911. *European Archives of Psychiatry and Clinical Neurosciences, 248*, 111–122.
4. Citron, M. (2004). Strategies for disease modification in Alzheimer's disease. *Nature Reviews, 5*, 677–685.
5. Matagne, A., & Dobson, C. M. (1998). The folding process of hen lysozyme: A perspective from the 'new view'. *Cellular and Molecular Life Sciences, 54*, 363–371.
6. Dobson, C. M. (2003). Protein folding and misfolding. *Nature, 426*, 884–890.
7. Stefani, M. (2004). Protein misfolding and aggregation: New examples in medicine and biology of the dark side of the protein world. *Biochimica et Biophysics Acta, 1739*, 5–25.
8. Findeis, M. A., Musso, G. M., Arico-Muendel, C. C., Benjamin, H. W., Hundal, A. M., Lee, J.-J., Chin, J., Kelley, M., Wakefield, J., Hayward, N. J., & Molineaux, S. M. (1999). Modified-peptide inhibitors of amyloid β-peptide polymerization. *Biochemistry, 38*, 6791–6800.

9. Nielsen, L., Khurana, R., Coats, A., Frokjaer, S., Brange, J., Vyas, S., Uversky, V. N., & Fink, A. L. (2001). Effect of environmental factors on the kinetics of insulin fibril formation: Elucidation of the molecular mechanism. *Biochemistry, 40*, 6036–6046.
10. Morozova-Rochea, L. A., Zurdo, J., Spencera, A., Noppe, W., Receveur, V., Archer, D. B., Joniau, M., & Dobson, C. M. (2000). Amyloid fibril formation and seeding by wild-type human lysozyme and its disease-related mutational variants. *Journal of Structural Biology, 130*, 339–351.
11. Pepys, M. B. (2006) Amyloidosis. *Annual Review of Medicine, 57*, 223–241.
12. Dubois, B., Feldman, H. H., Jacova, C., DeKosky, S. T., Barberger-Gateau, P., Cummings, W. Jr., Delacourte, A., Galasko, D., Gauthier, S., Jicha, G., Meguro, K., O'Brien, J., Pasquier, F., Robert, P., Rossor, M., Salloway, S., Stern, Y., Visser, P. J., & Scheltens, P. (2007). Research criteria for the diagnosis of Alzheimer's disease: Revising the NINCDS–ADRDA criteria. *The Lancet Neurology, 6*, 734–746.
13. Hamley, I. W. (2012). The amyloid beta peptide: A chemist's perspective. Role in Alzheimer's and fibrillization. *Chemical Reviews, 112*, 5147–5192.
14. Pepeu, G. (2001). Overview and perspective on the therapy of Alzheimer's disease from a preclinical viewpoint. *Progress in Neuro-Psychopharmacology and Biological Psychiatry, 25*(1), 193–209.
15. Fisher, A., & Hanin, I. (1986). Potential animal models for senile dementia of Alzheimer's type, with emphasis on AF64A-induced cholinotoxicity. *Annual Review of Pharmacology and Toxicology, 26*(1), 161–181.
16. Loring, J. F., Paszty, C., Rose, A., McIntosh, T. K., Murai, H., Pierce, J. E., ... & Peterson, K. R. (1996). Rational design of an animal model for Alzheimer's disease: Introduction of multiple human genomic transgenes to reproduce AD pathology in a rodent. *Neurobiology of Aging, 17*(2), 173–182.
17. Hsiao, K. (1998). Transgenic mice expressing Alzheimer amyloid precursor proteins. *Experimental Gerontology, 33*(7-8), 883–889.
18. Quon, D., Wang, Y., Catalano, R., Scardina, J. M., Murakami, K., & Cordell, B. (1991). Formation of β-amyloid protein deposits in brains of transgenic mice. *Nature, 352*(6332), 239–241.
19. Codita, A., Winblad, B., & Mohammed, A. H. (2006). Of mice andmen: More neurobiology in dementia. *Current Opinion in Psychiatry, 19*, 555–563.
20. Games, D., Buttini, M., Kobayashi, D., et al. (2006). Miceas models: Transgenic approaches and Alzheimer's disease. *Journal of Alzheimers Diseases, 9*(suppl), 133–149.
21. McGowan, E., Eriksen, J., & Hutton, M. (2006). A decade of modeling Alzheimer's disease in transgenic mice. *Trends in Genetics, 22*, 281–289.
22. Guo, Q., Sebastian, L., Sopher, B. L., et al. (1999). Increased vulnerability of hippocampal neurons from presenilin-1 mutant knock-in mice to amyloidbeta-peptide toxicity: Central roles of superoxide production and caspase activation. *Journal of Neurochemical, 72*, 1019–1029.
23. Flood, D. G., Reaume, A. G., Dorfman, K. S., et al. (2002). FAD mutant PS-1 gene-targeted mice: Increased A beta 42 and A beta deposition without APP over production. *Neurobiology of Aging, 23*, 335–348.

24. Goedert, M., Spillantini, M. G., Jakes, R., Rutherford, D., & Crowther, R. A. (1989). Multiple isoforms of human microtubule-associated protein tau: Sequences and localization in neurofibrillary tangles of Alzheimer's disease. *Neuron, 3*, 519–526.
25. Himmler, A., Drechsel, D., Kirschner, M. W., & Martin, D. W. Jr. (1989). Tau. consists of a set of proteins with repeated C-terminal microtubule-binding domains and variable N-terminal domains. *Molecular Cell Biology, 9*, 1381–1388.
26. Weingarten, M. D., Lockwood, A. H., Hwo, S. Y., & Kirschner, M. W. (1975). A protein factor essential for microtubule assembly. *Proceedings of the National Academy of Science of the United States of America, 72*, 1858–1862.
27. Grundke-Iqbal, I., Iqbal, K., Tung, Y. C., Quinlan, M., Wisniewski, H. M., & Binder, L. I. (1986b). Abnormal phosphorylation of the microtubule-associated protein tau (tau) in Alzheimer cytoskeletal pathology. *Proceedings of the National Academy of Sciences of the United States of America, 83*, 4913–4917.
28. Braak, H., Braak, E., Grundke-Iqbal, I., & Iqbal, K. (1986). Occurrence of neuropil threads in the senile human brain and in Alzheimer's disease: A third location of paired helical filaments outside of neurofibrillary tangles and neuritic plaques. *Neuroscience Letters, 65*, 351–355.
29. Arriagada, P. V., Growdon, J. H., Hedley-Whyte, E. T., & Hyman, B. T. (1992). Neurofibrillary tangles but not senile plaques parallel duration and severity of Alzheimer's disease. *Neurology, 42*, 631–639.
30. Avila, J., & Hernandez, F. (2007). GSK-3 inhibitors for Alzheimer's disease. *Expert Review of Neurotherapeutics, 7*, 1527–1533.
31. Alonso, A. D., Zaidi, T., Grundke-Iqbal, I., & Iqbal, K. (1994). Role of abnormally phosphorylated tau in the breakdown of microtubules in Alzheimer disease. *Proceedings of the National Academy of Sciences of the United States of America, 91*, 5562–5566.
32. Alonso, A. D., Mederlyova, A., Novak, M., Grundke-Iqbal, I., & Iqbal, K. (2004). Promotion of hyperphosphorylation by frontotemporal dementia tau mutations. *Journal of Biological Chemistry, 279*, 34873–34881.
33. Alonso, A. D., Zaidi, T., Novak, M., Barra, H. S., Grundke-Iqbal, I., & Iqbal, K. (2001). Interaction of tau isoforms with Alzheimer's disease abnormally hyperphosphorylated tau and in vitro phosphorylation into the disease-like protein. *Journal of Biological Chemistry, 276*, 37967–37973.
34. Khatoon, S., Grundke-Iqbal, I., & Iqbal, K. (1992). Brain levels of microtubule-associated protein tau are elevated in Alzheimer's disease: A radioimmuno-slot-blot assay for nanograms of the protein. *Journal of Neurochemistry, 59*, 750–753.
35. Ackmann, M., Wiech, H., & Mandelkow, E. (2000). Nonsaturable binding indicates clustering of tau on the microtubule surface in a paired helical filament-like conformation. *Journal of Neurochemistry, 275*, 30335–30343.
36. Poorkaj, P., Bird, T. D., Wijsman, E., Nemens, E., Garruto, R. M., Anderson, L., et al. (1998). Tau is a candidate gene for chromosome 17 frontotemporal dementia. *Annals of Neurology, 43*, 815–825.
37. Grundke-Iqbal, I., Iqbal, K., Quinlan, M., Tung, Y. C., Zaidi, M. S., & Wisniewski, H. M. (1986a). Microtubule-associated protein tau. A component of Alzheimer paired helical filaments. *Journal of Biological Chemistry, 261*, 6084–6089.

38. Novak, M., Jakes, R., Edwards, P. C., Milstein, C., & Wischik, C. M. (1991). Difference between the tau protein of Alzheimer paired helical filament core and normal tau revealed by epitope analysis of monoclonal antibodies 423 and 7.51. *Proceedings of the National Academy of Sciences of the United States of America, 88*, 5837–5841.
39. Bancher, C., Brunner, C., Lassmann, H., Budka, H., Jellinger, K., Wiche, G., et al. (1989). Accumulation of abnormally phosphorylated tau precedes the formation of neurofibrillary tangles in Alzheimer's disease. *Brain Research, 477,* 90–99.
40. Iqbal, K., Grundke-Iqbal, I., Zaidi, T., Merz, P. A., Wen, G. Y., Shaikh, S. S., et al. (1986). Defective brain microtubule assembly in Alzheimer's disease. *Lancet, 2,* 421–426
41. Lonso, A. D., Grundke-Iqbal, I., Barra, H. S., & Iqbal, K. (1997). Abnormal phosphorylation of tau and the mechanism of Alzheimer neurofibrillary degeneration: sequestration of microtubule-associated proteins 1 and 2 and the disassembly of microtubules by the abnormal tau. *Proceedings of the National Academy of Sciences of the United States of America, 94*, 298–303.
42. Alonso, A. D., Li, B., Grundke-Iqbal, I., & Iqbal, K. (2006). Polymerization of hyperphosphorylated tau into filaments eliminates its inhibitory activity. *Proceedings of the National Academy of Sciences of the United States of America, 23*, 8864–8869.
43. Li, B., Chohan, M. O., Grundke-Iqbal, I., & Iqbal, K. (2007). Disruption of microtubule network by Alzheimer abnormally hyperphosphorylated tau. *Acta Neuropathologica (Berlin), 113*, 501–511.
44. Iqbal, K., Liu, F., Gong, C. X., & Grundke-Iqbal, I. (2010). Tau in Alzheimer disease and related tauopathies. *Current Alzheimer Research, 7*(8), 656–664.
45. Lewis, J., Dickson, D. W., Lin, W. L., Chisholm, L., Corral, A., Jones, G., ... & McGowan, E. (2001). Enhanced neurofibrillary degeneration in transgenic mice expressing mutant tau and APP. *Science, 293*(5534), 1487–1491.
46. Rosario, E. R., Carroll, J. C., Oddo, S., LaFerla, F. M., & Pike, C. J. (2006). Androgens regulate the development of neuropathology in a triple transgenic mouse model of Alzheimer's disease. *Journal of Neuroscience, 26*(51), 13384–13389.
47. Nishimura, N., & Schaffer, C. B. (2013). Big effects from tiny vessels: imaging the impact of microvascular clots and hemorrhages on the brain. *Stroke, 44*(6_suppl_1), S90–S92.
48. Alzheimer's Association. (2019). Alzheimer's disease facts and figures. *Alzheimers Dement, 15*, 321–387.
49. Long, J. M., & Holtzman, D. M. (2019). Alzheimer disease: An update on pathobiology and treatment strategies. *Cell, 179*, 312–339.
50. Kosik, K. S., Joachim, C. L., & Selkoe, D. J. (1986). Microtubule-associated protein tau (tau) is a major antigenic component of paired helical filaments in Alzheimer disease. *Proceedings of the National Academy of Sciences of the United States of America, 83*, 4044–4048.
51. Hong, S., Beja-Glasser, V. F., Nfonoyim, B. M., Frouin, A., Li, S., Ramakrishnan, S., et al. (2016). Complement and microglia mediate early synapse loss in Alzheimer mouse models. *Science, 352*, 712–716.

52. Walsh, D. M., Klyubin, I., Fadeeva, J. V., Cullen, W. K., Anwyl, R., Wolfe, M. S., et al. (2002). Naturally secreted oligomers of amyloid beta protein potently inhibit hippocampal long-term potentiation in vivo. *Nature, 416*, 535–539.
53. Penninkilampi, R., Brothers, H. M., & Eslick, G. D. (2017). Safety and efficacy of anti-amyloid-β immunotherapy in Alzheimer's disease: A systematic review and meta-analysis. *Journal of Neuroimmune Pharmacology, 12*, 194–203.
54. Savarin-Vuaillat, C., & Ransohoff, R. M. (2007). Chemokines and chemokine receptors in neurological disease: Raise, retain, or reduce? *Neurotherapeutics, 4*, 590–601.
55. Xia, M. Q., Qin, S. X., Wu, L. J., Mackay, C. R., & Hyman, B. T. (1998). Immunohistochemical study of the beta-chemokine receptors CCR3 and CCR5 and their ligands in normal and Alzheimer's disease brains. *American Journal of Pathology, 153*, 31–37
56. El Khoury, J., Toft, M., Hickman, S. E., et al. (2007). Ccr2 deficiency impairs microglial accumulation and accelerates progression of Alzheimer-like disease. *Nature Medicine, 13*, 432–438.
57. Schroder, K., & Tschopp, J. (2010). The inflammasomes. *Cell, 140*, 821–832.
58. Halle, A., Hornung, V., Petzold, G. C., et al. (2008). The NALP3 inflammasome is involved in the innate immune response to amyloid-beta. *Nature Immunology, 9*, 857–865
59. Heneka, M. T., Kummer, M. P., Stutz, A., et al. (2013). NLRP3 is activated in Alzheimer's disease and contributes to pathology in APP/PS1 mice. *Nature, 493*, 674–678.
60. Rohn, T. T., Kokoulina, P., Eaton, C. R., & Poon, W. W. (2009). Caspase activation in transgenic mice with Alzheimer-like pathology: results from a pilot study utilizing the caspase inhibitor, Q-VD-OPh. *International Journal of Clinical and Experimental Medicine, 2*, 300–308.
61. Choi, S. H., Aid, S., Caracciolo, L., et al. (2013). Cyclooxygenase-1 inhibition reduces amyloid pathology and improves memory deficits in a mouse model of Alzheimer's disease. *Journal of Neurochemistry, 124*, 59–68.
62. Woodling, N. S., Wang, Q., Priyam, P. G., et al. (2014). Suppression of Alzheimer-associated inflammation by microglial prostaglandin-E2 EP4 receptor signaling. *Journal of Neuroscience, 34*, 5882–5894.
63. Christen, Y. (2000). Oxidative stress and Alzheimer disease. *American Journal of Clinical Nutrition, 71*, 621S–629S.
64. Butterfield, D. A., Hensley, K., Cole, P., Subramaniam, R., Aksenov, M., Aksenova, M., Bummer, P. M., Haley, B. E., & Carney, J. M. (1997). Oxidatively induced structural alteration of glutamine synthetase assessed by analysis of spin label incorporation kinetics: Relevance to Alzheimer's disease. *Journal of Neurochemistry, 68*, 2451–2457.
65. Cooke, M. S., Evans, M. D., Dizdaroglu, M., & Lunec, J. (2003). Oxidative DNA damage: Mechanisms, mutation, and disease. *FASEB Journal, 17*, 1195–1214.
66. van der Bliek, A. M., Shen, Q., & Kawajiri, S. (2013). Mechanisms of mitochondrial fission and fusion. *Cold Spring Harbor Perspectives in Biology, 5*(6), a011072.
67. Calkins, M. J., Manczak, M., Mao, P., Shirendeb, U., & Reddy, P. H. (2011). Impaired mitochondrial biogenesis, defective axonal transport ofmitochondria, abnormal

mitochondrial dynamics and synaptic degeneration in a mouse model of Alzheimer's disease. *Human Molecular Genetics, 20*, 4515–4529.

68. Trushina, E., Nemutlu, E., Zhang, S., Christensen, T., Camp, J., Mesa, J., Siddiqui, A., Tamura, Y., Sesaki, H., Wengenack, T. M., & Dzeja, P. P. (2012). Defects in mitochondrial dynamics and metabolomic signatures of evolving energetic stress in mouse models of familial Alzheimer's disease. *PLoS One, 7*(2), e32737.
69. Djordjevic, J., Chowdhury, S. R., Snow, W. M., Perez, C., Cadonic, C., Fernyhough, P., & Albensi, B. C. (2020). Early onset of sex-dependent mitochondrial deficits in the cortex of 3xTg Alzheimer's mice. *Cells, 9*, 1541.
70. Li, X. C., Hu, Y., Wang, Z. H., Luo, Y., Zhang, Y., Liu, X. P., Feng, Q., Wang, Q., Ye, K., Liu, G. P., et al. (2016). Human wild-type full-lengthtau accumulation disrupts mitochondrial dynamics and the functions via increasing mitofusins. *Scientific Report, 6*.
71. Wang, X., Su, B., Fujioka, H., & Zhu, X. (2008). Dynamin-like protein 1 reduction underlies mitochondrial morphology and distributionabnormalities in fibroblasts from sporadic Alzheimer's disease patients. *The American Journal of Pathology, 173*, 470–482.
72. Birnbaum, J. H., Wanner, D., Gietl, A. F., Saake, A., Kündig, T. M., Hock, C., Nitsch, R. M., & Tackenberg, C. (2018). Oxidative stress andaltered mitochondrial protein expression in the absence of amyloid-β and tau pathology in iPSC-derived neurons from sporadicAlzheimer's disease patients. *Stem Cell Research, 27*, 121–130.
73. Jucker, M. (2010). The benefits and limitations of animal models for translational research in neurodegenerative diseases. *Nature Medicine, 16*(11), 1210–1214.
74. Kokjohn, T. A., & Roher, A. E. (2009). Amyloid precursor protein transgenic mouse models and Alzheimer's disease: understanding the paradigms, limitations, and contributions. *Alzheimer's & Dementia, 5*(4), 340–347.
75. Mantile, F., & Prisco, A. (2020). Vaccination against β-Amyloid as a strategy for the prevention of Alzheimer's disease. *Biology, 9*, 425.
76. Loureiro, J. C., & Pais, M. V. (2020). Passive antiamyloid immunotherapy for Alzheimer's disease. *Current Opinion in Psychiatry, 33*, 284–291.
77. Nelson, P. T., & Alafuzoff, I. (2012). Correlation of Alzheimer disease neuropathologic changes with cognitive status: A review of the literature. *Journal of Neuropathology and Experimental Neurology, 71*, 362–381.
78. Cardoso, B. R., & Roberts, B. R. (2019). Supranutritional sodium selenate supplementation delivers selenium to the central nervous system: Results from a randomized controlled pilot trial in Alzheimer's disease. *Neurotherapeutics, 16*, 192–202.
79. Cummings, J. L., Morstorf, T., & Zhong, K. (2014). Alzheimer's disease drug-development pipeline: Few candidates, frequent failures. *Alzheimers Research Therapy, 6*(4), 37. https://www.clinicaltrials.gov/search?cond=Alzheimer%27s%20 Disease
80. Salloway, S., Sperling, R., Fox, N. C., Blennow, K., Klunk, W., Raskind, M., ... & Brashear, H. R. (2014). Two phase 3 trials of bapineuzumab in mild-to-moderate Alzheimer's disease. *New England Journal of Medicine, 370*(4), 322–333.

81. Egan, M. F., Kost, J., Tariot, P. N., Aisen, P. S., Cummings, J. L., Vellas, B., ... & Michelson, D. (2018). Randomized trial of verubecestat for mild-to-moderate Alzheimer's disease. *New England Journal of Medicine, 378*(18), 1691–1703.
82. Finn, R. S., Martin, M., Rugo, H. S., Jones, S., Im, S. A., Gelmon, K., ... & Slamon, D. J. (2016). Palbociclib and letrozole in advanced breast cancer. *New England Journal of Medicine, 375*(20), 1925–1936.
83. Gauthier, S., Feldman, H. H., Schneider, L. S., Wilcock, G. K., Frisoni, G. B., Hardlund, J. H., Moebius, H. J., Bentham, P., Kook, K. A., Wischik, D. J., & Schelter, B. O. (2016). Efficacy and safety of tau-aggregation inhibitor therapy in patients with mild or moderate Alzheimer's disease: A randomised, controlled, double-blind, parallel-arm, phase 3 trial. *The Lancet, 388*(10062), 2873-2884.

CHAPTER 5

Methods for Tailoring Medicines in Alzheimer's Disease: Customizing Therapies According to Individual Genetic and Molecular Characteristics

P. VEERESH BABU[1,*] and M. SRI RAMACHANDRA[2]

[1]*Department of Pharmacology, Gokaraju Rangaraju College of Pharmacy, Bachupally, Hyderabad, Telangana, India*

[2]*Dr. K. V. Subba Reddy Institute of Pharmacy, Dupapdu, Kurnool, Andhra Pradesh, India*

*Corresponding author

ABSTRACT

Alzheimer's disease (AD) presents a formidable challenge to modern medicine due to its complex etiology and heterogeneous nature across patients. Recent advancements in genomic and molecular research have shed light on the intricate mechanisms underlying AD pathogenesis, revealing significant inter-individual variability in genetic and molecular profiles. This variability underscores the necessity for personalized therapeutic approaches to effectively manage AD. Tailoring medicines for AD involves integrating patient-specific genetic, molecular, and clinical data to design individualized treatment strategies aimed at targeting the underlying disease mechanisms. In this review, we discuss cutting-edge methods for tailoring medicines in AD, focusing on the customization of therapies according to individual genetic and molecular characteristics. We first explore the role of genetic factors in AD susceptibility and progression,

Cellular and Molecular Mechanisms in Alzheimer's Disease Therapeutics, Volume 2: Molecular Mechanisms.
Prashant Tiwari, Sunil Kumar Kadiri, and Kunjbihari Sulakhiya (Eds)

DOI: 10.1201/9781779641816-5

highlighting the potential of genetic testing to identify actionable variants for personalized treatment selection. Next, we delve into molecular profiling techniques such as biomarker analysis, transcriptomics, proteomics, and metabolomics, which offer insights into disease mechanisms and aid in the identification of novel therapeutic targets. Furthermore, we examine the emerging field of pharmacogenomics, which elucidates how genetic variations influence drug response and toxicity, enabling the optimization of medication regimens for AD patients. Additionally, we discuss the integration of multidimensional data through bioinformatics and machine learning approaches to develop predictive models for treatment outcomes and disease progression. Moreover, we explore innovative therapeutic strategies, including gene therapy, immunotherapy, and precision drug delivery systems, which hold promise for targeted interventions tailored to individual molecular profiles. Finally, we concentrate on the challenges and limitations associated with implementing personalized medicine approaches in AD, including ethical considerations, regulatory hurdles, and healthcare system implications. In summary, personalized medicine provides a paradigm change in the treatment of AD by directing care toward individualized therapy that take into consideration each patient's particular genetic and molecular makeup. Personalized therapy approaches, which make use of cutting-edge technologies and interdisciplinary collaboration, have the potential to enhance treatment outcomes and open the door to more successful disease-modifying strategies for AD.

5.1 INTRODUCTION TO TAILORING MEDICINES IN ALZHEIMER'S DISEASE

5.1.1 AN OVERVIEW OF THE COMPLEXITY OF ALZHEIMER'S DISEASE

Aberrant protein aggregation buildup in the brain is a hallmark of Alzheimer's disease (AD). Beta-amyloid plaques and tau tangles are the main culprits. Tau tangles form inside neurons, whereas beta-amyloid plaques form outside of them. These protein buildups cause cell death by interfering with neuronal communication [1]. AD symptoms usually appear gradually and get worse over time. Forgetfulness, trouble focusing, and difficulties solving problems are some early indicators [2]. As the illness worsens, people may become confused, disoriented, suffer changes in their mood or demeanor, and eventually lose their capacity to carry out daily chores. Despite the fact that the precise etiology of AD is unknown, a number of risk factors have been found. These include ageing (as risk rises with age), genetics (a family history of AD increases risk), specific genes (such as *apolipoprotein E* (*ApoE*)*4*), lifestyle choices (such as food and exercise), and illnesses (such as diabetes, cardiovascular disease) [3–5]. There is not yet a single test that can be used to conclusively diagnose AD. A patient's medical history, cognitive tests, neurological examinations, blood testing, brain imaging (magnetic resonance imaging (MRI) or positron emission tomography (PET) scans), and occasionally cerebrospinal fluid analysis to find disease-related biomarkers [6] are used to make the diagnosis. AD has no known cure; however, certain drugs can help control symptoms and halt the illness's progression, especially in the early stages. These drugs mainly enhance behavior and cognitive performance via controlling neurotransmitters in the brain [7–9]. Patients, carers, and healthcare systems all face a variety of difficulties as a result of AD. It can have a major negative influence on quality of life, cause stress and exhaustion in carers, and result in large financial burdens [10]. As the world's population ages, AD is expected to grow more prevalent, highlighting the urgent need for effective cures and treatments. Researchers are still looking at the fundamental causes of Alzheimer's and possible treatment approaches. Research aims to create new treatments, such as immunotherapies and targeted pharmacological therapies, and to comprehend the roles that inflammation, heredity, and other factors play in the genesis of disease [11–13].

5.1.2 IMPORTANCE OF TAILORED MEDICINE IN ALZHEIMER'S TREATMENT

Tailored medicine has emerged as a capable approach in the treatment of AD, offering tailored interventions based on individual variability in genetics, biomarkers, and lifestyle factors [14]. Here are some key aspects highlighting the importance of personalized medicine in AD treatment:

1. **Early Detection and Diagnosis:** Personalized medicine allows for the identification of individuals at risk of developing Alzheimer's at an earlier stage. By analyzing genetic predispositions, biomarkers, and lifestyle factors, healthcare providers can detect AD even before symptoms manifest. Early detection enables the implementation of interventions aimed at slowing or halting disease progression [15].

2. **Precision Treatment Selection:** AD is a complex disease with varying underlying mechanisms among patients. Personalized medicine enables the selection of treatment strategies tailored to the specific molecular and genetic characteristics of each patient. This approach increases the likelihood of treatment efficacy while minimizing potential adverse effects [16].
3. **Genetic Profiling:** Genetic factors play a significant role in AD risk and progression. Personalized medicine utilizes genetic profiling to identify specific genetic variants associated with AD susceptibility, such as the *ApoE ε4* allele. Understanding an individual's genetic risk profile helps in devising personalized prevention and treatment strategies [17].
4. **Biomarker Assessment:** Biomarkers offer important information on the underlying pathology of AD. Examples of biomarkers include the proteins tau and amyloid-beta in cerebrospinal fluid or through imaging methods such as PET scans. Biomarker assessment is integrated into personalized medicine to support early diagnosis, track the course of disease, and evaluate therapy response [18–20].
5. **Tailored Lifestyle Interventions:** Lifestyle factors, including physical activity, cognitive stimulation, social engagement and diet, influence Alzheimer's risk and progression. Personalized medicine incorporates these factors into treatment plans, offering tailored lifestyle interventions to optimize brain health and potentially mitigate disease progression [21].
6. **Clinical Trial Design and Patient Stratification:** Personalized medicine enhances the design and carrying out of clinical trials for Alzheimer's treatments. By stratifying patients based on their genetic profiles, biomarker status, and other relevant characteristics, researchers can identify subpopulations most likely to benefit from specific interventions, leading to more efficient and successful clinical trials [22].
7. **Personalized Medicine:** It prioritizes patients' unique characteristics, preferences, and values, involving them in decision-making processes and tailoring treatment plans to improve outcomes and quality of life in Alzheimer's care [23–25].

5.1.3 HISTORICAL BACKGROUND AND ADVANCEMENTS IN PERSONALIZED MEDICINE APPROACHES

Personalized medicine, or precision medicine, is a healthcare approach that tailors treatment to an individual's unique genetic makeup, lifestyle, and environment, gaining momentum in recent decades due to genomics, technology, and data analytics [26].

1. **Historical Background:**
   a. Early Beginnings: The idea of personalized medicine traces back to ancient times when physicians recognized that individual differences influence disease susceptibility and response to treatments [27].
   b. Blood Typing: One of the earliest examples of personalized medicine was the discovery of blood types by Karl Landsteiner in the early 20th century. This laid the foundation for personalized blood transfusions [28].
   c. Pharmacogenetics: In the mid-20th century, researchers began to investigate how genetic variations affect responses to drugs, leading to the field of pharmacogenetics.
   d. Human Genome Project (HGP): One of the most significant milestones was the completion of the HGP in 2003, which provided a comprehensive map of the human genome and laid the groundwork for understanding genetic variations underlying health and disease [29, 30].
2. **Advancements in Personalized Medicine:**
   a. **Genomic Sequencing:** The cost of genomic sequencing has decreased significantly since the HGP, making it more accessible for research and clinical applications. Next-generation sequencing technologies have enabled rapid and affordable whole-genome or whole-exome sequencing [31–33].
   b. **Biobanks and Large-Scale Data:** The establishment of biobanks and large-scale genomic databases has facilitated research into the genetic basis of diseases and drug responses. Projects like the UK Biobank and All of Us in the US collect

vast amounts of health and genetic data from diverse populations [34].

c. **Bioinformatics and Data Analytics:** Advances in bioinformatics and data analytics have enabled the analysis of large genomic datasets to identify disease-associated genetic variants, predict treatment responses, and stratify patient populations [35].

d. **Targeted Therapies:** Personalized medicine has led to the development of targeted therapies tailored to specific genetic mutations or molecular biomarkers. Examples include targeted cancer therapies like Herceptin for HER2-positive breast cancer and kinase inhibitors for various cancers [36–38].

e. **Companion Diagnostics:** The integration of diagnostic tests with specific therapies, known as companion diagnostics, allows clinicians to identify patients who are most likely to benefit from a particular treatment. This approach improves treatment outcomes and reduces adverse effects [39].

d. **Immunotherapy:** In oncology, immunotherapy drugs harness the body's immune system to target and destroy cancer cells. Personalized approaches are being explored to identify patients who are most likely to respond to immunotherapy based on their tumor's genetic profile.

f. **Gene Editing Technologies:** Emerging gene editing technologies like CRISPR-Cas9 offer the potential to correct disease-causing genetic mutations directly. While still in early stages, these technologies hold promise for personalized treatments for genetic disorders [40].

Overall, personalized medicine continues to evolve rapidly, driven by advances in genomics, technology, and interdisciplinary collaborations between researchers, clinicians, and industry partners. As our understanding of the genetic basis of diseases deepens, personalized approaches are expected to play an increasingly prominent role in healthcare, leading to more precise diagnoses, targeted treatments, and improved patient outcomes.

5.2 GENETIC BASIS OF ALZHEIMER'S DISEASE

5.2.1 ALZHEIMER'S DISEASE-RELATED GENETIC RISK FACTORS

AD is a multifactorial condition that is impacted by both hereditary and environmental factors [41]. Age is still the biggest risk factor; however, some genetic differences can greatly raise the likelihood of developing the condition. The following are some important genetic risk factors for AD:

1. ***ApoE* Gene:** The *ApoE* gene, found in three variants, is a significant genetic risk factor for late-onset AD, with individuals with one or two copies of the *ApoE ε4* allele more likely to develop AD [42–45].
2. **Rare Mutations** in *amyloid precursor protein* (*APP*), *presenilin* (*PSEN)1*, and *PSEN2* genes are linked to early-onset familial AD, causing increased amyloid-beta protein synthesis. Additional genes, including *CLU*, *CR1*, *PICALM*, and *BIN1*, are identified as risk factors for late-onset AD, affecting lipid metabolism, immunological response, and synaptic function [46–48].
3. ***TREM2* Gene**: Rare variations in the *TREM2* gene, involved in immunological response and phagocytosis of cellular debris, have been linked to an increased risk of late-onset AD [49].
4. ***SORL1* Gene**: There is additional evidence linking variations in the *sortilin-related receptor 1 (SORL1)* gene to a higher risk of AD. The processing and trafficking of *APP* are associated with *SORL1*, which has an impact on amyloid-beta [50] synthesis and clearance. It is crucial to remember that while these genetic variables raise a person's risk of developing Alzheimer's, they do not guarantee that a person will get the illness. Significant contributions are also played by environmental variables, lifestyle decisions, and other genetic differences. Furthermore, genetic testing for the likelihood of developing AD is available; however, because of the disease's complexity and the scarcity of existing treatments, it is generally not advised for screening the general public.

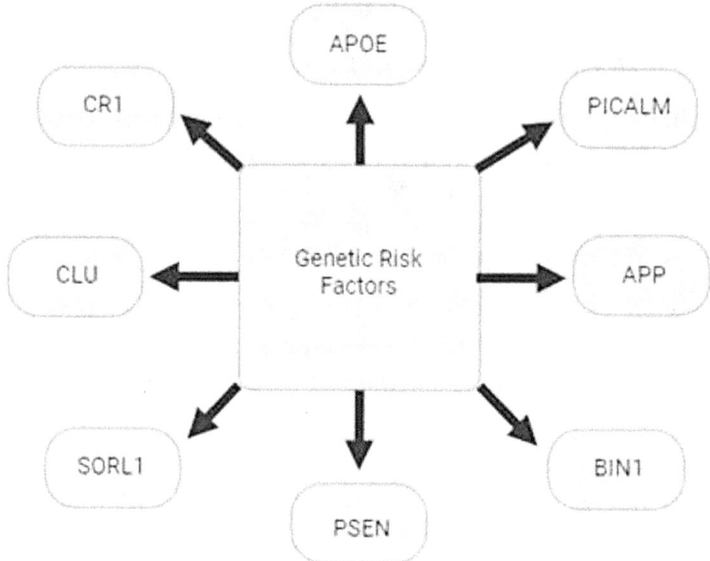

FIGURE 5.1 Genetic players of AD.

5.2.2 ALZHEIMER'S PATHOLOGY AND THE ROLE OF GENES LIKE APP, PSEN, AND ApoE

AD pathogenesis is significantly influenced by genes such as *PSEN*, *APP*, and *ApoE*.

1. ***ApoE:*** *ApoE*, a gene involved in cholesterol metabolism, has three primary alleles. The ε4 allele is the most significant genetic risk factor for late-onset Alzheimer's disease, with carriers developing Alzheimer's earlier than noncarriers [51].
2. **APP:** The APP, which is produced by enzyme cleavage to yield amyloid-beta (Aβ) peptides, is encoded by the *APP* gene. Aβ peptides build up in the brain during AD, causing plaques that impair neuronal function and hasten neurodegeneration. Aβ can be overproduced or processed differently as a result of *APP* gene mutations, which increases the risk of plaque formation and hastens the onset of AD. The processing of *APP* and the buildup of Aβ are essential to the pathophysiology of both familial and sporadic types of AD, even though mutations in APP are

comparatively uncommon and usually linked to familial early-onset AD [52, 53].

3. **PSEN:** Proteins called presenilins are engaged in the APP processing pathway. In contrast to sporadic occurrences, family variants of AD are caused by mutations in the *PSEN1* and *PSEN2* genes, and symptoms usually manifest earlier in life. Presenilins are a component of the complex known as γ-secretase, which cleaves *APP* to yield Aβ peptides. *PSEN* gene mutations can change how *APP* is processed, which increases the synthesis of Aβ42 and other more hazardous forms of the protein. The neuro-degenerative cascade observed in AD [54] and the development of amyloid plaques are both influenced by this increased production of Aβ.

5.2.3 GENETIC VARIATIONS' EFFECTS ON THE COURSE OF DISEASE AND THE RESPONSE TO TREATMENT

Genetic variations are important for how different medical problems progress and how well treatments work.

TABLE 5.1 Role of Genes in Pathophysiology of AD.

S.No.	Genes	Role
1	*Amyloid Precursor Protein (APP)*	Increased synthesis of amyloid—beta (Aβ) is caused by the mutations of these genes and forms fragments. These fragements forms plaques in the brain.
2	*Presenilin 1 (PSEN1)*	
3	*Presenilin 2 (PSEN2)*	
4	***Apolipoprotein E (ApoE)***	
4.1	*Apolipoprotein E ε2 (ApoE ε2)*	Provide some protection against the disease.
4.2	*Apolipoprotein E ε3 (ApoE ε3)*	The most common allele is believed to have a neutral effect on Alzheimer's disease, neither decreasing nor increasing the risk.
4.3	*Apolipoprotein E ε4 (ApoE ε4)*	The risk of Alzheimer's increases in certain populations and is linked to an earlier age of disease onset.
	Other genes	
	CLU, CR1, PICALM, and BIN1 genes	Identified as risk factors for late-onset Alzheimer's disease.

This is a summary of their influence:

1. **Disease Vulnerability:** A person's vulnerability to contracting particular diseases may be increased by certain genetic differences. For instance, there is a considerable increase in the risk of ovarian and breast cancers due to mutations in the *BRCA1* and *BRCA2* genes. Strategies for early detection and prevention can benefit from an understanding of these genetic predispositions [55].
2. **Illness Advancement:** Genetic variants can alter parameters including illness severity, rate of advancement, and risk of complications, which can all have an impact on how a disease develops. For example, certain genetic variations are linked to an earlier onset and faster progression of symptoms in illnesses such as AD [56].
3. **Response to Treatment:** Genetic differences can affect an individual's reaction to drugs and other forms of therapy. Treatment regimens are becoming more and more individualized thanks to pharmacogenomics, the study of how a person's genes impact how they respond to medications. For instance, differences in the genes that encode drug-metabolizing enzymes can affect the efficacy and adverse effects of pharmaceuticals [57].
4. **Precision Medicine:** The development of genomic technologies has made it possible to customize therapies based on a patient's genetic composition. Healthcare professionals can identify targeted medicines with a higher chance of success and fewer side effects by analyzing a patient's genetic profile [58].
5. **Risk Assessment:** Information from genetic testing can be very helpful in risk assessment and stratification. Genetic testing, for instance, can identify people with a higher genetic risk for cardiovascular illnesses, enabling early interventions and lifestyle changes to lower the chance of unfavorable outcomes
6. **Prognostic Markers**: Genetic differences can act as prognostic markers, influencing therapy choices and assisting in the prediction of disease outcomes. For example, certain genetic mutations in cancer may signal a worse prognosis or a greater chance of recurrence, which would lead to more aggressive treatment strategies [59].

7. **Drug Development:** It is crucial to comprehend how genetic variations affect disease biology and treatment response in order to develop effective new drugs. Through the use of genetic insights to identify therapeutic targets, researchers can create more precise and effective treatments with fewer side effects [60].

5.3 MOLECULAR CHARACETRISTICS OF ALZHEIMER'S DISEASE

5.3.1 AMYLOID-BETA AND TAU PROTEINS: KEY PLAYERS IN ALZHEIMER'S PATHOLOGY

AD is primarily caused by the abnormal folding of amyloid-beta (Aβ) protein, leading to the formation of amyloid plaques. Tau protein, crucial for neuronal microtubules, undergoes abnormal changes, including hyper-phosphorylation, resulting in tau tangles, which impair normal neuronal function and lead to neuronal degeneration [61, 62].

Amyloid plaques and tau tangles are believed to be involved in the neurodegeneration and cognitive loss associated with AD. In addition to developing new treatments that target these proteins, research is continuing to determine the precise processes by which these proteins contribute to the disease process [63].

5.3.2 OTHER MOLECULAR PATHWAYS IMPLICATED IN DISEASE DEVELOPMENT

Some additional pathways beyond the well-known amyloid-beta and tau pathologies:

1. **Neuroinflammation:** The development of AD is significantly influenced by chronic neuroinflammation. Pro-inflammatory cytokines are released upon activation of microglia and astrocytes, contributing to the degeneration and death of neurons [64].

2. **Oxidative Stress:** Elevated levels of reactive oxygen species (ROS) and diminished defenses against antioxidants cause oxidative stress to neurons, which in turn aggravates AD-related neurodegeneration [65].

3. **Cholesterol Metabolism:** AD pathology has been linked to dysregulation of cholesterol metabolism, specifically changes in

cholesterol transport and metabolism inside the brain. Amyloid plaque development may be facilitated by elevated cholesterol levels [66].

4. **Mitochondrial Dysfunction:** In AD, impaired mitochondrial function, including flaws in energy metabolism and elevated ROS generation, leads to neuronal malfunction and cell death [67].

5. **Synaptic Dysfunction:** A hallmark of AD, early synaptic dysfunction, including decreased neurotransmitter release and synapse loss, contributes to cognitive decline [68].

6. **Glutamate Toxicity:** AD can cause neuronal damage and death due to dysregulation of glutamate neurotransmission and excitotoxicity.

7. **Vascular Factors:** AD risk has been associated with vascular risk factors, including diabetes, hypertension, and atherosclerosis. Neurodegeneration can be exacerbated by vascular dysfunction, which can result in decreased cerebral blood flow, hypoxia, and nutritional deprivation [69].

8. **Insulin Signaling:** The pathophysiology of AD has been linked to insulin resistance and compromised insulin signaling in the brain, which results in dysregulated glucose metabolism and heightened neuronal vulnerability.

9. **Tau Phosphorylation:** Aberrant tau protein phosphorylation impairs microtubule stability and neuronal function, even though tau pathology is mostly linked to intracellular neurofibrillary tangles in AD.

10. **Epigenetic Modifications:** Variations in noncoding RNA expression, histone modifications, and DNA methylation can affect gene expression patterns linked to the pathophysiology of AD [70].

5.3.3 BIOMARKERS FOR DIAGNOSING AND MONITORING ALZHEIMER'S PROGRESSION

Biomarkers are essential for the diagnosis and tracking of AD. Some of the most important biomarkers in the field are as follows:

1. **Levels of the Protein Fragment Beta-Amyloid (Aβ):** People with AD develop plaques in their brains made of beta-amyloid. Increased beta-amyloid concentrations, especially in cerebrospinal fluid (CSF), may be a sign of AD pathology [71].
2. **Tau Protein:** Tau protein helps keep neuronal microtubules stable. Neurofibrillary tangles are caused by hyperphosphorylation of tau in AD. AD is indicated by elevated tau levels in CSF or alterations in tau protein in imaging studies [72].
3. **Neuroimaging Techniques** like PET scans can detect beta-amyloid plaques and neurofibrillary tangles in the brain, which are linked to AD and brain structure and function changes [73].
4. **Genetic Markers:** The likelihood of having AD is significantly influenced by genetic variables. Early-onset familial AD is linked to mutations in genes including *APP*, *PSEN1*, and *PSEN2*. One of the main risk factors for late-onset AD is the *ApoE ε4* allele [74].
5. **Neuroinflammation Markers:** One of the main characteristics of AD is inflammatory in the brain. Markers in CSF or blood, such as inflammatory mediators, chemokines, and cytokines, can show neuroinflammation linked to the pathophysiology of AD [75].
6. **Markers of Synaptic Dysfunction:** The early stages of AD involve synapse loss and dysfunction. CSF contains biomarkers that are linked to synaptic dysfunction, including SNAP-25 and neurogranin.
7. **Blood-Based Biomarkers:** Levels of beta-amyloid, tau, and other proteins linked to neurodegeneration are among the blood-based biomarkers for AD that researchers are now examining. Blood testing have the potential to be more accessible and less intrusive diagnostic techniques [76].
8. **Cognitive Exams**: Cognitive assessments like the Montreal Cognitive Assessment (MoCA) and the Mini-Mental State Examination (MMSE) are valuable diagnostic and tracking tools for AD progression [77].

In clinical studies, these indicators are employed singly or in combination to monitor the course of AD and gauge the patient's reaction to treatment. It is crucial to remember that there is no one biomarker that can be used to diagnose AD; instead, a thorough evaluation that includes imaging, clinical assessment, and biomarker analysis is frequently required for a precise diagnosis.

5.4 PHARMACOGENOMICS IN THE MANAGEMENT OF AD

5.4.1 SYNOPSIS OF PHARMACOGENOMICS AND ITS USE IN ALZHEIMER'S TREATMENT

Pharmacogenomics is the study of how a person's genetic composition affects how they react to medications. It blends the fields of genomics—the study of genes and their functions—with pharmacology, the study of pharmaceuticals. Pharmacogenomics seeks to optimize pharmacological therapy for people by comprehending how genetic variations impact drug metabolism, efficacy, and adverse responses. Pharmacogenomics is a promising field in the treatment of AD. Its uses include:

1. **Predicting Drug Reaction**: Genetic differences can affect how people react to medications used to treat AD, including memantine and cholinesterase inhibitors (donepezil, rivastigmine, etc.). Physicians can anticipate a patient's response to a certain medication by examining their genetic profile. For instance, different genetic variants may impact how certain medications are metabolized, resulting in differences in their effectiveness or susceptibility to adverse effects [78].
2. **Pharmacogenomics** makes personalized medicine strategies for Alzheimer's treatment possible. Healthcare practitioners can customize drug regimens based on an individual's genetic profile, as an alternative to using a one-size-fits-all therapy approach. This may result in less adverse effects and better treatment outcomes [79].
3. **Drug Creation**: Knowledge of the hereditary causes of AD can help in the creation of new medications. Pharmacogenomic research has the ability to pinpoint possible therapeutic targets by examining genetic risk factors linked to the illness. Furthermore, by designing clinical trials with enriched patient populations that are likely to respond to experimental medications, researchers might benefit from genetic insights [80].

4. **Risk Assessment and Prevention:** There is a genetic mutation that has been linked to a higher chance of AD. The application of pharmacogenomics to the identification of high-risk patients can facilitate the implementation of early intervention or tailored preventive measures. Furthermore, knowledge of genetic risk factors can guide therapies and lifestyle changes to slow the progression of disease [81].
5. **Prediction of Adverse Drug Reaction:** Certain genetic differences may put people at risk for negative drug reactions. These genetic markers can be found using pharmacogenomic testing, which gives medical professionals the ability to foresee and stop certain negative drug reactions to Alzheimer's treatments.
6. **Pharmacogenomics** presents ethical questions about genetic privacy, consent, and the fair allocation of healthcare resources in spite of its potential advantages. The proper integration of pharmacogenomics into clinical practice requires addressing concerns of access and affordability, safeguarding genetic data, and ensuring informed consent [82].

All things considered, pharmacogenomics presents a promising future for personalized medicine methods, better treatment outcomes, and direction for drug discovery in the context of Alzheimer's medication optimization. However, in order to fully realize its promise in managing this intricate neurodegenerative disease, more research and clinical validation are required.

5.4.2 GENETIC VARIATIONS INFLUENCING DRUG METABOLISM AND RESPONSE

Genetic differences are important in determining how drugs are metabolized and how people respond to them. The following are some salient points:

1. **Enzymes in Drug Metabolism:** The bulk of pharmaceuticals are broken down by liver-based enzymes, mainly cytochrome P450 (CYP) family members. Different drug metabolism rates can result from changes in the activity of these enzymes caused by genetic polymorphisms. For instance, differences in the *CYP2D6* gene might lead to the classification of individuals as ultrarapid, extensive, or poor metabolizers, which can impact the toxicity and effectiveness of medications metabolized by this enzyme [83].

2. **Pharmacokinetics:** Genetic differences can affect the absorption, distribution, metabolism, and excretion (ADME) of drugs. Drug levels in the body can be impacted by variations in drug transporters, such as P-glycoprotein, which can influence drug absorption and distribution [84].
3. **Drug Targets:** Receptors and enzymes involved in the drug's mechanism of action are examples of drug targets that can be impacted by genetic differences. Changes in these targets may have an impact on the toxicity or efficacy of drugs. For instance, the responsiveness to beta-blockers used in the treatment of cardiovascular disease may be impacted by certain genetic variants in the gene encoding the β_1-adrenergic receptor [85].
4. **Adverse Effects and Drug Response:** Genetic differences may play a role in the interindividual variability of adverse effect susceptibility and drug response. For instance, changes in the area of the *HLA* gene have been linked to hypersensitive reactions to specific medications, including abacavir used to treat HIV or carbamazepine used to treat epilepsy [86].
5. **Pharmacogenomics:** This field of study, which aims to personalize drug therapy, is gaining traction in clinical practice. It examines how genetic differences affect drug response. Healthcare professionals can improve medicine selection, dosage, and monitoring for individual patients, improving treatment outcomes and lowering the risk of side effects, by discovering genetic variants that alter drug metabolism and response [87].

In general, individualized medicine approaches for enhancing the effectiveness and safety of medication therapy depend on an understanding of how genetic variations impact drug metabolism and response.

5.4.3 CUSTOMIZING MEDICATION REGIMENS ACCORDING TO UNIQUE GENETIC PROFILES

The study of customizing medication regimens according to unique genetic profiles is called pharmacogenomics. This method considers genetic variants that impact an individual's reaction to drugs.

Gene variations among individuals can impact the way their bodies metabolize drugs. The metabolism, effectiveness, and possible side effects of drugs can all be impacted by these genetic variations. Using DNA analysis, pharmacogenomic testing determines particular genetic variants that may affect an individual's response to a certain medicine. There are several ways to accomplish this testing, such as taking saliva samples or doing blood tests. Healthcare professionals can select medications that are most likely to be safe and effective for a certain patient based on the results of genetic testing. For instance, a patient's doctor might recommend a different prescription or change the dosage if they have a genetic mutation that impacts how they metabolize a certain drug. Healthcare professionals can also benefit from genetic information in determining the best way to administer a drug to a specific patient. Based on their genetic makeup, some individuals may need higher or lower doses of specific medications. Healthcare professionals can reduce the likelihood of negative responses and side effects by customizing medication therapies based on each patient's genetic profile. Treatment outcomes overall and patient safety can both be enhanced by this individualized approach [88]. We could witness even more advanced personalized treatment strategies as pharmacogenomics research progresses. This might involve creating brand-new medications created especially for people with particular genetic variants.

5.5 TAILORED THERAPEUTIC METHODS

5.5.1 SPECIFIC MOLECULAR PATHWAYS TARGETED BY TARGETED THERAPIES

Research on targeted therapeutics for AD is ongoing, with the goal of tackling the underlying molecular mechanisms that contribute to the onset and progression of the illness. The following are some of the main molecular pathways that currently under development medications target:

1. **Aggregation of Beta-Amyloid:** The protein beta-amyloid causes plaques to develop in the brains of Alzheimer's patients. The goal of beta-amyloid-targeting therapies is to either stop it from forming or encourage its removal from the brain. Among them are monoclonal antibodies made to attach to beta-amyloid and aid in its elimination [89].

2. **Anomalies of the Tau Protein:** Alzheimer's sufferers' brains tangle with tau, another protein. Treatments for tau are intended

to stop its aberrant aggregation or encourage its excretion. Among the strategies are immunotherapies that specifically target tau or tiny compounds that prevent tau from aggregating [90].

3. **Neuroinflammation:** It is believed that inflammation in the brain has a role in the development of AD. In order to lessen inflammation, therapies that target neuroinflammation try to alter the immune response in the brain. Targeting particular inflammatory pathways or immune cells implicated in the inflammatory response may be one method to achieve this [91].

4. **Synaptic Dysfunction:** The loss of synapses, or the connections between neurons, is a hallmark of AD. Treatments for synaptic dysfunction seek to preserve already-existing synapses or encourage the growth of new ones. Targeting chemicals involved in synaptic function or signaling pathways that control synaptic plasticity may be one method to do this [92].

5. **Mitochondrial Dysfunction**: The energy-producing organelles found in cells, mitochondria, are involved in the pathophysiology of AD. The goal of therapies aimed at mitochondrial dysfunction is to enhance cellular energy metabolism and restore mitochondrial function. Utilizing antioxidants that target mitochondria or other compounds that support mitochondrial health may be one way to do this [93].

6. **Abnormalities in the Metabolism of Cholesterol Have Been Linked to AD**. Treatments that target cholesterol metabolism seek to lower the synthesis of harmful cholesterol metabolites or return normal levels of cholesterol to the brain. The use of statins or other cholesterol-lowering medications may be necessary for this [94].

7. **Cognitive Enhancement**: Certain treatments target neurotransmitter systems involved in memory and learning, such as the cholinergic system, in an effort to improve cognitive performance in Alzheimer's patients. Cholinesterase inhibitors and other medications that increase cholinergic neurotransmission may be used in these treatments [95].

By focusing on the particular molecular pathways implicated in the pathophysiology of AD, these targeted medicines provide intriguing new avenues for treating the condition. It's crucial to remember that a lot of these treatments are still in the experimental phase and need additional clinical research to ascertain their efficacy and safety in people. Comprehending the intricate interactions among these biological pathways is essential for creating efficacious therapy approaches for AD.

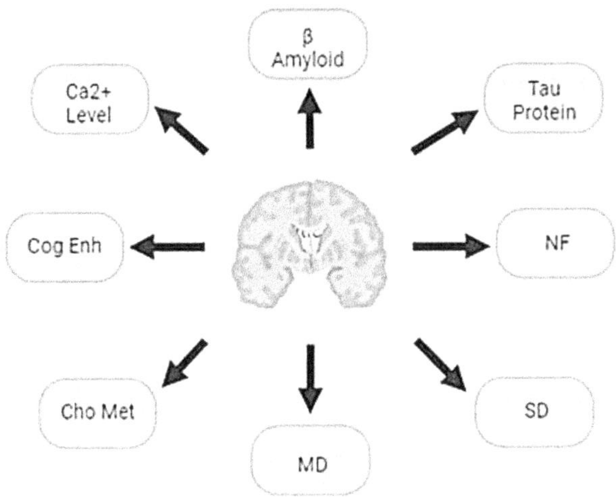

FIGURE 5.2 Personalized approaches of AD.

5.5.2 IMMUNOTHERAPY AND ITS POTENTIAL FOR PERSONALIZED TREATMENT

Immunotherapy is a novel method to cancer care, utilizing the body's immune system to combat cancerous cells.

1. **Targeted Therapy:** Immunotherapy can be designed to specifically target chemicals on cancer cells or immune system checkpoints. Immune checkpoint drugs, such as CTLA-4 or PD-1 inhibitors, for example, function by obstructing the signals that cancer cells employ to avoid the immune system [96].

2. **Biomarker Identification:** Personalized immunotherapy relies heavily on identifying biomarkers—indicators that can predict how a patient will respond to a particular treatment. Biomarkers might include genetic mutations, protein expression levels, or other factors. For example, the expression of PD-L1 in tumors can predict response to PD-1/PD-L1 inhibitors.

3. **Genetic Profiling:** Advances in genomic sequencing allow for a deeper understanding of the genetic makeup of tumors. By analyzing the genetic mutations driving cancer growth, doctors can identify targets for immunotherapy drugs or other targeted therapies [97].
4. **Tumor Microenvironment Analysis:** The tumor microenvironment plays a crucial role in cancer progression and response to treatment. Immunotherapy can be personalized by analyzing the composition of immune cells, stromal cells, and other components within the tumor microenvironment [98].
5. **Combination Therapies:** Personalized treatment often involves combining different therapies to target multiple aspects of cancer biology. Immunotherapy can be combined with chemotherapy, radiation therapy, targeted therapy, or other immunotherapies to enhance treatment efficacy [99].
6. **Predictive Models:** Machine learning and other computational approaches are being used to develop predictive models that can help identify which patients are most likely to benefit from immunotherapy. These models integrate various clinical, molecular, and imaging data to make personalized treatment recommendations.
7. **Clinical Trials:** Participation in clinical trials is essential for advancing personalized immunotherapy. Clinical trials often incorporate biomarker testing and other personalized approaches to identify the most effective treatments for individual patients [100].

All things considered, the promise of personalized immunotherapy is its capacity to customize treatment plans according to the particulars of each patient's malignancy. Personalized immunotherapy has the ability to decrease needless side effects and enhance outcomes by determining which medicines are most beneficial for certain patients.

5.5.3 DRUG REPURPOSING AND COMBINATION THERAPIES FOR PERSONALIZED CARE

Drug repurposing and combination therapies hold promising avenues for personalized care in AD, a complex neurodegenerative disorder characterized by cognitive decline and memory loss. Following are some strategies and considerations:

1. **Drug Repurposing:** This involves identifying existing drugs approved for other conditions that may have therapeutic effects on Alzheimer's. Several drugs originally developed for unrelated conditions have shown potential benefits in AD treatment. For instance:

    a. **Memantine:** Originally used for moderate-to-severe Alzheimer's, it's now also being explored for its potential in treating mild cases.

    b. **Anti-Inflammatory Medications:** Nonsteroidal anti-inflammatory medications (NSAIDs) are one type of anti-inflammatory medication that has demonstrated promise in lowering inflammation linked to AD pathology.

    c. **Diabetes Medications:** Drugs used to manage diabetes, such as insulin-sensitizing agents, are being investigated for their neuroprotective effects in AD [101–103].

2. **Combination Therapies:** Given the complexity of AD, a combination of drugs targeting multiple pathways may be more effective than single-drug therapies. Combining drugs with complementary mechanisms of action can potentially enhance treatment outcomes and slow disease progression. For example:

    a. **Cholinesterase Inhibitors (ChEIs) + Memantine:** This combination is sometimes used to target both cholinergic dysfunction and glutamatergic excitotoxicity, two key mechanisms implicated in AD.

    b. **Anti-Amyloid Therapies + Anti-Inflammatory Drugs:** Combining drugs targeting amyloid-beta accumulation with anti-inflammatory agents may provide synergistic effects in mitigating neuroinflammation and reducing amyloid plaque burden [104, 105].

3. **Personalized Medicine Approaches**: AD is heterogeneous, with various underlying pathologies and genetic factors contributing to its development and progression. Personalized medicine approaches aim to tailor treatment strategies to individual patients based on

their unique genetic makeup, biomarker profiles, and disease stage. Key elements of personalized care in AD may include:

a. **Genetic Testing**: Identifying genetic risk factors such as the presence of *ApoE* alleles can help guide treatment decisions and predict disease progression.

b. **Biomarker Profiling**: Utilizing biomarkers such as amyloid-beta and tau levels in cerebrospinal fluid or through neuroimaging techniques (e.g., PET scans) can aid in diagnosing AD and monitoring treatment response.

c. **Cognitive and Functional Assessments:** Regular assessments of cognitive function and daily living activities help clinicians track disease progression and adjust treatment strategies accordingly [106–108].

4. **Emerging Therapeutic Targets:** Ongoing research continues to uncover novel therapeutic targets and treatment approaches for AD. These include targeting neuroinflammation, synaptic dysfunction, mitochondrial dysfunction, and epigenetic modifications. Investigational therapies such as monoclonal antibodies targeting specific protein aggregates (e.g., amyloid-beta, tau) and small molecules modulating synaptic function are being actively studied in clinical trials [109].

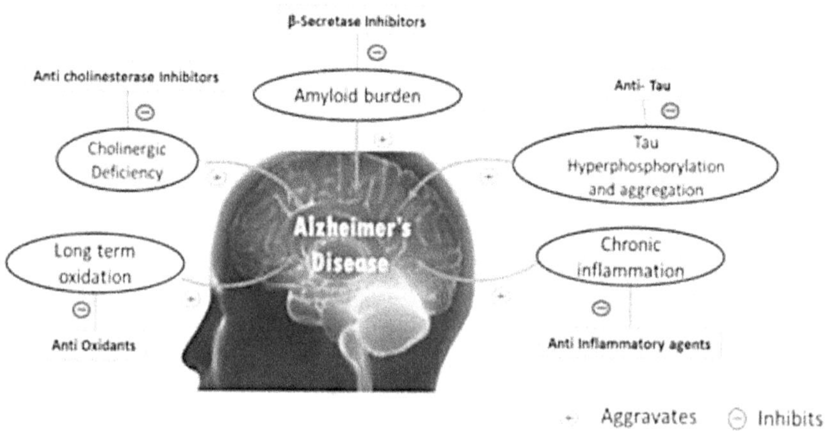

FIGURE 5.3 Various treatment options of AD.

5.6 CHALLENGES AND OPPORTUNITIES IN PERSONALIZED MEDICINE FOR AD

5.6.1 ETHICAL CONSIDERATIONS IN GENETIC TESTING AND PERSONALIZED TREATMENT

Genetic testing and personalized treatment for AD present a range of ethical considerations that need careful attention. Following are some key points:

1. **Informed Consent:** Patients undergoing genetic testing for AD should fully understand the implications of the test results, together with the potential psychological impact and limitations of predictive testing. Informed consent ensures that individuals are aware of the potential consequences before deciding to undergo testing [110].
2. **Privacy and Confidentiality:** Genetic information is highly sensitive and raises concerns about privacy and confidentiality. There is a risk of discrimination in employment, insurance, and social contexts based on genetic predispositions. Safeguards should be in place to defend individuals from unauthorized access to their genetic data [111].
3. **Psychological Impact**: Learning about one's genetic risk for AD can have profound psychological effects on individuals and their families. It may lead to anxiety, depression, or altered family dynamics. Support systems should be available to help individuals cope with the emotional consequences of genetic testing [112].
4. **Stigmatization:** There is a risk of stigmatization associated with genetic predispositions to AD. Patients may face discrimination or social isolation based on their genetic risk status. Efforts to educate the public and combat stigma are essential to promote understanding and acceptance [113].
5. **Equitable Access:** To guarantee that everyone has the chance to profit from advancements in genomic medicine, regardless of socioeconomic status or location, access to genetic testing and customized therapy should be equitable. To avoid more disparities in Alzheimer's care, disparities in access to genetic testing and healthcare services need to be addressed [114].

6. **Beneficence and Nonmaleficence:** Medical professionals need to weigh the possible advantages of genetic testing and individualized AD treatment against the associated risks and uncertainty. While the nonmaleficence principle stresses avoiding needless risks to patients, the beneficence principle calls for optimizing benefits and minimizing harm [115].

7. **Data Sharing and Research Ethics:** Determining the fundamental causes of AD and creating efficient treatments depend heavily on genetic research. To safeguard the rights and welfare of study participants, however, researchers must abide by stringent ethical guidelines with regard to data sharing, permission, and participant confidentiality [116].

8. **End-of-Life Decision-Making:** Alzheimer's patients and their families may find that the results of genetic testing have an impact on their final decisions. To guarantee that patients' desires are honored and their quality of life is maximized, advance care planning and conversations regarding treatment preferences—including palliative and hospice care—are crucial [117]. Healthcare professionals, researchers, legislators, and patient advocacy organizations must work together to address these ethical issues and create standards and laws that support the values of autonomy, beneficence, justice, and respect for people's rights and dignity.

5.6.2 ACCESSIBILITY AND AFFORDABILITY OF PERSONALIZED THERAPIES

Addressing the increasing prevalence of AD requires improving the accessibility and cost of personalized therapies. The following tactics could be useful:

1. **Early Diagnosis and Detection:** Regular cognitive screenings for people over a specific age may aid in the early diagnosis of AD. Timely action and customized treatment approaches are made possible by early diagnosis [118].

2. **Developments in Biomarker Research:** Further investigation into AD biomarkers may result in more precise diagnosis and

individualized treatment plans. Personalized therapies can be made possible by the use of biomarkers to identify people who are at risk or in the early stages of the disease [119].

3. **Precision Medicine Approaches:** Customizing treatment regimens for each patient according to their genetic composition, biomarker profile, and other characteristics may result in more potent medications. By considering each patient's distinct qualities, precision medicine techniques may improve therapy efficacy [120].
4. **Clinical Trials and Research Financing:** Increasing financing for research on AD can hasten the discovery of novel treatments and make them more accessible. This is especially true in the field of personalized medicine. Patients may be able to obtain cutting-edge therapies by taking part in clinical trials [121].
5. **Regulatory Policies:** Reducing development costs and increasing patient access to personalized therapeutics can be achieved by streamlining regulatory procedures. Pharmacies and researchers can collaborate with regulatory bodies to hasten the approval of customized therapies that have proven to be safe and effective [122].
6. **Support from the Healthcare System**: By guaranteeing that certain treatments are covered by insurance plans, healthcare systems may significantly contribute to increasing the affordability and accessibility of customized therapies. Personalized medicine approaches can also be taught to healthcare professionals so they can better serve patients with AD [123].
7. **Patient Education and Support:** Raising knowledge and acceptance of these treatments can be achieved by teaching patients and their carers about personalized medicine and its possible advantages. For those impacted by AD, support groups and instructional materials can offer invaluable knowledge and resources [124]

By putting these tactics into practice and encouraging cooperation between researchers, medical professionals, legislators, and patients, we can endeavor to increase the affordability and accessibility of tailored therapy for Alzheimer's patients.

5.6.3 FUTURE PATHS AND POSSIBLE ADVANCES IN PERSONALIZED MEDICINE FOR AD

Since AD is a multifaceted neurodegenerative condition, personalized medicine presents a promising avenue for enhancing diagnosis, treatment, and care. The following are some possible paths and developments in customized care for AD in the future:

1. **Genetic Screening and Risk Assessment**: People who are more susceptible to AD can be identified thanks to advancements in genetic testing, especially with genome-wide association studies (GWAS). Targeted therapies and preventative measures can be made possible by personalized risk assessments based on genetic markers [125].
2. **Biomarker-Based Diagnostics:** Biomarkers are essential for tracking the advancement of disease and for making early diagnoses. Research on imaging biomarkers, including tau and amyloid-beta proteins, and blood-based CSF will increase the sensitivity and accuracy of identifying Alzheimer's pathology [126].
3. **Precision Imaging Methods:** New developments in neuroimaging technology, such as MRI and PET, provide improved specificity and resolution for identifying brain abnormalities linked to AD. When imaging data is combined with additional indicators, it can offer a thorough understanding of disease causes and customized treatment recommendations.
4. **Machine Learning and Data Analytics**: To identify disease subtypes, forecast progression trajectories, and customize treatment plans, machine learning algorithms can examine huge datasets including genetic, clinical, imaging, and omics data. Therapeutic optimization and personalized risk prediction are possible applications of these computational approaches [127].
5. **Targeted Therapies:** The goal of personalized medicine is to match patients with therapies that are specific to their disease characteristics and distinct molecular profiles. There is hope for more tailored and efficient interventions thanks to emerging pharmacotherapies that target particular pathways linked to the

etiology of AD, including tau hyperphosphorylation, neuroinflammation, amyloid-beta aggregation, and synaptic dysfunction

6. **Lifestyle Interventions**: Based on a person's unique risk factors and illness stage, personalized therapies include lifestyle changes such as nutrition, exercise, mental stimulation, and social interaction. Aside from reducing the risk of AD, interventions aimed at vascular risk factors like obesity, diabetes, and hypertension may also enhance general brain health.
7. **Regenerative Medicine and Gene Therapy:** Cutting-edge approaches, such gene editing and stem cell-based treatments, show promise in altering disease-causing genes, regaining neuronal function, and encouraging Alzheimer's patients' brain repair. Personalized methods that take into account individual response profiles and genetic variability can maximize therapy efficacy and safety.
8. **Patient-Centered Care:** When making decisions, personalized medicine places a strong emphasis on the aims, values, and preferences of the patient. For people with AD and their families, integrated care approaches that include multidisciplinary teams, carer support, and community services can improve patient-centered outcomes and quality of life [128].

We can better understand the variability of AD, customize interventions to meet the needs of each patient, and ultimately enhance clinical outcomes and disease management by utilizing the power of personalized medicine. But to fully utilize personalized medicine in Alzheimer's, multidisciplinary cooperation, long-term research, government backing, and broad acceptance of cutting-edge tools and procedures are needed.

5.7 FINAL THOUGHTS AND UPCOMING PROSPECTS

5.7.1 KEY DISCOVERIES AND TAKEAWAYS FROM RESEARCH ON PERSONALIZED MEDICINE AND AD

Personalized medicine in AD refers to adjusting care plans and treatments to each patient according to their particular genetic, molecular, and clinical traits. The main conclusions and revelations from research on personalized therapy in AD are as follows:

1. **Genetic Variants and Risk Assessment:** Research has linked a number of genetic variations, including *ApoE* gene variants, to an elevated risk of AD. Genetic testing is used in personalized medicine methods to determine a person's genetic risk profile, which can guide lifestyle changes and early intervention plans.

2. **Biomarkers for Early Detection:** Biomarkers that indicate AD pathology before clinical symptoms appear, such as tau and amyloid-beta proteins in cerebrospinal fluid or through neuroimaging methods like PET scans, can aid in the early identification of the disease. By integrating these biomarkers, personalized medicine enables targeted medicines at preclinical stages and facilitates early diagnosis and intervention [129].

3. **Treatment Response Prediction:** The need for individualized treatment strategies is highlighted by Alzheimer's patients' variability in treatment response. To maximize therapeutic outcomes and reduce side effects, pharmacogenomic studies seek to find genetic markers linked to responsiveness to particular drugs, such as memantine or cholinesterase inhibitors. **4. Precision Drug Development:** Targeted treatments that target certain pathogenic pathways, such as tau protein pathology or amyloid-beta aggregation, have been developed as a result of advances in our understanding of the molecular mechanisms behind AD. Precision medications that are suited to specific individuals according to their individual disease characteristics are made possible by personalized medicine.

5. **Lifestyle and Environmental Considerations:** Personalized medicine takes into account socioeconomic, environmental, and lifestyle determinants of health in addition to genetic considerations. A customized treatment plan should include lifestyle interventions such as exercise, nutrition, social interaction, and cognitive stimulation in order to lower the risk of Alzheimer's and improve brain health in general.

6. **Clinical Trial Design and Recruitment:** To improve trial populations and increase the chance of discovering treatment benefits, personalized medicine approaches are used to stratify patients based on biomarker profiles, genetic risk factors, and illness stage.

By focusing on those who are most likely to benefit from experimental therapies, precision recruitment tactics help to expedite drug development efforts.

7. **Carer Support and Education**: The patient is not the only target of personalized treatment; family members and carers are also included. Encouraging carers to get individualized education, counseling, and support services that are catered to their particular needs and preferences might help them better manage the difficulties of providing care for a loved one suffering with AD [130].

5.7.2 POSSIBLE EFFECTS OF PERSONALIZED THERAPIES ON PATIENT OUTCOMES AND DISEASE MANAGEMENT

Personalized therapies have the power to significantly enhance patient outcomes and disease management in the following ways:

1. **Customized Treatment Strategies:** AD is a varied and complex disease that affects people in different ways. Personalized treatments consider the distinct genetic, molecular, and environmental elements that impact each patient's illness. Therapies that are more specifically designed to address these characteristics can be more successful in treating symptoms or delaying the advancement of the disease.
2. **Early Identification and Intervention:** AD biomarkers can be found and treated even before symptoms appear thanks to personalized medicine. This makes it possible for medical professionals to take action sooner in the course of the illness, which may reduce its progression and prolong the preservation of cognitive function.
3. **Precision Medication:** Pharmacogenomics is a crucial aspect of personalized medicine, enabling healthcare professionals to prescribe drugs based on an individual's genetic makeup, thereby reducing the likelihood of adverse reactions [131].
4. **Targeted Medicines:** Thanks to personalized medicine, certain biological pathways linked to AD can be addressed by targeted medicines. Compared to traditional methods, which frequently have poor efficacy and sometimes have unfavorable side effects, these therapies have the potential to be more accurate and efficient.

5. **Better Clinical Trials:** Targeted and effective clinical trial design is made easier by personalized medicine. Researchers can find patient subgroups that are more likely to respond to a specific treatment by stratifying patients based on their genetic or molecular profiles. This approach can result in more fruitful clinical trials and quicker drug development.
6. **Improved Care Planning**: Information from personalized medicine is very helpful for managing and organizing care. Healthcare professionals can create individualized care plans that cater to each patient's unique requirements and preferences by knowing their genetic risk factors and the course of their condition.
7. **Improved Patient Adherence and Engagement:** Patients are more engaged and stick to therapy when therapies are customized for them specifically. Patients are more inclined to follow treatment plans and take an active role in their own care when they see that personalized treatments are producing noticeable results [132].

5.7.3 SUGGESTIONS FOR MORE RESEARCH AND THE INCORPORATION OF PERSONALIZED MEDICINE INTO CLINICAL PRACTICE

The advancement of treatment and care for AD depends on the investigation and application of personalized medicine in clinical practice. The ensuing are suggestions for additional investigation and execution:

1. **Genomic Profiling**: Keep up the genomic research to find genetic variations linked to the onset and risk of AD. By classifying patients according to their genetic susceptibility, genomic information can be incorporated into clinical practice to enable individualized risk assessment and focused therapies.
2. **The development of biomarkers** is important for early detection, disease staging, and treatment response monitoring in AD. Research should be done to identify and validate trustworthy biomarkers. Neuroimaging indicators or the amounts of tau and amyloid-beta proteins in cerebrospinal fluid are examples of biomarkers that might help customize treatment plans for specific individuals.

3. **Machine Learning and AI:** Examine how to analyze complicated datasets, such as genetic, imaging, and clinical data, using machine learning algorithms and artificial intelligence approaches. These instruments can help find patterns and connections that could inform more precise illness progression prediction and personalized therapy choices [133].

4. **Environmental and Lifestyle Factors:** Examine how food, exercise, social interaction, cognitive stimulation, and other lifestyle choices affect the development and risk of AD. Individual risk profiles can be used to tailor interventions that target particular lifestyle changes in order to either postpone the development of disease or decrease its course.

5. **Drug Development and Targeted Therapies:** Give top priority to the creation of disease-modifying treatments that target certain biochemical pathways, including as tau phosphorylation, amyloid-beta aggregation, neuroinflammation, and synaptic dysfunction, that are linked to the pathogenesis of AD. One way to approach personalized treatment is to choose medicines according to the underlying disease mechanisms that are more prevalent in each individual patient [134].

6. **Clinical Trials Design:** Utilize biomarker-driven patient classification, adaptive trial designs, and outcome measures sensitive to the impacts of tailored treatments when designing clinical trials. These are just a few of the personalized medicine ideas to keep in mind. Facilitating the use of personalized medicine approaches in AD research and medication development requires cooperation between academia, industry, regulatory bodies, and patient advocacy groups.

7. **Ethical and Regulatory Considerations**: Talk about the ethical and legal issues around patient privacy, data security, informed consent, and fair access to individualized diagnosis and treatments when it comes to personalized medicine in AD. To protect patient safety and ethical integrity, rules and guidelines for the appropriate application of personalized medicine in clinical practice must be developed.

8. **Patient and Carers Education:** Inform patients, carers, and medical professionals about the possible advantages and restrictions of customized treatment for AD. Treatment adherence, contentment, and clinical outcomes can all be improved by providing people with information about individualized treatment alternatives and include them in joint decision-making processes.
9. **Healthcare Infrastructure and Reimbursement Policies:** Assess the state of healthcare systems' preparedness, taking into account the availability of diagnostic tests, provider training, and reimbursement mechanisms, to incorporate personalized medicine into standard clinical practice. Ensuring that all patients with AD have equitable access to personalized medicine services requires addressing both financial and practical hurdles.
10. **Longitudinal Research and Real-World Evidence:** To evaluate the long-term efficacy, safety, and cost-effectiveness of personalized medicine therapies in AD, conduct longitudinal research and real-world evidence analyses. Optimizing patient care and health system resource allocation requires ongoing assessment and improvement of personalized treatment algorithms based on real-world outcomes data [135].

We can improve outcomes for AD patients and progress the integration of personalized medicine into clinical practice by following these research areas and encouraging cross-disciplinary collaboration.

KEYWORDS

- **Alzheimer's disease**
- **biomarkers**
- **genetic factors**
- **personalized therapy**
- **pharmacogenomics**
- **tailored interventions**

REFERENCES

1. Sloane, P. D., Zimmerman, S., Suchindran, C., Reed, P., Wang, L., Boustani, M., & Sudha, S. (2002). The public health impact of Alzheimer's disease, 2000–2050: Potential implication of treatment advances *Annual Review of Public Health, 23*, 213–231.
2. Siemieniuk, R. A. C., & Guyatt, G. H. (2015). Corticosteroids in the treatment of community-acquired pneumonia: An evidence summary. *Polskie Archiwum Medycyny Wewnetrznej*, *125*, 570–575.
3. Panza, F., Lozupone, M., Logroscino, G., & Imbimbo, B. P. (2019). A critical appraisal of Amyloid-β-Targeting therapies for Alzheimer disease. *Nature Reviews Neurology*, *15*, 73–88.
4. Hampel, H., Caraci, F., Cuello, A. C., Caruso, G., Nisticò, R., Corbo, M., Baldacci, F., Toschi, N., Garaci, F., Chiesa, P. A., et al. (2020). A path toward precision medicine for neuroinflammatory mechanisms in Alzheimer's disease. *Frontiers in Immunology*, *11*, 456.
5. Ohm, D. T., Fought, A. J., Martersteck, A., Coventry, C., Sridhar, J., Gefen, T., Weintraub, S., Bigio, E., Mesulam, M.-M., Rogalski, E., et al. (2021). Accumulation of neurofibrillary tangles and activated microglia Is associated with lower neuron densities in the aphasic variant of Alzheimer's disease. *Brain Pathology*, *31*, 189–204.
6. Heneka, M. T., Carson, M. J., El Khoury, J., Landreth, G. E., Brosseron, F., Feinstein, D. L., Jacobs, A. H., Wyss-Coray, T., Vitorica, J., Ransohoff, R. M., et al. (2015). Neuroinflammation in Alzheimer's disease. *Lancet Neurology, 14*, 388–405.
7. Nordengen, K., Kirsebom, B.-E., Henjum, K., Selnes, P., Gísladóttir, B., Wettergreen, M., Torsetnes, S. B., Grøntvedt, G. R., Waterloo, K. K.; Aarsland, D., et al. (2019). Glial activation and inflammation along the Alzheimer's disease Continuum. *Journal of Neuroinflammation*, *16*, 46.
8. Hampel, H., Frank, R., Broich, K., Teipel, S. J., Katz, R. G., Hardy, J., Herholz, K., Bokde, A. L. W., Jessen, F., Hoessler, Y. C., et al. (2010). Biomarkers for Alzheimer's disease: Academic, industry and regulatory perspectives. *Nature Reviews Drug Discovery, 9*, 560–574.
9. Chen, R., Mias, G. I., Li-Pook-Than, J., Jiang, L., Lam, H. Y. K., Chen, R., Miriami, E., Karczewski, K. J., Hariharan, M., Dewey, F. E., et al. (2012). Personal omics profiling reveals dynamic molecular and medical phenotypes. *Cell, 148*, 1293–1307.
10. Hampel, H., O'Bryant, S. E., Durrleman, S., Younesi, E., Rojkova, K., Escott-Price, V., Corvol, J.-C., Broich, K., Dubois, B., Lista, S., et al. (2017). A precision medicine initiative for Alzheimer's disease: The road ahead to biomarker-guided integrative disease modeling. *Climacteric Journal of the International Menopause Society*, *20*, 107–118.
11. Berkowitz, C., Mosconi, L., Scheyer, O., Rahman, A., Hristov, H., & Isaacson, R. (2018) Precision medicine for Alzheimer's disease prevention. *Healthcare*, *6*, 82.
12. Antman, E. M., & Loscalzo, J. (2016). Precision medicine in cardiology. *Nature Reviews Cardiology, 13*, 591–602.
13. Shin, S. H., Bode, A. M., & Dong, Z. (2017). Precision medicine: The foundation of future cancer therapeutics. *NPJ Precision Oncology*, *1*, 12.

14. Reitz, C. (2016). Toward precision medicine in Alzheimer's disease. *Annals of Translational Medicine*, *4*, 107.
15. Ridge, P. G., Hoyt, K. B., Boehme, K., Mukherjee, S., Crane, P. K., Haines, J. L., Mayeux, R., Farrer, L. A., Pericak-Vance, M. A., Schellenberg, G. D., et al. (2016). Assessment of the genetic variance of late-onset Alzheimer's disease. *Neurobiology of Aging*, *41*, 200.e13–200.e20.
16. Bisht, K., Sharma, K., & Tremblay, M.-È. (2018). Chronic stress as a risk factor for Alzheimer's disease: Roles of microglia-mediated synaptic remodeling, inflammation, and oxidative stress. *Neurobiology of Stress, 9*, 9–21.
17. Decourt, B., Lahiri, D. K., & Sabbagh, M. N. (2017). Targeting tumor necrosis factor Alpha for Alzheimer's disease. *Current Alzheimer Research, 14*, 412–425.
18. Brosseron, F., Krauthausen, M., Kummer, M., & Heneka, M. T. (2014). Body fluid Cytokine levels in mild cognitive impairment and Alzheimer's disease: A comparative overview. *Molecular Neurobiology*, *50*, 534–544.
19. Eikelenboom, P., Zhan, S. S., van Gool, W. A., & Allsop, D. (1994). Inflammatory mechanisms in Alzheimer's disease. *Trends in Pharmacological Sciences, 15*, 447–450.
20. Caraci, F., Gulisano, W., Guida, C. A., Impellizzeri, A. A. R., Drago, F., Puzzo, D., & Palmeri, A. (2015). A key role for TGF-B1 in Hippocampal synaptic plasticity and memory. *Scientific Reports, 5*, 11252.
21. Tong, L., Balazs, R., Soiampornkul, R., Thangnipon, W., & Cotman, C. W. (2008). Interleukin-1 beta impairs brain derived neurotrophic factor-induced signal transduction. *Neurobiology of Aging*, *29*, 1380–1393.
22. Bruno, M. A., Leon, W. C., Fragoso, G., Mushynski, W. E., Almazan, G., & Cuello, A. C. (2009). Amyloid beta-induced nerve growth factor dysmetabolism in Alzheimer disease. *Journal of Neuropathology & Experimental Neurology, 68*, 857–869.
23. Iulita, M. F., do Carmo, S., Ower, A. K., Fortress, A. M., Flores Aguilar, L., Hanna, M., Wisniewski, T., Granholm, A.-C., Buhusi, M., Busciglio, J., et al. (2014). Nerve growth factor metabolic dysfunction in Down's Syndrome brains. *Brain Journal of Neurology, 137*, 860–872.
24. Agid, Y. (1998). Levodopa: Is toxicity a myth? *Neurology*, *50*, 858–863.
25. Dickson, D. W., Braak, H., Duda, J. E., Duyckaerts, C., Gasser, T., Halliday, G. M., Hardy, J., Leverenz, J. B., del Tredici, K., Wszolek, Z. K., et al. (2009). Neuropathological assessment of Parkinson's disease: Refining the diagnostic criteria. *Lancet Neurology, 8*, 1150–1157.
26. Houlden, H., & Singleton, A. B. (2012). The genetics and neuropathology of Parkinson's disease. *Acta Neuropathologica*, *124*, 325–338.
27. Riederer, P., Sofic, E., Rausch, W. D., Schmidt, B., Reynolds, G. P., Jellinger, K., & Youdim, M. B. (1989). Transition Metals, Ferritin, Glutathione, and Ascorbic Acid in Parkinsonian brains. *Journal of Neurochemistry*, *52*, 515–520.
28. Schapira, A. H., & Jenner, P. (2011). Etiology and pathogenesis of Parkinson's disease. *Movement Disorders Official Journal of Movement Disorders Society, 26*, 1049–1055.
29. Tofaris, G. K., Spillantini, M. G. (2005). Alpha-Synuclein dysfunction in Lewy body diseases. *Movement Disorders Official Journal of Movement Disorders Society*, *20*(Suppl. S12), S37–S44.

30. Hernández-Romero, M. C., Delgado-Cortés, M. J., Sarmiento, M., de Pablos, R. M., Espinosa-Oliva, A. M., Argüelles, S., Bández, M. J., Villarán, R. F., Mauriño, R., Santiago, M., et al. (2012). Peripheral inflammation increases the deleterious effect of CNS inflammation on the nigrostriatal dopaminergic system. *Neurotoxicology, 33*, 347–360.
31. Armstrong, M. J. (2021). Advances in Dementia with Lewy bodies. *Therapeutic Advances in Neurological Disorders, 14*, 17562864211057666.
32. Chen, X., Kordich, J. K., Williams, E. T., Levine, N., Cole-Strauss, A., Marshall, L., Labrie, V., Ma, J., Lipton, J. W., & Moore, D. J. (2019). Parkinson's disease-linked D620N VPS35 Knockin Mice manifest Tau neuropathology and dopaminergic neurodegeneration. *Proceedings National Academy Sciences USA*, 116, 5765–5774.
33. Boillée, S., Vande Velde, C., & Cleveland, D. W. (2006). ALS: A disease of motor neurons and their nonneuronal neighbors. *Neuron, 52*, 39–59.
34. Sasaki, S., & Iwata, M. (2007). Mitochondrial alterations in the spinal cord of patients with sporadic amyotrophic lateral sclerosis. *Journal of Neuropathology & Experimental Neurology, 66*, 10–16.
35. Magrané, J., & Manfredi, G. (2009). Mitochondrial function, morphology, and axonal transport in amyotrophic lateral sclerosis. *Antioxidants & Redox Signalling, 11*, 1615–1626.
36. Higgins, C. M. J., Jung, C., & Xu, Z. (2003). ALS-associated mutant SOD1G93A causes mitochondrial vacuolation by expansion of the intermembrane space and by involvement of SOD1 aggregation and peroxisomes. *BMC Neuroscience, 4*, 16.
37. Bacman, S. R., Bradley, W. G., Moraes, C. T. (2006). Mitochondrial involvement in amyotrophic lateral sclerosis: Trigger or target? *Molecular Neurobiology, 33*, 113–131.
38. Tohgi, H., Abe, T., Yamazaki, K., Murata, T., Ishizaki, E., & Isobe, C. (1999). Remarkable increase in cerebrospinal fluid 3-Nitrotyrosine in patients with sporadic amyotrophic lateral sclerosis. *Annals of Neurology, 46*, 129–131.
39. Zarei, S., Carr, K., Reiley, L., Diaz, K., Guerra, O., Altamirano, P. F., Pagani, W., Lodin, D., Orozco, G., & Chinea, A. (2015). A comprehensive review of amyotrophic lateral sclerosis. *Surgical Neurology International, 6*, 171.
40. Zhao, W., Beers, D. R., & Appel, S. H. (2013).Immune-mediated mechanisms in the pathoprogression of amyotrophic lateral Sclerosis. *Journal Neuroimmune Pharmacology Official Journal Society NeuroImmune Pharmacology, 8*, 888–899.
41. Hampel, H., O'Bryant, S. E., Castrillo, J. I., Ritchie, C., Rojkova, K., Broich, K., Benda, N., Nisticò, R., Frank, R. A., Dubois, B., et al. (2016). Precision medicine—The golden gate for detection, treatment and prevention of Alzheimer's disease. *Journal of Prevention Alzheimer's Disease, 3*, 243–259.
42. *National Research Council (US) Committee on A Framework for Developing a New Taxonomy of Disease toward Precision Medicine: Building a Knowledge Network for Biomedical Research and a New Taxonomy of Disease; The National Academies Collection: Reports Funded by National Institutes of Health.* National Academies Press (2011).
43. Hampel, H., Lista, S., & Khachaturian, Z. S. (2012). Development of biomarkers to chart all Alzheimer's disease stages: The royal road to cutting the therapeutic Gordian Knot. *Alzheimer's & Dementia Journal of Alzheimer's Association, 8*, 312–336.

44. Lista, S., Khachaturian, Z. S., Rujescu, D., Garaci, F., Dubois, B., & Hampel, H. (2016). Application of systems theory in longitudinal studies on the origin and progression of Alzheimer's disease. *Methods in Molecular Biology (Clifton NJ), 1303*, 49–67.
45. Dubois, B., Feldman, H. H., Jacova, C., Hampel, H., Molinuevo, J. L., Blennow, K., DeKosky, S. T., Gauthier, S., Selkoe, D., Bateman, R., et al. (2014). Advancing research diagnostic criteria for Alzheimer's disease: The IWG-2 criteria. *Lancet Neurology, 13*, 614–629.
46. Hampel, H., Lista, S., Teipel, S. J., Garaci, F., Nisticò, R., Blennow, K., Zetterberg, H., Bertram, L., Duyckaerts, C., Bakardjian, H., et al. (2014). Perspective on future role of biological markers in clinical therapy trials of Alzheimer's disease: A long-range point of view beyond 2020. *Biochemical Pharmacology, 88*, 426–449.
47. Hampel, H., Lista, S. (2013). Use of biomarkers and imaging to assess pathophysiology, mechanisms of action and target engagement. *Journal of Nutrition Health Aging, 17*, 54–63.
48. Zetzsche, T., Rujescu, D., Hardy, J., & Hampel, H. (2010). Advances and perspectives from genetic research: Development of biological markers in Alzheimer's disease. *Expert Review Molecular Diagnostics, 10*, 667–690.
49. Hampel, H., & Lista, S. (2012). Alzheimer disease: From inherited to sporadic AD-crossing the biomarker bridge. *Nature Reviews Neurology, 8*, 598–600.
50. Lista, S., O'Bryant, S. E., Blennow, K., Dubois, B., Hugon, J., Zetterberg, H., & Hampel, H. (2015). Biomarkers in sporadic and familial Alzheimer's disease. *Journal of Alzheimer's Disease, 47*, 291–317.
51. Blennow, K., Hampel, H., Weiner, M., & Zetterberg, H. (2010). Cerebrospinal fluid and plasma biomarkers in Alzheimer disease. *Nature Reviews Neurology, 6*, 131–144.
52. Hampel, H., Shen, Y., Walsh, D. M., Aisen, P., Shaw, L. M., Zetterberg, H., Trojanowski, J. Q., & Blennow, K. (2010). Biological markers of Amyloid beta-related mechanisms in Alzheimer's disease. *Experimental Neurology, 223*, 334–346.
53. Henriksen, K., O'Bryant, S. E., Hampel, H., Trojanowski, J. Q., Montine, T. J., Jeromin, A., Blennow, K., Lönneborg, A., Wyss-Coray, T., Soares, H., et al. (2014). The future of blood-based biomarkers for Alzheimer's disease. *Alzheimer's & Dementia Journal of Alzheimer's Association, 10*, 115–131.
54. O'Bryant, S. E., Lista, S., Rissman, R. A., Edwards, M., Zhang, F., Hall, J., Zetterberg, H., Lovestone, S., Gupta, V., Graff-Radford, N., et al. (2016). Comparing biological markers of Alzheimer's disease across blood fraction and platforms: Comparing apples to oranges. *Alzheimer's and Dementia Diagnosis Assessment & Disease Monitoring, 3*, 27–34.
55. Ewers, M., Sperling, R. A., Klunk, W. E., Weiner, M. W., & Hampel, H. (2011). Neuroimaging markers for the prediction and early diagnosis of Alzheimer's disease dementia. *Trends in Neuroscience, 34*, 430–442.
56. Teipel, S. J., Grothe, M., Lista, S., Toschi, N., Garaci, F. G., & Hampel, H. (2013). Relevance of magnetic resonance imaging for early detection and diagnosis of Alzheimer disease. *Medical Clinics of North America, 97*, 399–424.
57. Poil, S.-S., de Haan, W., van der Flier, W. M., Mansvelder, H. D., Scheltens, P., & Linkenkaer-Hansen, K. (2013). Integrative EEG biomarkers predict progression to Alzheimer's disease at the MCI stage. *Frontiers in Aging Neuroscience, 5*, 58.

58. Jørgensen, J. T. (2015). Companion diagnostics: The key to personalized medicine. *Expert Reviews of Molecular Diagnostics*, *15*, 153–156.
59. Lyman, G. H., & Moses, H. L. (2016). Biomarker tests for molecularly targeted therapies—The key to unlocking precision medicine. *New England Journal of Medicine*, *375*, 4–6.
60. Ballard, C., Gauthier, S., Corbett, A., Brayne, C., Aarsland, D., & Jones, E. (2011). Alzheimer's disease. *Lancet London England*, *377*, 1019–1031.
61. Mahley, R. W. (1988). Apolipoprotein E: Cholesterol transport protein with expanding role in cell biology. *Science*, *240*, 622–630.
62. Corder, E. H., Saunders, A. M., Strittmatter, W. J., Schmechel, D. E., Gaskell, P. C., Small, G. W., Roses, A. D., Haines, J. L., & Pericak-Vance, M. A. (1993). Gene dose of Apolipoprotein E Type 4 Allele and the risk of Alzheimer's disease in late onset families. *Science*, *261*, 921–923.
63. Talbot, C., Lendon, C., Craddock, N., Shears, S., Morris, J. C., & Goate, A. (1994). Protection against Alzheimer's disease with ApoE Epsilon 2. *Lancet London England*, *343*, 1432–1433.
64. Farrer, L. A., Cupples, L. A., Haines, J. L., Hyman, B., Kukull, W. A., Mayeux, R., Myers, R. H., Pericak-Vance, M. A., Risch, N., van Duijn, C. M. (1997). Effects of age, sex, and ethnicity on the association between Apolipoprotein E Genotype and Alzheimer disease. A meta-analysis. APOE and Alzheimer disease meta analysis consortium. *JAMA*, *278*, 1349–1356.
65. Strittmatter, W. J., Weisgraber, K. H., Huang, D. Y., Dong, L. M., Salvesen, G. S., Pericak-Vance, M., Schmechel, D., Saunders, A. M., Goldgaber, D., & Roses, A. D. (1993). Binding of human Apolipoprotein E to synthetic amyloid beta peptide: Isoform-specific effects and implications for late-onset Alzheimer disease. *Proceedings of the National Academy of Sciences of the USA*, *90*, 8098–8102.
66. Van Cauwenberghe, C., van Broeckhoven, C., & Sleegers, K. (2016). The genetic landscape of Alzheimer disease: Clinical implications and perspectives. *Genetics in Medicine: Official Journal of the American College of Medical Genetics*, *18*, 421–430.
67. Solomon, A., Turunen, H., Ngandu, T., Peltonen, M., Levälahti, E., Helisalmi, S., Antikainen, R., Bäckman, L., Hänninen, T., Jula, A., et al. (2018). Effect of the Apolipoprotein E Genotype on cognitive change during a multidomain lifestyle intervention: A subgroup analysis of a randomized clinical trial. *JAMA Neurology*, *75*, 462–470.
68. Masson, L. F., McNeill, G., & Avenell, A. (2003). Genetic variation and the lipid response to dietary intervention: A systematic review *American Journal of Clinical Nutrition*, *77*, 1098–1111.
69. Martínez-Lapiscina, E. H., Galbete, C., Corella, D., Toledo, E., Buil-Cosiales, P., Salas-Salvado J., Ros, E., & Martinez-Gonzalez, M. A. (2014). Genotype patterns at CLU, CR1, PICALM and APOE, cognition and mediterranean diet: The PREDIMED-NAVARRA trial. *Genes and Nutrition*, *9*, 393.
70. Etnier, J. L., Caselli, R. J., Reiman, E. M., Alexander, G. E., Sibley, B. A., Tessier, D., & McLemore, E. C. (2007). Cognitive performance in older women relative to ApoE-Epsilon4 genotype and aerobic fitness. *Medicine & Science in Sports & Exercise*, *39*, 199–207.

71. Yassine, H. N., Braskie, M. N., Mack, W. J., Castor, K. J., Fonteh, A. N., Schneider, L. S., Harrington, M. G., & Chui, H. C. (2017). Association of Docosahexaenoic Acid supplementation with Alzheimer disease stage in Apolipoprotein E E4 carriers: A review. *JAMA Neurology*, *74*, 339–347.
72. Sibani, S., Christensen, B., O'Ferrall, E., Saadi, I., Hiou-Tim, F., Rosenblatt, D. S., & Rozen, R. (2000). Characterization of six novel mutations in the Methylenetetrahydrofolate Reductase (MTHFR) gene in patients with homocystinuria. *Human Mutation*, *15*, 280–287.
73. Román, G. C. (2015). MTHFR gene mutations: A potential marker of late-onset Alzheimer's disease? *Journal of Alzheimer's Disease*, *47*, 323–327.
74. Romero-Sánchez, C., Gómez-Gutierrez, A., Gómez, P. E., Casas-Gomez, M. C., & Briceño, I. (2015). C677T (RS1801133) MTHFR Gene Polymorphism frequency in a Colombian population. *Colombia Médica (Cali, Colombia)*, *46*, 75–79.
75. McCaddon, A., Hudson, P., Davies, G., Hughes, A., Williams, J. H., & Wilkinson, C. (2001). Homocysteine and cognitive decline in healthy elderly. *Dementia and Geriatric Cognitive Disorders*, *12*, 309–313.
76. Seshadri, S., Beiser, A., Selhub, J., Jacques, P. F., Rosenberg, I. H., D'Agostino, R. B., Wilson, P. W. F., & Wolf, P. A. (2002). Plasma homocysteine as a risk factor for dementia and Alzheimer's disease. *New England Journal of Medicine*, *346*, 476–483.
77. Clarke, R., Birks, J., Nexo, E., Ueland, P. M., Schneede, J., Scott, J., Molloy, A., & Evans, J. G. (2007). Low Vitamin B-12 status and risk of cognitive decline in older adults. *American Journal of Clinical Nutrition*, 86, 1384–1391.
78. Weisberg, I., Tran, P., Christensen, B., Sibani, S., & Rozen, R. (1998). A second genetic Polymorphism in Methylenetetrahydrofolate Reductase (MTHFR) associated with decreased enzyme activity. *Molecular Genetics and Metabolism*, *64*, 169–172.
79. Mansouri, L., Fekih-Mrissa, N., Klai, S., Mansour, M., Gritli, N., & Mrissa, R. (2013). Association of Methylenetetrahydrofolate Reductase Polymorphisms with susceptibility to Alzheimer's disease. *Clinical Neurology and Neurosurgery*, *115*, 1693–1696.
80. Wakutani, Y., Kowa, H., Kusumi, M., Nakaso, K., Yasui, K.-I., Isoe-Wada, K., Yano, H., Urakami, K., Takeshima, T., & Nakashima, K. (2004). A Haplotype of the Methylenetetrahydrofolate reductase gene is protective against late-onset Alzheimer's disease. *Neurobiology of Aging*, *25*, 291–294.
81. Smith, A. D., Smith, S. M., de Jager, C. A., Whitbread, P., Johnston, C., Agacinski, G., Oulhaj, A., Bradley, K. M., Jacoby, R., & Refsum, H. (2010). Homocysteine-lowering by B Vitamins slows the rate of accelerated brain atrophy in mild cognitive impairment: A randomized controlled trial. *PLoS ONE*, *5*, e12244.
82. Douaud, G., Refsum, H., de Jager, C. A., Jacoby, R., Nichols, T. E., Smith, S. M., & Smith, A. D. (2013). Preventing Alzheimer's disease-related gray matter atrophy by B-Vitamin treatment. *Proceedings of the National Academy of Sciences USA*, *110*, 9523–9528.
83. Hekmatdoost, A., Vahid, F., Yari, Z., Sadeghi, M., Eini-Zinab, H., Lakpour, N., & Arefi, S. (2015). Methyltetrahydrofolate vs. Folic Acid supplementation in idiopathic recurrent miscarriage with respect to methylenetetrahydrofolate reductase C677T and A1298C polymorphisms: A randomized controlled trial. *PLoS ONE*, *10*, e0143569.

84. Schellenberg, G. D., Bird T. D., Wijsman, E. M., Orr, H. T., Anderson, L., Nemens, E., White, J. A., Bonnycastle, L., Weber, J. L., & Alonso, M. E. (1992). Genetic linkage evidence for a familial Alzheimer's disease locus on Chromosome 14. *Science*, *258*, 668–671.
85. Van Broeckhoven, C., Backhovens, H., Cruts, M., de Winter, G., Bruyland, M., Cras, P., & Martin, J. J. (1992). Mapping of a gene predisposing to early-onset Alzheimer's disease to Chromosome 14q24.3. *Nature Genetics*, *2*, 335–339.
86. Sherrington, R., Rogaev, E. I., Liang, Y., Rogaeva, E. A., Levesque, G., Ikeda, M., Chi, H., Lin, C., Li, G., Holman, K., et al. (1995). Cloning of a gene bearing missense mutations in early-onset familial Alzheimer's disease. *Nature*, *375*, 754–760.
87. Steiner, H., Fluhrer, R., & Haass, C. (2008). Intramembrane proteolysis by gamma-secretase. *Journal of Biological Chemistry*, *283*, 29627–29631.
88. Scheuner, D., Eckman, C., Jensen, M., Song, X., Citron, M., Suzuki, N., Bird, T. D., Hardy, J., Hutton, M., Kukull, W., et al. (1996). Secreted amyloid beta-protein similar to that in the senile plaques of Alzheimer's disease is increased in Vivo by the Presenilin 1 and 2 and APP mutations linked to familial Alzheimer's disease. *Nature Medicine*, *2*, 864–870.
89. Levy-Lahad, E., Wasco, W., Poorkaj, P., Romano, D. M., Oshima, J., Pettingell, W. H., Yu, C. E., Jondro, P. D., Schmidt, S. D., & Wang, K. (1995). Candidate gene for the Chromosome 1 Familial Alzheimer's disease Locus. *Science*, *269*, 973–977.
90. Campion, D., Dumanchin, C., Hannequin, D., Dubois, B., Belliard, S., Puel, M., Thomas-Anterion, C., Michon, A., Martin, C., Charbonnier, F., et al. (1999). Early-onset autosomal dominant Alzheimer disease: Prevalence, genetic heterogeneity, and mutation spectrum. *American Journal of Human Genetics*, *65*, 664–670.
91. Jiang, H., Jayadev, S., Lardelli, M., & Newman, M. (2018). A review of the familial Alzheimer's disease Locus PRESENILIN 2 and Its relationship to PRESENILIN 1. *Journal of Alzheimer's Disease*, *66*, 1323–1339.
92. De Strooper, B., Iwatsubo, T., & Wolfe, M. S. (2012). Presenilins and -Secretase: Structure, function, and role in Alzheimer disease. *Cold Spring Harbor Perspectives Medicine*, *2*, a006304.
93. Jayadev, S., Leverenz, J. B., Steinbart, E., Stahl, J., Klunk, W., Yu, C.-E., & Bird, T. D. (2010). Alzheimer's disease phenotypes and genotypes associated with mutations in Presenilin 2. *Brain Journal of Neurology*, *133*, 1143–1154.
94. Finckh, U., Alberici, A., Antoniazzi, M., Benussi, L., Fedi, V., Giannini, C., Gal, A., Nitsch, R. M., & Binetti, G. (2000). Variable expression of familial Alzheimer disease associated with Presenilin 2 Mutation M239I. *Neurology*, *54*, 2006–2008.
95. Blauwendraat, C., Wilke, C., Jansen, I. E., Schulte, C., Simón-Sánchez, J., Metzger, F. G., Bender, B., Gasser, T., Maetzler, W., Rizzu, P., et al. (2016). Pilot whole-exome sequencing of a German early-onset Alzheimer's disease Cohort reveals a substantial frequency of PSEN2 variants. *Neurobiology Aging*, *37*, 08.e11–208.e17.
96. Yan, L., Li, L., Han, W., Pan, B., Xue, X., & Mei, B. (2013). Age-related neuropsychiatric symptoms in Presenilins conditional double Knockout mice. *Brain Research Bulletin*, *97*, 104–111.
97. Jayadev, S., Case, A., Alajajian, B., Eastman, A. J., Möller, T., & Garden, G. A. (2013). Presenilin 2 Influences MiR146 level and activity in Microglia. *Journal of Neurochemistry*, *127*, 592–599.

98. Jayadev, S., Case, A., Eastman, A. J., Nguyen, H., Pollak, J., Wiley, J. C., Möller, T., Morrison, R. S., & Garden, G. A. (2010). Presenilin 2 Is the Predominant γ-Secretase in Microglia and Modulates Cytokine release. *PLoS ONE, 5*, e15743.
99. Jun, G. R., Chung, J., Mez, J., Barber, R., Beecham, G. W., Bennett, D. A., Buxbaum, J. D., Byrd, G. S., Carrasquillo, M. M., Crane, P. K., et al. (2017). Transethnic genome-wide scan identifies novel Alzheimer's disease Loci. *Alzheimer's & Dementia Journal of the Alzheimer's Association, 13*, 727–738.
100. Patel, D., Zhang, X., Farrell, J. J., Chung, J., Stein, T. D., Lunetta, K. L., & Farrer, L. A. (2021). Cell-type-specific expression quantitative Trait Loci associated with Alzheimer disease in blood and brain tissue. *Translational Psychiatry, 11*, 250.
101. Moingeon, P. (2016). Biomarkers for allergen immunotherapy: A "panoromic" view. *Immunology and Allergy Clinics of North America, 36*, 161–179.
102. Currie, G., & Delles, C. (2016). Use of biomarkers in the evaluation and treatment of hypertensive patients. *Current Hypertension Reports, 18*, 54.
103. Comabella, M., Sastre-Garriga, J., & Montalban, X. (2016). Precision medicine in multiple sclerosis: Biomarkers for diagnosis, prognosis, and treatment response. *Current Opinion in Neurology, 29*, 254–262.
104. Hambly, N., Shimbori, C., & Kolb, M. (2015). Molecular classification of idiopathic pulmonary fibrosis: Personalized medicine, genetics and biomarkers. *Respirology Carlton Vic, 20*, 1010–1022.
105. Pearson, E. R. (2016). Personalized medicine in diabetes: The role of "omics" and biomarkers. *Diabetic Medikcine Journal of British. Diabetic Association, 33*, 712–717.
106. O'Bryant, S. E., Xiao, G., Zhang, F., Edwards, M., German, D. C., Yin, X., Como, T., Reisch, J., Huebinger, R. M., Graff-Radford, N., et al. (2014). Validation of a serum screen for Alzheimer's disease across Assay platforms, species, and tissues. *Journal of Alzheimer's Disease, 42*, 1325–1335.
107. DeMarshall, C. A., Nagele, E. P., Sarkar, A., Acharya, N. K., Godsey, G., Goldwaser, E. L., Kosciuk, M., Thayasivam, U., Han, M., Belinka, B., et al. (2016). Detection of Alzheimer's disease at mild cognitive impairment and disease progression using autoantibodies as blood-based biomarkers. *Alzheimer's & Dementia Diagnosis, Assessment & Disease Monitoring, 3*, 51–62.
108. Winston, C. N., Goetzl, E. J., Akers, J. C., Carter, B. S., Rockenstein, E. M., Galasko, D., Masliah, E., & Rissman, R. A. (2016). Prediction of conversion from mild cognitive impairment to dementia with neuronally derived blood exosome protein profile. *Alzheimer's & Dementia Diagnosis, Assessment & Disease Monitoring, 3*, 63–72.
109. Guedes, J. R., Santana, I., Cunha, C., Duro, D., Almeida, M. R., Cardoso, A. M., de Lima M. C. P., & Cardoso, A. L. (2016). MicroRNA deregulation and chemotaxis and phagocytosis impairment in Alzheimer's disease. *Alzheimer's & Dementia Diagnosis, Assessment & Disease Monitoring, 3*, 7–17.
110. Tampubolon, G. (2016). Repeated systemic inflammation was associated with cognitive deficits in Older Britons. *Alzheimer's & Dementia Diagnosis, Assessment & Disease Monitoring, 3*, 1–6.
111. Hu, W. T., Chen-Plotkin, A., Arnold, S. E., Grossman, M., Clark, C. M., Shaw, L. M., McCluskey, L., Elman, L., Karlawish, J., Hurtig, H. I., et al. (2010). Biomarker

discovery for Alzheimer's disease, frontotemporal lobar degeneration, and Parkinson's disease. *Acta Neuropathologica, 120*, 385–399.

112. Henchcliffe, C., Dodel, R., & Beal, M. F. (2011). Biomarkers of Parkinson's disease and dementia with Lewy bodies. *Progress in Neurobiology, 95*, 601–613.
113. Heneka, M. T., Kummer, M. P., & Latz, E. (2014). Innate immune activation in neurodegenerative disease. *Nature Reviews Immunology, 14*, 463–477.
114. Durrenberger, P. F., Fernando, F. S., Kashefi, S. N., Bonnert, T. P., Seilhean, D., Nait-Oumesmar, B., Schmitt, A., Gebicke-Haerter, P. J., Falkai, P., Grünblatt, E., et al. (2015). Common mechanisms in neurodegeneration and neuroinflammation: A brain net Europe Gene expression microarray study. *Journal of Neural Transmission, 122*, 1055–1068.
115. Duong, T., Nikolaeva, M., & Acton, P. J. (1997). C-Reactive protein-like immunoreactivity in the neurofibrillary tangles of Alzheimer's disease. *Brain Research, 749*, 152–156.
116. O'Bryant, S. E., Johnson, L., Edwards, M., Soares, H., Devous, M. D., Ross, S., Rohlfing, G., & Hall, J. (2013). Texas Alzheimer's research & care consortium the link between C-Reactive protein and Alzheimer's disease among Mexican Americans. *Journal of Alzheimer's Disease, 34*, 701–706.
117. Glass, C. K., Saijo, K., Winner, B., Marchetto, M. C., & Gage, F. H. (2010). Mechanisms underlying inflammation in neurodegeneration. *Cell, 140*, 918–934.
118. Wyss-Coray, T., & Rogers, J. (2012). Inflammation in Alzheimer disease—A brief review of the basic science and clinical literature. *Cold Spring Harbor Perspectives in Medicine, 2*, a006346.
119. In'T Veld, B. A., Ruitenberg, A., Hofman, A., Launer, L. J., van Duijn, C. M., Stijnen, T., Breteler, M. M., & Stricker, B. H. (2001). Nonsteroidal anti inflammatory drugs and the risk of Alzheimer's disease. *New England Journal of Medicine, 345*, 1515–1521.
120. Aisen, P. S., Schafer, K. A., Grundman, M., Pfeiffer, E., Sano, M., Davis, K. L., Farlow, M. R., Jin, S., Thomas, R. G., Thal, L. J., et al. (2003). Effects of Rofecoxib or Naproxen vs. Placebo on Alzheimer disease progression: A randomized controlled trial. *JAMA, 289*, 2819–2826.
121. Thal, L. J., Ferris, S. H., Kirby, L., Block, G. A., Lines, C. R., Yuen, E., Assaid, C., Nessly, M. L., Norman, B. A., Baranak, C. C., et al. (2005). A randomized, double-blind, study of Rofecoxib in patients with mild cognitive impairment. *Neuropsychopharmacology Official Publication of the American College of Neuropsychopharmacology, 30*, 1204–1215.
122. Jones, L., Holmans, P. A., Hamshere, M. L., Harold, D., Moskvina, V., Ivanov, D., Pocklington, A., Abraham, R., Hollingworth, P., Sims, R., et al. (2010). Genetic evidence implicates the immune system and cholesterol metabolism in the aetiology of Alzheimer's disease. *PLoS ONE, 5*, e13950.
123. O'Bryant, S. E., Gupta, V., Henriksen, K., Edwards, M., Jeromin, A., Lista, S., Bazenet, C., Soares, H., Lovestone, S., Hampel, H., et al. (2015). Guidelines for the standardization of preanalytic variables for blood-based biomarker studies in Alzheimer's disease research. *Alzheimer's & Dementia Journal of the Alzheimer's Association, 11*, 549–560.

124. Li, M., Yu, T., & Hu, Y.-F. (2015). The impact of companion diagnostic device measurement performance on clinical validation of personalized medicine. *Statistics Medicine*, *34*, 2222–2234.
125. Li, M. (2015). Statistical consideration and challenges in bridging study of personalized medicine. *Journal of Biopharmceutical Statistics*, *25*, 397–407.
126. Hampel, H., Vassar, R., de Strooper, B., Hardy, J., Willem, M., Singh, N., Zhou, J., Yan, R., Vanmechelen, E., de Vos A., et al. (2021). The β-Secretase BACE1 in Alzheimer's disease. *Biological Psychiatry*, *89*, 745–756.
127. Janeiro, M. H., Ardanaz, C. G., Sola-Sevilla, N., Dong, J., Cortés-Erice, M., Solas, M., Puerta, E., & Ramírez, M. J. (2021). Biomarkers in Alzheimer's disease. *Advances in Laboratory Medicine/ Avances en Medicina de Laboratorio*, *2*, 27–37.
128. Urbanelli, L., Magini, A., Ciccarone, V., Trivelli, F., Polidoro, M., Tancini, B., & Emiliani, C. (2009). New perspectives for the diagnosis of Alzheimer's disease. *Recent Patents CNS Drug Discovery*, *4*, 160–181.
129. Brier, M. R., Gordon, B., Friedrichsen, K., McCarthy, J., Stern, A., Christensen, J., Owen, C., Aldea, P., Su, Y., Hassenstab, J., et al. (2016). Tau and Aβ Imaging, CSF measures, and cognition in Alzheimer's disease. *Science Translational Medicine*, *8*, 338ra66.
130. Ferreira, L. K., & Busatto, G. F. (2011). Neuroimaging in Alzheimer's disease: Current role in clinical practice and potential future applications. *Clinics Sao Paulo Brazil*, *66*(Suppl. S1), 19–24.
131. Herholz, K., & Ebmeier, K. (2011). Clinical amyloid imaging in Alzheimer's disease. *Lancet Neurology*, *10*, 667–670.
132. Isaac, M., Vamvakas, S., Abadie, E., Jonsson, B., Gispen, C., & Pani, L. (2011). Qualification opinion of novel methodologies in the predementia stage of Alzheimer's disease: Cerebro-spinal-fluid related biomarkers for drugs affecting amyloid burden—Regulatory considerations by European medicines agency focusing in improving benefit/risk in regulatory trials European Neuropsychopharmacology. *Journal of the European College of Neuropsychopharmacology*, *21*, 781–788.
133. Shaw, L. M., Vanderstichele, H., Knapik-Czajka, M., Clark, C. M., Aisen, P. S., Petersen, R. C., Blennow, K., Soares, H., Simon, A., Lewczuk, P., et al. (2009). Cerebrospinal fluid biomarker signature in Alzheimer's disease neuroimaging initiative subjects. *Annals of Neurology*, *65*, 403–413.
134. Mulder, C., Verwey, N. A., van der Flier, W. M., Bouwman, F. H., Kok, A., van Elk, E. J., Scheltens, P., & Blankenstein. M. A. (2010). Amyloid-Beta (1-42), Total Tau, and Phosphorylated Tau as cerebrospinal fluid biomarkers for the diagnosis of Alzheimer disease. *Clinical Chemistry*, *56*, 248–253.
135. Buchhave, P., Minthon, L., Zetterberg, H., Wallin, A. K., Blennow, K., & Hansson, O. (2012). Cerebrospinal fluid levels of β-Amyloid 1-42, but not of Tau, are fully changed already 5 to 10 years before the onset of Alzheimer dementia. *Archives of General Psychiatry*, *69*, 98–106.

CHAPTER 6

Ethical Considerations in the Field of Alzheimer's Research and Overcoming Obstacles in Clinical Trials and Ethical of Research

YASHRAJ YADAV[1,*], PAWAN KUMAR DUBEY[1], DISHANT GUPTA[1], JAYA SHARMA[1], DEEPAK KUMAR GUPTA[2], and ABHISHEK AGARWAL[3]

[1]*Department of Pharmacology, Swami Vivekananda College of Pharmacy, Indore, Madhya Pradesh, India*

[2]*Bengal College of Pharmaceutical Technology, Dubrajpur, West Bengal, India*

[3]*H&H Healthcare and Cosmetics Private Limited, Indore, Madhya Pradesh, India*

*Corresponding author

ABSTRACT

Alzheimer's disease (AD) is distinguished by cognitive decline along with significant forgetfulness, although it can also appear with nonamnestic cognitive impairment. AD often causes short-term memory difficulties, but it can also affect expressing speech, visual information analyzing, and executive function. AD stands as one of the most pressing public health challenges of our time, affecting millions worldwide and posing profound social, economic, and ethical dilemmas. Cognitive impairments have extremely detrimental effects on symptoms, leading to moral dilemmas. This chapter covers every ethical consideration that arises during AD clinical trials. Various scientific search engines, such as Scopus, WoS, PubMed, Medline,

Cellular and Molecular Mechanisms in Alzheimer's Disease Therapeutics, Volume 2: Molecular Mechanisms. Prashant Tiwari, Sunil Kumar Kadiri, and Kunjbihari Sulakhiya (Eds)

DOI: 10.1201/9781779641816-6

Embase, and Cochrane Library, are utilized in the development of chapters. It is difficult to achieve a compromise between these ethical concerns and the need to conduct major and impactful research. Nonetheless, it is vital to protect study participants' privacy and health, as well as the integrity of the research process. Whether intended or not, it is always necessary to safeguard individuals from serious or irreparable harm. Breaking research ethics will also harm research credibility because people will be less likely to trust your findings if your practices are unethical. The investigators, medical research coordinators, pharmacists, patients, and their caregivers play critical roles in ensuring the success of an AD ethical issue during clinical trials. The information gathered from all stakeholders should be used to perform Alzheimer's ethical clinical trials and remove obstacles.

6.1 INTRODUCTION

Alzheimer's disease (AD) is a dangerous nervous system ailment that usually manifests in middle to late life. It causes a steady and permanent impairment in memory, as well as degradation of other cognitive capacities. The illness is defined by the death of nerve cells along with neural connections in the cerebral cortex of the brain, as well as a severe loss of brain mass [1]. German scientist and neurophysiologist first define about the AD in 1906. Alzheimer's disorder is a neurodegenerative condition characterized by tau-containing tangles of neurofibrillary fibers and beta-amyloid (Aβ) plaques. It appears in the form of dementia and is characterized by substantial amnestic cognitive impairment, although it can also appear with nonamnestic cognitive impairment. AD often causes short-term memory difficulties, but it can also affect expressing speech, visual information analyzing, and executive function. As researchers strive to unravel the complexities of AD and develop effective interventions, they confront a myriad of ethical considerations that permeate every aspect of the research process. This introduction serves as a gateway into the ethical dimensions of Alzheimer's research, drawing insights from scholarly works to shed light on the multifaceted ethical landscape that researchers navigate [2].

6.1.1 THE IMPACT OF ALZHEIMER'S DISEASE

The burden that AD places on people is substantial and growing for families, health care systems, and society, offering significant difficulties that go well beyond medical concerns [3]. AD, a particularly prevalent type of dementia, gradually robs people of their mental capacity, sense of freedom, and quality of life, which eventually results in severe disability and reliance. To mitigate its effects and improve outcomes for impacted individuals and communities, public health programs, legislative initiatives, and research endeavors must consider the complex burden of AD [4]. AD exacts a substantial toll on healthcare systems and economies worldwide, placing immense strain on healthcare resources, long-term care facilities, and caregiver networks [5]. The direct costs of medical care, including diagnosis, treatment, and long-term care, are staggering, whereas indirect costs associated with caregiver burden, productivity loss, and reduced quality of life further exacerbate the economic impact of the disease.

1. **The Caregiver's Concern and Lifestyle Quality:** Alzheimer's places a profound emotional, physical, and financial burden on caregivers, who often shoulder the responsibility of providing round-the-clock care and support for affected individuals [6]. Caregivers frequently experience high levels of stress, anxiety, depression, and social isolation, compromising their own health and well-being. Moreover, the demands of caregiving can disrupt employment, strain family relationships, and erode the caregiver's own quality of life, perpetuating a cycle of caregiving-related stress and burden [7].

2. **Social and Psychological Impact:** AD impacts cognition, autonomy, social connections, identity, and sense of self, leading to personality changes, isolation, and stigma. Family members and caregivers may struggle to cope with the emotional toll of witnessing their loved one's decline and adapting to the challenges of caregiving, further exacerbating the psychological impact of the disease on affected individuals and families [8].

3. **Public Health and Policy Challenges:** AD poses formidable public health and policy challenges, requiring a comprehensive and coordinated response to address its multifaceted impact on individuals, families, and efforts to improve early detection and diagnosis, enhance access to quality care and support services, and promote dementia-friendly communities that are essential for mitigating the burden of investments in Alzheimer's research

which are crucial for improving lives of patients and caregivers, understanding the disease, and finding a cure to reduce its global burden [9].

6.1.2 UNDERSTANDING THE ETHICAL CONTEXT IN ALZHEIMER'S RESEARCH

Delving into the ethical landscape of Alzheimer's research reveals a complex interplay of principles, regulations, and societal considerations that shape the conduct and outcomes of scientific inquiry in this field [10]. At its core, the ethical landscape of Alzheimer's research revolves around the principles of respect for persons, beneficence, and justice. Respecting people with Alzheimer's means valuing their dignity and autonomy, ensuring informed consent (IC) and protection of those unable to make decisions [11]. Researchers must prioritize maximizing benefits and minimizing harm in Alzheimer's research. This includes developing interventions to alleviate suffering, improve quality of life, and advance knowledge while prioritizing participant safety. Justice requires a fair distribution of research benefits and burdens, as well as equal access to participation opportunities. Given the impact of Alzheimer's on certain communities, inclusivity and diversity in research participation are crucial for addressing health disparities and promoting health equity [12]. Practical ethical problems in Alzheimer's research include IC, privacy, and developing technology. Researchers must offer thorough information about the study and get participants' voluntary, informed, and continuing permission. Privacy and confidentiality are critical considerations in Alzheimer's research, given the sensitive nature of health information and the potential for stigmatization and discrimination. Researchers must implement robust data security measures, anonymize data whenever possible, and obtain explicit consent for data sharing to uphold participant privacy and confidentiality. Emerging technologies like biomarker assays, neuroimaging, and genetic testing could revolutionize Alzheimer's research and interventions [13]. Researchers must follow high ethical standards and legal limits, such as the Declaration of Helsinki, the Belmont Report, and national and international laws governing human subject research. The ethics committee evaluates research protocols, ensuring compliance with ethical standards and regulations to maintain institutional integrity [14]. Moreover, fostering ethical research practices requires fostering a culture of integrity, transparency, and accountability within the scientific community. This entails promoting open communication, collaboration, and ethical reflection among researchers, clinicians, policymakers, and other stakeholders involved in Alzheimer's research.

6.1.3 CLINICAL TRIALS AND ETHICS

Clinical trials are vital for developing new Alzheimer's treatments, providing hope for those affected. However, Alzheimer's research faces numerous challenges, such as recruitment and participant safety. To tackle these challenges, ethical considerations including beneficence, respect for persons, justice, and adherence to regulations and standards are necessary [15]. Recruiting an adequate number of participants who meet stringent inclusion criteria while ensuring diversity and representativeness of the population is essential for the generalizability and validity of study findings. Moreover, retaining participants over the course of the trial presents logistical and ethical challenges, particularly in long-term studies where attrition rates may be high due to disease progression, caregiver burden, or participant withdrawal [16]. Ethically addressing recruitment and retention challenges requires proactive strategies to engage potential participants and caregivers, build trust and rapport, and address concerns about trial participation. This may involve collaborating with community organizations, leveraging social and digital media platforms, and study details procedures, purpose risks, and potential benefits provided comprehensively. Additionally, offering support services, such as transportation assistance, respite care, and counseling, can help alleviate logistical and emotional barriers to participation and retention [17]. Ensuring participant's safety and well-being is paramount in Alzheimer's clinical trials, given the vulnerability of the population and the potential risks associated with experimental interventions. Ethical considerations in ensuring participant's safety encompass rigorous risk-benefit assessment, IC processes, monitoring of adverse events, and provision of appropriate medical care and support services. Researchers must carefully balance the potential benefits of participation with the risks of harm and prioritize participant's welfare throughout the trial [18]. Maintaining scientific integrity and data quality is essential for ethical conduct in Alzheimer's clinical trials, as accurate and reliable data are critical for evaluating the safety and efficacy of interventions. Ethical challenges such as data integrity, bias, and conflicts of interest may arise, necessitating transparency, accountability, and adherence to methodological standards [10]. Researchers must work closely with IRBs, regulatory agencies, and stakeholders to meet ethical standards and guidelines. Ethical review processes should include evaluating study protocols, risk mitigation, and participant protections, upholding the principles of beneficence, justice, and respect [19].

6.1.4 ENSURING ETHICAL CONDUCT: REGULATORY FRAMEWORKS AND GUIDELINES

A strong system of regulations and guidelines ensures ethical practices in Alzheimer's research, safety of participants, and maintaining scientific integrity. These frameworks outline principles, standards, and procedures that researchers must follow in the design, conduct, and oversight of studies to safeguard the rights, welfare, and dignity of participants [20]. Central to ensuring ethical conduct in Alzheimer's research is adherence to national and international regulations governing human subject's research. These regulations establish legal needs and ethical principal for the protection of research participants, including requirements for IC, ethical review, and data privacy and confidentiality. The Common Rule in the US (Title 45 CFR Part 46) establishes ethical guidelines to protect human research subjects, including those with AD. It covers IC, review by Institutional Review Boards (IRBs), and safeguarding vulnerable populations [21]. Various regulations and guidelines exist around the world to ensure ethical and standardized practices in clinical trials for Alzheimer's research. In the US, the Food and Drug Administration (FDA) has specific regulations (Title 21 CFR Parts 50, 56, and 312) that cover the design, conduct, and reporting of clinical trials. The International Conference on Harmonization (ICH) also provides guidelines in this regard. Europe has the European Union Clinical Trials Regulation (Regulation (EU) No 536/2014), which establishes harmonized rules for clinical trials within EU member states. The Council of Europe's Oviedo Convention and Additional Protocol offer ethical standards for biomedical research. Additionally, international ethical guidelines like the Declaration of Helsinki, the Belmont Report, and the World Medical Association's Declaration of Taipei provide ethical principles for research involving human subjects. These guidelines emphasize respect for persons, beneficence, and justice [22]. IRBs are responsible for safeguarding the rights and welfare of research participants by assessing the risks and benefits of the study, the qualifications of the research team, and the adequacy of the IC process. They are also tasked with monitoring ongoing studies, evaluating protocol deviations, and ensuring that any changes to the study are ethically justified. In addition to IRBs, regulatory agencies such as the FDA and the National Institutes of Health (NIH) set the guidelines and standards that must be followed in research studies. Sponsors, who provide funding and resources for the study, also play a key role in ensuring compliance by overseeing and supporting the research process. Effective collaboration between all these parties is essential to ensure that research studies are conducted in an ethical and legally compliant manner. By working together, researchers, sponsors, regulatory agencies, and ethical review bodies can uphold the highest standards of research integrity and protect the rights and well-being of research participants [23].

TABLE 6.1 Regulatory Frameworks and Guidelines.

Regulatory Framework	Key Ethical Principles	Application in Alzheimer's Research
Common Rule (45 CFR Part 46)	Respect for Persons, Beneficence, Justice	Informed Consent, IRB Review
The FDA Regulations, specifically Parts 50, 56, and 312 of Title 21 of the Code of Federal Regulations	Participant Safety, Data Integrity	Clinical Trial Conduct, Safety Monitoring
Europe Union CTR (Regulation (EU) No. 536/2014)	Informed Consent, Participant Rights	Clinical Trial Conduct in EU Member States
Declaration of Helsinki	Respect for Autonomy, Beneficence	Ethical Conduct of Research, Informed Consent
Belmont Report ,1979	Ethics values respect, kindness, and fairness	Ethical priorities: Participant protection, consent, privacy. Adherence to regulations, research integrity

6.2 CLINICAL FEATURES AND ETHICAL CONSIDERATIONS

AD is defined by a range of clinical symptoms that progress over time, drastically impacting people's cognitive, emotional, and functional abilities. Addressing these symptoms ethically necessitates careful consideration of the concepts of autonomy, beneficence, and justice, as well as a dedication to protecting the dignity and rights of Alzheimer's patients and caregivers [24]. Common clinical indicators include language impairments, confusion, memory loss, difficulty solving problems, disorientation, mood swing disorders, and a lack of initiative. AD symptoms include cognitive decline, behavioral changes, functional impairments, and memory loss, posing ethical challenges. As individuals progress through the stages of Alzheimer's, their capacity for decision-making and autonomy may become compromised, raising questions about IC, patient rights, and the balance between beneficence and respect for persons [25]. As Alzheimer's progresses, individuals may experience difficulties in decision-making and understanding complex information. Ethical considerations surrounding IC involve assessing individuals' decision-making capacity, ensuring comprehension and voluntariness, and respecting their right to participate in decisions that affect their health and well-being [26]. Behavioral and psychological symptoms of Alzheimer's, such as agitation, aggression, and wandering, present ethical challenges for caregivers and healthcare professionals, who must balance individuals' autonomy and dignity with the need to ensure their safety and the safety of others. Ethical approaches to managing these symptoms involve individualized care plans, nonpharmacological interventions, and strategies that prioritize least restrictive measures while promoting individuals' well-being and preserving their autonomy and dignity [27].

Ethical care for individuals with Alzheimer's requires sensitivity to their preferences, values, and life experiences, recognizing their unique identity and personhood beyond the limitations imposed by the disease. Caregivers and healthcare professionals must strive to engage individuals in decisions about their care, respect their cultural and spiritual beliefs [28]. Ethical considerations in Alzheimer's care encompass a commitment to enhancing individuals' quality of life and preserving their dignity throughout the course of the disease. This involves providing comprehensive support services, promoting meaningful social engagement and activities, and addressing physical and emotional needs in a compassionate and person-centered manner. Ethical palliative care approaches focus on symptom management, comfort, and dignity at the end of life, prioritizing individuals' preferences and ensuring a peaceful and dignified transition [29].

TABLE 6.2 Clinical Symptom and Ethical Considerations.

Clinical Symptom	Ethical Considerations
Cognitive decline	Impact on decision-making capacity, challenges in informed consent
Memory impairment	Preservation of identity and personhood, informed consent challenges
Behavioral symptoms	Balancing autonomy and safety, respect for individual preferences
Caregiver burden	Supporting caregivers ethically, promoting caregiver's well-being
Care at the end of life	Hospice care ethics: Honoring patients' choices for treatment

Ethical considerations include addressing caregivers' needs for education, respite, and support services, respecting their autonomy and rights as decision-makers, and recognizing the importance of their well-being in providing high-quality care to dementia subjects [30]. Various clinical manifestations of Alzheimer's, including agitation, aggression, psychosis, and wandering, pose ethical challenges for caregivers and healthcare providers, who must balance the need for compassionate care and support with ensuring safeguard of individuals with Alzheimer's and those around them. Ethical approaches to managing behavioral symptoms involve comprehensive assessment, personalized interventions, and multidisciplinary collaboration, prioritizing nonpharmacological approaches, caregiver education, and individualized care plans that respect individuals' preferences, values, and dignity [8]. As Alzheimer's progresses, maintaining individuals' life with quality and ensuring compassionate care of life become paramount ethical imperatives. Ethical considerations in supportive and care of life for individuals with Alzheimer's include respecting their wishes regarding medical interventions, symptom management, and advanced care planning, as well as supporting families and caregivers in navigating complex decisions and providing holistic, person-centered care that promotes comfort, dignity, and meaning in the final stages of life [31].

6.3 ETHICAL PRINCIPLES DURING TREATMENT OF ALZHEIMER'S DISEASE

Early detection techniques are necessary since the World Health Organization (WHO) projects a threefold increase in Alzheimer's incidence by 2050. A two-step machine learning (ML) technique is being investigated to solve the problems associated with large-scale assessments. Finding high-risk individuals based on medication or sociodemographic data is the first step. To choose appropriate participants for clinical research, the identified high-risk people will undergo individual biomarker tests in the second stage. Delaying or slowing down AD's progression is the aim. Fighting the stigma attached to dementia requires compassionate care that emphasizes the individual strengths and connections of the patient. To address ethical concerns regarding illness development, it is important to avoid imposing invasive medical procedures on individuals who may not understand the objective of the therapy, since this might be considered an assault. Humane medical therapy in advanced stages of sickness should prioritize comfort measures, such as behavioral or pain drugs, above prolonging morbidity. A hospice-like approach to caring for individuals with AD involves evaluating their burdens and recognizing that death is not an enemy. For those with severe AD, the most effective way to show care and compassion is by attentive "being with" rather than technical interventions. The ethics of intervention in AD are intricate and often entail questions of autonomy, beneficence, nonmaleficence, and fairness [32]. To keep patients, we must respect their independence, be transparent and discreet, fair, and loyal, and truthful. Pharmacological treatments for Alzheimer's illness must be carefully studied considering ethical issues. Following are a few of the important ethical considerations:

6.3.1 PATIENT INDEPENDENCE

Whenever feasible, get the patient involved in treatment decisions. If they are unable to engage, evaluate their previously expressed desires and ideals. Autonomy is further defined by its degree of purposeful activities and freedom from dominating influences. Appreciation for liberty is seen as a core concept of medical ethics. Simplicity of essentials will bring together what is needed for daily existence in a simpler style. Communicating through spoken words on what must be done next: talk to the individual and advise them regarding "what they should do next" in the context of caring for persons with daily living difficulties. Verbal communication provides opportunities to complete a task and offer possibilities for the individual to execute the tasks of daily living on their own [33]. To maintain functional autonomy, which is crucial for the planning of care needs, a predictive system would foresee an unfavorable evolution of functional autonomy and optimize the management of patients with AD, particularly by assisting in the creation of tailored interventions, such as new disease-modifying medications. In the framework of anti-Alzheimer's medications, autonomy refers to patients' ability to choose among a suitable range of treatment alternatives. This involves having the right decision to embrace or decline therapies based on a thorough understanding of the prospective advantages and hazards. It is critical for medical professionals to make sure that folks, or their authorized those making decisions, have accurate information and can agree to the treatment regimen. Furthermore, the moral framework for pharmacy leadership argues that efforts to limit pharmaceutical prices can influence which treatments are covered, compromising patient autonomy. Limitations in prescription coverage can limit patient and provider options, potentially causing disputes between individual requirements and payer policies [34]. Some individuals may have made advanced directives or chosen healthcare proxy to make choices for them when they are unable to do so. Respecting these directions guarantees that patients' choices are upheld even when their cognitive abilities deteriorate. In practice, the change from medical patronage to care that prioritizes patients has raised the importance of patient autonomy. This means that healthcare practitioners must engage in collaborative decision-making with patients while prescribing pharmaceuticals, considering their beliefs and preferences. Overall, the association between prescription of anti-Alzheimer's medications and patient autonomy highlights the need of balancing ethical values such as beneficence, benevolence, and justice while respecting patients' autonomy in making healthcare decisions [35].

6.3.2 BENEFICENCE

Beneficence is a way of doing good for others, which includes generosity, compassion, and mercy. It also conveys a powerful moral message. Beneficence is a vital principle ensuring that doctors prioritize patients' well-being and provide optimal care. This involves safeguarding the patient, assisting those with impairments, and eliminating circumstances that might be harmful. There is a moral duty to behave in the best interests of others and to avoid harm, and it is the most important commitment in Alzheimer's care. A beneficial option is the one that occurs regardless of its decision and includes providing the best treatment for the patient. In Alzheimer's care, kindness can appear in a variety of ways, such as providing proper medical therapy, including prescription of anti-Alzheimer's medicines to assist control symptoms and perhaps delay the course of the illness; maintaining

Alzheimer's people's respect, convenience, and civic engagement is necessary to ensure their quality of life [36]. AI technology supports caregivers and loved ones in the care of people with Alzheimer's, while maintaining respect for patient autonomy and recognizing their choices and principles as cognitive abilities decline. This frequently leads to ethical quandaries, particularly when a patient's preferences collide with what healthcare personnel feel is in their greatest interests. For example, a patient may refuse some treatments or choose to remain inside despite safety concerns. To negotiate these issues, healthcare practitioners should engage in collaborative decision-making with clients and their families, evaluate their previous and current preferences, and involve them in care choices whenever possible. In circumstances when patients are unable to arrive at informed choices, healthcare proxy or legal guardians can identify the appropriate course of action based on their beliefs and best interests [37]. Anti-Alzheimer's drugs treat symptoms, slow Alzheimer's development, and enhance cognitive function. They help patients maintain everyday tasks and independence, improving their quality of life [38].

6.3.3 NONMALEFICENCE

Nonmaleficence, or avoiding harm to the patient, is crucial in treating AD. Clinicians must carefully consider adverse drug reactions and potential risks of therapy to minimize negative effects on the patient. The principle of nonmaleficence, also known as "first, do no harm," is particularly important when using psychopharmacological agents with cognitively challenged individuals. Clinicians should assess potential side effects that may further harm the patient, such as disorientation, abnormal metabolism, cerebrovascular events, or even death. By balancing nonmaleficence and beneficence, healthcare practitioners should aim to use pharmaceuticals that enhance the patient's well-being while minimizing side effects. Additionally, healthcare providers must carefully consider the potential for drug interactions and how introducing new medications may affect the patient's overall health, especially considering the presence of coexisting mental and physical illnesses [39]. "Reducing social and economic conditions that are detrimental to people's and communities" health is what justice refers to. There could be a few factors related to injustice that lead to increased rates of illness, damage, disability, and mortality in Alzheimer's patients. Among the contributing factors include low income, lack of education, lack of health insurance, and shortage of medical care [40].

6.3.4 JUSTICE AND EQUALITY

Ensuring justice, fairness, and fair treatment for people with AD entails numerous critical components, including their privileges, worth, and equal access to adequate care and support. Here are some considerations. Individuals with Alzheimer's condition are entitled to legal protection and respect of their human rights. This involves making certain that individuals are treated properly within the law, given access to legal counsel, and are not discriminated against because of their condition. People with AD should have equal access to healthcare services, regardless of race, background, or financial level. Caregivers require proper training, support resources, and acknowledgment for their vital job. Promoting public knowledge of AD helps to overcome stigma and ensures that those affected are treated with empathy and compassion in society. Justice is making financial resources and social support available to people with Alzheimer's and their caregivers [41]. This might include financial assistance, social services, and community initiatives that encourage inclusion and well-being. Advocacy activities should center on policies that promote justice and fairness in Alzheimer's care, such as healthcare, employment, housing, and disability rights. Ensuring justice, fairness, and fair treatment for people with AD entails numerous critical components, including their rights, dignity, and access to appropriate care and support. Individuals with AD are entitled to legal protection and respect of their human rights. This involves making certain that individuals are treated properly under the law, have access to legal counsel, and are not discriminated against because of their condition. Patient with AD should have equal access to healthcare services, regardless of race, background, or financial level. This includes access to diagnostic, therapeutic, and long-term care services. A medical system that reflects cultural competency is critical for providing fair Alzheimer's care. This involves having a diverse workforce, overcoming language difficulties, and offering culturally sensitive patient handouts and practice statements [42].

6.3.5 EDUCATION AND AWARENESS

Education and awareness are critical in comprehending ethical concepts in Alzheimer's therapy. Understanding moral implications: Ethical awareness entails understanding the ethical implications of every activity. To successfully resolve issues and satisfy patient's needs, healthcare practitioners must first identify the potential ethical consequences of their actions.

Understanding cultural and ethnic distinctions: Research indicates that a person's culture, preferred speech, country of origin, ethnic background, and ethnic all have a substantial impact on cognitive test performance. Higher levels of schooling are also linked to enhanced cognitive reserve, which protects against cognitive deterioration in the context of neuro-degeneration. As a result, education can aid in recognizing cultural and ethnic distinctions, allowing for more individualized and effective care. Education and awareness are critical for understanding and addressing the ethical considerations behind Alzheimer's treatment. They contribute to the promotion of ethical conduct, the protection of patients' rights and well-being, and the provision of effective and tailored treatment and care [43].

6.4 ETHICAL CONCERNS IN ALZHEIMER'S DISEASE CLINICAL TRIALS

AD is a type of dementia that presents various ethical dilemmas for patients, family members, caregivers, physicians, and researchers. These concerns vary and alter with the disease's stage and severity. Ethicists have addressed challenges in AD research, including managing IC for patients with limited decision capacity, using placebos in this group of people, and worries about privacy and public image. As the field of Alzheimer's research has advanced, ethical questions have arisen regarding the dangers of developing the illness [44]. Major ethical issues raised in AD clinical research are displayed in Figure 6.1.

In the pursuit of an Alzheimer's cure, ethical challenges arise. If handled carefully, these obstacles are not insurmountable and will not hinder the investigation process. Neglecting them, however, could harm subjects, researchers, and impede progress [45].

1. Research with Alzheimer's patients raises significant ethical concerns, including the need for large study cohorts that last until the participants' death.
2. Cohorts often include individuals who initially consented but later developed cognitive impairment.
3. Adaptive trial designs further complicate obtaining IC, maintaining equipoise, and ensuring fairness.
4. Additionally, biomarkers and predictive tests indicating risk, rather than certainty, could potentially harm participants if accessed by insurance firms or long-term care providers.

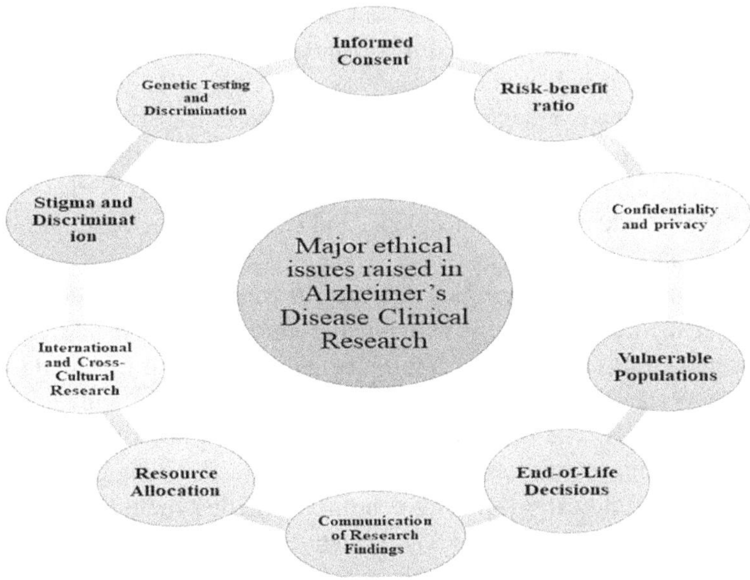

FIGURE 6.1 Major ethical issues raised in Alzheimer's disease clinical research.

5. The involvement of study partners also carries risks to the subject-study partner relationship.
6. To address these complex ethical challenges, increased attention and support are needed at all levels, including federal and local studies, with a centralized regulatory structure such as genetic research.
7. The principle of "respect for persons" is particularly relevant in Alzheimer's research, recognizing that some patients may lack autonomy [43].

6.4.1 VOLUNTARY PARTICIPATION

The voluntary nature of participation is a fundamental ethical aspect of research. AD research can only proceed if patients are willing to take part in research studies. However, enrolling and retaining participants in research studies for AD is difficult. Furthermore, because individuals with AD have reduced decision-making capacities, clinical trials must include caregivers. The distinct problems are experienced throughout the recruitment process for Alzheimer's illness clinical studies. Human subject participation in clinical trials is crucial for increasing our understanding of AD and developing more effective treatments. Medical research would come to a halt without voluntary participants. Delays in subject recruiting can have a negative impact on clinical trial outcomes while also raising trial expenditures. AD clinical studies have distinct hurdles when recruiting patients. To enhance subject accrual into clinical trials, it is crucial to identify these problems and execute effective measures to address them [46, 47].

Alzheimer targets persons who are 60 years old or older. Thus, comorbid medical issues are unavoidable among research participants. Learning, cultural upbringing, culture, neighborhood, and financial status are all important factors in determining who qualifies and volunteers for AD clinical trials. The recruitment process begins with a dialog between potential volunteers and researchers [48]. A subject's living environment and the role of the caregiver may determine whether they are eligible to take part in a clinical study for AD. Effective recruiting requires teaching potential participants about the trials and raising their awareness of clinical trials. During routine clinic visits, the lead investigator can update the patients about new research while also assisting them in comprehending the specifics of the trials. Regarding any new experiment, prospective volunteers could have a range of queries [49]. In many respects, the caregiver serves as both a key participant and stakeholder at every stage of the clinical research, thereby making them a "co-subject" in the study. Researchers who address caregivers' needs, worries, and unique situations will have more success recruiting participants [50].

6.4.2 RIGHT TO WITHDRAW

Trial subjects have absolute right to quite clinical trials as participant at any time without any reason. A participant or the caregiver for a participant should have the option to withdraw permission and end the research at any point throughout the trial [51]. Clinical trial participants are frequently the subject of voluntary or involuntary withdrawals. The following are some justifications for withdrawal: decline in health, study violation, investigator's choice, ineffectiveness of the medication under test, unfavorable and unbearable side effects occurring, inadequate patient follow-up, and patient's refusal to participate in the clinical study. Inquire about the withdrawal procedure. To officially withdraw from some studies, you may need to follow certain procedures, such as filling out a withdrawal form or passing through final evaluations. They could suggest more clinical trials or tried-and-true therapies that might be helpful. Ask about any postwithdrawal follow-up care that may be required. This might entail switching to an alternative medical treatment or keeping an eye out for any side effects.

6.4.3 PROTECTION OF PARTICIPANTS

Research involving Alzheimer's patients raises important ethical concerns, such as how one safeguards mentally ill people from exploitation. Protecting participants is vital in Alzheimer's clinical research to ensure ethical behavior and safety. The various crucial elements of participant's safety in clinical studies for Alzheimer's illness: Participants must be given thorough information about the goals, methods, possible dangers, and advantages of a research before signing up for it. Participants freely provide IC, whereas IRBs conduct ethical reviews of studies on AD. These committees assess the study protocol to guarantee ethical behavior and participant's safety. Safety-monitoring procedures are put into place for clinical studies to quickly detect and handle any unfavorable occurrences or participant's side effects. Regular check-ins, medical evaluations, and reporting systems are a few examples of monitoring [43]. The experiment may be supervised by independent data safety monitoring boards (DSMBs), who will assess safety data and trial progress on a regular basis. The highest ethical standards in medical studies are upheld, participant's safety is given priority, and participant's autonomy is respected by investigators and groups conducting AD clinical trials [52].

6.4.4 INFORMED CONSENT AND ALZHEIMER'S DISEASE

Designing IC procedures with proxies raises significant ethical concerns. IC is necessary for ethical research on AD, but individuals with AD may find it difficult to give it due to cognitive impairment. The IC procedure can be burdensome and may not satisfy participants' requests for information. There is disagreement over the most accurate way to evaluate the decision-making skills of AD research participants. IC ethics in AD research is a complicated topic [11]. The following are some salient points:

1. **Impairment of Cognitive Function:** Significant cognitive impairment is common in AD patients. (Learning, language, and

visual-spatial skills. Having difficulty finding the proper term. Difficulty interpreting visual imagery and spatial connections. Impaired reasoning or judgment. Difficulty focusing and thinking, particularly concerning complex topics like numbers. Performing many tasks simultaneously is very tough.), which may make it more difficult for them to give informed permission. This turns into a serious moral conundrum, particularly if the study is intrusive, time-consuming, and fraught with danger.

2. **Policy Uncertainty:** The absence of established policies regarding the inclusion of individuals with decision-making deficits in clinical studies persists across different nations, despite prolonged deliberation. There is policy uncertainty elsewhere besides the India and USA. For instance, the location and kind of study with dementia subjects may give rise to three distinct legislations in the UK [53].
3. **Authenticity and Independence**: Both are two crucial ethical notions in this context: autonomy (the patient's freedom to make decisions) and authenticity (research involvement that is consistent with the patient's ideals). Authenticity in the context of research participation relates to the idea that individuals should be genuinely expressing their own values and beliefs when deciding whether to participate in research studies. This principle emphasizes that research participants should not be manipulated or influenced in ways that compromise their true desires or interests [11, 54].
4. **Informed Consent:** In some cases, a person's capacity to provide permission may change, and family members may be involved in assisting them in making choice. Obtain permission to see the person who has Alzheimer's from staff, family, or other designated individuals. The information sheet should include a concise title, goals, background, participation requirements, potential risks, and the right to withdraw at any time. Determine if the individual with Alzheimer's has the capacity to agree to participating in a certain activity. If they do, provide information to the person who has Alzheimer's about how to get involved. Document the consent process thoroughly, including who provided consent (the individual or surrogate), what information was provided, and how understanding was assessed. Use written or audio-visual materials to supplement verbal explanations. Surrogate decision

is circumstances when individuals are unable to offer informed approval; alternative decision-makers (such as authorized agents or family members) may be consulted. The surrogate's decisions should be based on the person's previously stated tastes, values, and greatest interests. Assessing the capacity of Alzheimer's patients to provide IC is crucial. Researchers need to use appropriate methods to evaluate participants' decision-making abilities and ensure that consent is truly voluntary and based on adequate understanding (Poppe et al., 2020). In conclusion, gaining informed permission for AD research necessitates sensitivity, adaptability, and a dedication to protecting the autonomy and well-being of people with cognitive impairment. Researchers must adjust communication tactics and decision-making procedures to meet the special demands of AD patients while maintaining ethical behavior throughout the study process [55].

5. **Advance Directives:** According to the Alzheimer's Association, involvement in research should need the agreement of an individual capable of coming to informed decisions. This informed permission might be supplied by advanced directives made by a fully competent individual who may later develop dementia and state their beliefs about study involvement.
6. **Long-Term Follow-Up:** Alzheimer's trials may involve long-term follow-up to assess treatment efficacy and safety. Ethical considerations include ensuring ongoing participant consent, providing continued access to care, and addressing any evolving risks or benefits [56].

6.4.5 SAMPLE SIZE

The objectives of the trial, the phase of Alzheimer's that is being investigated, and the kind of intervention being examined whether pharmacological, behavioral, or lifestyle-based collectively determine the study design, participant criteria, outcome measures, and ethical considerations for clinical evaluation. However, because of the intricacy and fluctuation of the condition, as well as the requirement for statistical power to detect meaningful changes among treatment groups, clinical trials for AD frequently require large sample sizes. Sample numbers might range from a few hundreds to thousands of people [57, 58]. This is owing to the lengthy preclinical period, high attrition rates among the elderly, and possible selection effects. To lower sample sizes in AD secondary prevention studies, use dynamic PET scans and/or limit enrolment to people with moderate amyloid load or APOE-ε4 carriers. Initial trials may require between 20 and 80 participants, but phase II trials that assess therapy effects seldom require more than 100 to 200 patients [59].

6.4.6 CONFIDENTIALITY AND PRIVACY

Confidentiality and privacy issues are also important ethical aspects of research. Absolutely, secrecy and privacy are critical in clinical studies, particularly those involving delicate illnesses such as AD. Here are how these ideals are upheld. While getting involved in a clinical experiment, participants must offer IC. This method ensures subjects comprehend the study's purpose, procedures, risks, benefits, and options, allowing informed decision-making and questioning [60]. Throughout the experiment, participants' confidentiality is safeguarded. Personal identifiers, such as names and contact details, are often replaced with codes or pseudonyms to make data anonymous. Access to participants' information has been restricted to authorized individuals, and data security measures are in place to prevent unauthorized usage or disclosure. Research institutions and sponsors must protect participants' data in accordance with applicable rules and regulations, such as HIPAA in the United States or GDPR in the European Union. To reduce the risk of breaches, data-handling processes guarantee that participants' information is stored, transmitted, and disposed of securely. These bodies assess the IC procedure, the risk–benefit ratio, and confidentiality and privacy precautions. Ongoing monitoring ensures that ethical standards are followed and that participants are safe. The research findings are published in a way that ensures participants' confidentiality. Individual participants' data are aggregated or anonymised in study papers and presentations to avoid identity. The findings are presented in aggregate form to add to scientific understanding while respecting participants' confidentiality. By adhering to confidentiality and privacy norms, clinical trial researchers and institutions demonstrate their commitment to protecting participants' rights and fostering confidence in the research process. This guarantees that people are confident and comfortable about their decision to join in AD clinical studies [61].

6.4.7 PLACEBO USE

The use of placebos in Alzheimer's trials can be ethically complex, especially when effective treatments are limited. Ethical guidelines recommend careful consideration of placebo use, such as offering alternative treatments after the trial or ensuring that participants receive the best standard of care. Randomized withdrawal design is used in Alzheimer's illness clinical research to assess how long a medication is beneficial. Following this time frame, each subject is swapped to a placebo, and both groups have ongoing evaluations. Presumably, the medicine was only symptomatic. It is considered that the medication affects the biology of the illness if the group that was moved from the active medication keeps some gains above the placebo group. The use of placebos in clinical trials raises several ethical issues. Placebo control might even be considered an ethical obligation, but some provisos should be kept in mind [62].

1. In situations where a medical condition is concerned, it is essential that there is a lack of satisfactory treatment available or if there is one, it is expected to come.
2. Appropriate duration for placebo intervention during trials.
3. The administration of a placebo must not pose unacceptable risks.
4. Ensuring comprehensive subject information and consent in research trials.

6.5 OVERCOMING OBSTACLES IN ALZHEIMER'S CLINICAL TRIALS

Overcoming hurdles in clinical research of Alzheimer's is critical for furthering research and discovering viable therapies for this complicated illness. Here are some typical challenges and techniques for overcoming them.

6.5.1 SUBJECT RECRUITMENT AND SCREENING

Strategies such as raising awareness, collaborating with advocacy groups, harnessing social media, and offering incentives might assist enhanced recruitment. Alzheimer's studies frequently have significant dropout rates due to the disease's degenerative nature and the duration of the trial period. Maintaining frequent communication with participants, supporting caregivers, reducing hardship through simple trial designs, and offering incentives for participation can all help to increase retention. Screening is costly and lengthy with many potential participants failing due to strict eligibility criteria [63].

6.5.2 LIMITED PATIENT AWARENESS

Patient knowledge is a major obstacle to Alzheimer's clinical studies. Doctors and patients lack understanding of the early stages of the disease and ongoing studies. Issues identified include doctors and patients being unaware of dementia's early stages, fear of an Alzheimer's diagnosis, overworked healthcare systems, lack of effective treatment options, absence of quick and affordable diagnostics, and limited knowledge about clinical trials. The key barriers preventing productive research on Alzheimer's are patient enrolment in studies. To overcome this, strategies for improving participant identification, recruitment, enrolment, and retention should be implemented in different phases of the disease. Promising solutions involve increasing public awareness of Alzheimer's, providing reliable biomarkers and tools for cognitive assessment to researchers and healthcare professionals, and facilitating trial participants' awareness, referral, and participation in research studies [64].

6.5.3 DISEASE COMPLEXITY AND BIOMARKERS

AD appears differently in everyone, resulting in variation among research populations. Using biomarkers for more accurate diagnosis and stratifying individuals based on illness features can help overcome this issue. Biomarkers that reliably predict disease progression or therapy response are critical for assessing the efficacy of proposed therapies. Even experienced investigators make false clinical diagnoses of AD in 10% to 15% of cases, and biomarkers may increase diagnostic accuracy. Importantly for the development of potential disease-modifying medications for AD, biomarkers may also serve as indirect measurements of disease severity. When utilized this way, clinical trial sample sizes may be lowered, and a change in biomarker may be interpreted as supporting evidence of illness alteration. Investing in biomarker research and confirming current biomarkers can lead to better trial outcomes [65].

6.5.4 LACK OF CLEAR DIAGNOSIS

The lack of a clear diagnosis can hinder trial success. Compared with studies for other diseases, AD research studies are slower, more expensive, and more complex. The screen failure rate is 44% for mild AD trials and 88% for preclinical trials. According to research, doctors list patients' comorbidities, lack of access to diagnostic clinical tools, worry about patients' risks associated with experimental protocols, and distance from a research center as obstacles to Alzheimer's clinical trial referral. As just fifty percent of AD patients receive a proper diagnosis based solely on a clinical evaluation, pathology testing would enable doctors to treat AD patients specifically and accurately while alerting them to potential causes of memory and cognitive impairments. Given the widespread nature of the illness and the scarcity of testing and support resources, providers understandably choose to treat overt ailments or screen for disorders that have widely recognized diagnostic processes and remedies [66, 67]. One small clinical trial noticed that physicians were unwilling to refer patients due to concerns about imagined harm (such as risk of surgery and prejudice of diagnosis in the absence of potent treatment) and fear of departing patients to different providers caring for patients in a clinical study setting. Screening and diagnosis can help clinicians better manage patients with impairments by providing patients and their loved ones with information about the causes of their impairment. In addition, recognizing cognitive impairment offers a chance to enrol those affected in studies that may improve our understanding of the condition and provide new medicines, particularly considering the current dearth of effective treatments. However, there is a significant financial cost associated with the procedure of testing for cognitive decline in people over 65 years. One researcher predicted that the entire cost of diagnostic assessment and screening for a typical primary care person might reach about $39,000 for each case that is discovered. This represents a substantial cost commitment that healthcare professionals will need to carefully weigh against the possible advantages of early identification and intervention [68].

6.5.5 REGULATORY CHALLENGES

Regulatory barriers, such as lengthy licensing processes and rigorous criteria, might cause clinical studies to start and end late. Collaborating with regulatory bodies' initial phase, taking use of accelerated routes where possible, and performing multicenter studies can all assist to overcome these hurdles. Alzheimer's medicine approval is subject to stringent conditions set by regulatory authorities such as the FDA and EMA, including clinical effectiveness and safety testing. Meeting these requirements demands rigorous trial design, data collecting, and analysis, which may be costly and time-consuming. Emerging Alzheimer's medicines, such as disease-modifying medications, may necessitate distinct regulatory systems due to their novel methods of action and ability to influence the progression of the disease rather than simple symptoms. To progress Alzheimer's medication development and improve outcomes for patients and families, researchers, regulatory agencies, healthcare providers, and patient advocacy groups must work together to address these issues [69]. Ensure GCP compliance with HIPAA rules when processing PHI in your study involving protected health information giving HIPAA-compliant services and resources; nevertheless, you must design your environment correctly. Tripartite guidelines that are harmonized have also been issued by the International Council for Harmonization (ICH). To make it easier for the regulatory bodies in the US, Japan, and the EU to accept clinical data from each other, the ICH E6(R1) guideline offers a common standard. An integrated addition to ICH E6(R1), ICH E6(R2) offers a more thorough framework for putting GCP into practice. The most recent iteration of the ICH GCP guideline is ICH E6(R3) [70].

6.5.6 USE OF DIGITAL HEALTH TOOLS AND ARTIFICIAL INTELLIGENCE

Some obstacles in Alzheimer's studies might be resolved with the help of digital health tools, such as remote recruitment and assessment. Recent advancements in numerous digital technologies, including wearables, discrete sensors, and passive data collection applications, coupled with augmented computational capacity, and refined analytical methodologies, offer significant potential for a more nuanced and in-depth comprehension of the initiation and progression of neurodegenerative conditions such as AD and similar disorders. With these technological breakthroughs, there is a possibility of not only identifying appropriate candidates for clinical trials but also detecting subtle alterations in cognition and functionality throughout the disease continuum. Moreover, these digital tools have the capability to enhance the evaluation of responses to treatment, consequently boosting both the accuracy and precision of diagnosis. By advocating for the concept of aging in place and supporting the implementation of personalized medical solutions, digital innovations have the capacity to ameliorate patient care standards and fortify overall safety within healthcare environments. In essence, the fusion of digital technologies with neurodegenerative disease research opens a realm of possibilities for improving the understanding, diagnosis, and treatment efficacy of conditions such as AD and cultivates a landscape where patient care and safety are markedly enhanced. Digital technologies do, however, also bring with them a host of new difficulties in the fields of science, medicine, technology, commerce, ethics, and regulations [71]. To solve these well-known difficulties, clinical studies use a variety of established digital technologies. Digital or direct data capturing is a stepping stone toward digital biomarkers. It enhances data quality and guidance during test administration by proactively reacting to errors, automatically calculating results, handling branching rules, ensuring answer consistency, and instantly responding to missing data. It also enables translations and integrates video and audio information collection into evaluations [72]. This method improves precision and accuracy, crucial for detecting effects beyond significant sizes. Passive surveillance can be conducted remotely in various settings using software, sensors, and mobile apps. These tools can detect behaviors such as sleep patterns, emotions, exercise routines, social interactions, and eating habits. By analyzing these behaviors, it is possible to infer cognitive and functional changes in individuals. To effectively monitor biomarkers of change, a network of sensors and the integration of data from electronic health records are necessary. The Oregon Center for Aging and Technology has developed a platform consisting of sensors and devices to collect data in real time from home environments. Empower Yu, another platform, utilizes multiple sensors to monitor mobility, appliance usage, door activity, and room entry, providing a comprehensive understanding of patients. The platform incorporates an inference engine and machine-learning techniques to assess factors such as gait, pacing, nocturnal movements, repetitive behaviors, and medication adherence [73].

Technology businesses are at the forefront of developing innovative screening methods tailored to track even subtle cognitive changes, particularly through the medium of gaming. One notable example of this is Akili Interactive Labs, which is actively refining the video game platform EVO to serve as both a screening tool and a potential indicator for amyloid positivity. EVO requires users to engage in multitasking activities, simultaneously challenging them with perceptual discrimination tasks and visuomotor-tracking exercises. Among these innovative approaches is NeuroRacer, an early version of EVO created by dedicated neuroscientists from the University of California, San Francisco. NeuroRacer specifically aims to pinpoint declines in cognitive function associated with aging, providing a promising opportunity for potential cognitive enhancement through targeted training interventions. This groundbreaking initiative could offer valuable insights into the malleability of cognitive decline and the potential effectiveness of cognitive training programs in addressing age-related cognitive changes [74]. Preliminary findings from research conducted on the 10-minute Altoida Neuro Motor Index (NMI) test revealed an impressive 94% accuracy rate in predicting the onset of AD dementia up to 6 years in advance. This innovative test, operated through an iPhone app, evaluates psychomotor processing speed, executive function, prospective memory, and spatial memory by presenting various levels of difficulty to the user. While a combination of established AD biomarkers such as fDG-PET, magnetic resonance imaging, electroencephalography/ ERPs, and phosphorylated tau protein/amyloid β42 in cerebrospinal fluid delivers superior performance compared with the NMI test, the latter shows significant correlation with these biomarkers. Notably, the NMI test emerges as a more reliable predictor than any single biomarker when used alone, highlighting its potential as a valuable tool for early AD dementia detection. The multifaceted nature of the NMI test enables a comprehensive assessment of cognitive domains, underscoring its importance in the realm of predictive medicine for neurodegenerative diseases such as Alzheimer's [75]. AI technology has the potential to address obstacles faced in clinical studies for AD, such as high screen failure and subject allocation issues. Recent research has focused on using AI for eligibility screening and randomization in clinical trials. Predicting protein and MRI biomarkers during the prescreening phase can reduce the high screen failure rate. Additionally, AI can stratify the AD population into quick and slow progressors, facilitating equitable allocation to intervention and control groups. Despite challenges with explainability and validation, integrating biological knowledge and rigorous validations could lead to new diagnostic biomarkers and precision medicine for Alzheimer's [76]. AI technologies can help address gaps and obstacles in medication discovery, development, and clinical trial design for AD. By forming open-research partnerships, AI has the potential to improve decision-making in academia, the pharmaceutical sector, and healthcare systems. It can expedite the development of Alzheimer's drugs by enhancing molecular design and synthesis, prediction of side effects and drug interactions, determining clinical trial sample sizes based on heterogeneity, analyzing trial data, and applying outcomes to patients who would benefit the most. AI plays a crucial role in accelerating medication research and tailored therapy for Alzheimer's patients.

6.6 CONCLUSION

AD is a major global health issue, affecting approximately 55 million people worldwide. However, this number is expected to almost double by 2030 and reach a staggering 139 million by 2050. The disease is particularly prevalent in poor- and middle-income countries, accounting for more than 60% of cases currently, and expected to reach 71% by 2050. In India, the incidence of AD is on the rise, projected to double by 2030 and quadruple by 2050 among individuals over the age of 60. AD is a neurodegenerative brain ailment that impairs memory and cognitive skills. Common symptoms include memory loss, poor comprehension, hearing, and visual impairment. These cognitive impairments pose ethical challenges in clinical studies, such as obtaining consent, ensuring quality of life and patient engagement, and addressing exclusion. Ethical considerations in clinical research for AD include respecting participants, independent evaluation, fair participant selection, and recruiting, maintaining a good risk?benefit ratio, social value, scientific validity, collaborative partnerships, and beneficence. Currently, only 1% of eligible volunteers are referred to or considered for Alzheimer's clinical trials. Overcoming participation obstacles is crucial to finding effective disease prevention and treatment. To better understand and overcome these hurdles, a survey and interviews were conducted with almost 900 stakeholders, including clinical trial workers, patients, caregivers, and clinicians. By following ethical norms and involving all stakeholders, successful and ethical clinical trials for AD can be carried out. Researchers, medical research coordinators, pharmacists, patients, and caregivers all play pivotal roles in ensuring the success of these trials. The information gathered from stakeholders should be utilized to improve the ethical conduct of clinical trials and remove obstacles.

KEYWORDS

- **clinical trials**
- **dementia**
- **ethical consideration**
- **informed consent form**
- **obstacles**

REFERENCES

1. The Editors of Encyclopaedia Britannica. *Alzheimer disease summary*. Encyclopaedia Britannica; 2021
2. Kumar, A., Sidhu, J., & Goyal, A. (2023). *Alzheimer disease*. Treasure Island (FL): Stat Pearls Publishing.
3. Dharmarajan, T. S., & Gunturu, S. G. (2009). Alzheimer's disease: A healthcare burden of epidemic proportion. *American Health & Drug Benefits, 2*(1), 39–47.
4. Livingston, G., & et al. (2020). Dementia prevention, intervention, and care: 2020 report of the Lancet Commission. *The Lancet, 396*(10248), 413–446.
5. Wong, W. (2020). Economic burden of Alzheimer disease and managed care considerations. *The American Journal of Managed Care, 26*(8 Suppl), S177–S183.
6. Koca, E., Taşkapilioğlu, Ö., & Bakar, M. (2017). Caregiver burden in different stages of Alzheimer's disease. *Archives of Neuropsychiatry, 54*(1), 82–86.
7. Schulz, R., & Sherwood, P. R. (2008). Physical and mental health effects of family caregiving. *The American Journal of Nursing*, *108*(9 Suppl), 23–27.
8. Li, X. L., & et al. (2014). Behavioural and psychological symptoms in Alzheimer's disease. *BioMed Research International, 2014*, 927804.
9. Macklin, K. (2021). On the frontlines of the Alzheimer's crisis: Advocacy organizations in Delaware and nationwide urge public health intervention to curb staggering disease trends. *Delaware Journal of Public Health, 7*(4), 20–23.
10. Post, S. G., & Whitehouse, P. J. (1995). Fairhill guidelines on ethics of the care of people with Alzheimer's disease: A clinical summary. *Journal of the American Geriatrics Society, 43*(12), 1423–1429.
11. Kim, S. Y. H. (2011). The ethics of informed consent in Alzheimer disease research. *Nature Reviews Neurology, 7*(7), 410–414.
12. Barrios, L. M., & et al. (2022). Rethinking benefits in health research, reflections of an ethics committee. *The Yale Journal of Biology and Medicine, 95*(3) 389–398.
13. Arafah, A., & et al. (2023). The future of precision medicine in the cure of Alzheimer's disease. *Biomedicines, 11*(2), 335.
14. Nagai, H., Nakazawa, E., & Akabayashi, A. (2022). The creation of the Belmont Report and its effect on ethical principles: A historical study. *Monash Bioethics Review, 40*(2), 157–170.

15. Watson, J. L., & et al. (2014). Obstacles and opportunities in Alzheimer's clinical trial recruitment. *Health Affairs, 33*(4), 574–579.
16. Chaudhari, N., & et al. (2020). Recruitment and retention of the participants in clinical trials: challenges and solutions. *Perspectives in Clinical Research, 11*(2), 64–69.
17. Probstfield, J. L., & Frye, R. L. (2011). Strategies for recruitment and retention of participants in clinical trials. *JAMA, 306*(16), 1798–1799.
18. Indorewalla, K. K., & et al. (2021). Modifiable barriers for recruitment and retention of older adults' participants from underrepresented minorities in Alzheimer's disease research. *Journal of Alzheimer's Disease, 80*(3), 927–940.
19. Onakomaiya, D., & et al. (2023). Challenges and recommendations to improve institutional review boards' review of community-engaged research proposals: A scoping review. *Journal of Clinical and Translational Science, 7*(1), e93.
20. Chandra, M., & et al. (2021). Ethical issues in dementia research. *Indian Journal of Psychological Medicine, 43*(5_suppl), S25–S30.
21. National Research Council, & et al. (2014). *Proposed revisions to the common rule for the protection of human subjects in the behavioural and social sciences.* Washington (DC): National Academies Press.
22. Tenti, E., & et al. (2018). Main changes in European clinical trials regulation (no 536/2014). *Contemporary Clinical Trials Communications, 11*, 99–101.
23. Kim, W. O. (2012). Institutional review board (IRB) and ethical issues in clinical research. *Korean Journal of Anesthesiology, 62*(1), 3–12.
24. Kumar, A., & et al. (2021). *Alzheimer disease (nursing).* Treasure Island (FL): StatPearls Publishing.
25. Bogdanovic, B., Eftimov, T., & Simjanoska, M. (2022). In-depth insights into Alzheimer's disease by using explainable machine learning approach. *Scientific Reports, 12*(1), 6508.
26. Howe, E. (2012). Informed consent, participation in research, and the Alzheimer's patient. *Innovations in Clinical Neuroscience, 9*(5–6), 47–51.
27. O'Connell, C., Sommer, B. R., & Dunn, L. B. (2017). Ethical challenges in geriatric psychiatry. *Focus, 15*(1), 59–64.
28. Howe, E. G. (2006). Ethical issues in diagnosing and treating Alzheimer disease. *Psychiatry (Edgmont), 3*(5), 43–53.
29. van der Geugten, W., & Goossensen, A. (2020). Dignifying and undignifying aspects of care for people with dementia: A narrative review. *Scandinavian Journal of Caring Sciences, 34*(4), 818–838.
30. Brodaty, H., & Donkin, M. (2009). Family caregivers of people with dementia. *Dialogues in Clinical Neuroscience, 11*(2), 217–228.
31. Hochwald, I. H., & et al. (2021). Ethical challenges in end-stage dementia: Perspectives of professionals and family care-givers. *Nursing Ethics, 28*(7-8), 1228–1243.
32. Morris, R., Caro, F. G., & Hansan, J. E. (1998). Personal assistance: The future of home care. Baltimore: Johns Hopkins University Press.
33. Suwa, S., Yumoto, A., Ueno, M., Yamabe, T., Hoshishiba, Y., & Sato, M. (2019). Characteristics of care methods for daily life disabilities in Alzheimer's type dementia that respect autonomy and independence. *Nursing Open, 6*(3), 930–941.

34. Arrieta Valero, I. (2019). Autonomies in interaction: Dimensions of patient autonomy and non-adherence to treatment. *Frontiers in Psychology, 10,* 1857. https://doi.org/10.3389/fpsyg.2019.01857.
35. Kim, H., & Lee, A. (2023). How patient autonomy drives the legal liabilities of medical practitioners and the practical ways to mitigate and resolve them. *Postgraduate Medical Journal, 99*(1168), 83– 88. https://doi.org/10.1093/postmj/qgad003.
36. Premeal. Medical ethics: Beneficence. *The Medic Portal.* https://www.themedicportal.com/application-guide/medical-school-interview/medical-ethics/medical-ethics-beneficence/ (accessed May 2, 2024).
37. Strech, D., Mertz, M., Knüppel, H., Neitzke, G., & Schmidhuber, M. (2013). The full spectrum of ethical issues in dementia care: Systematic qualitative review. *The British Journal of Psychiatry, 202,* 400–406. https://doi.org/10.1192/bjp.bp.112.116335.
38. Majidazar, R., Rezazadeh-Gavgani, E., Sadigh-Eteghad, S., & Naseri, A. (2022). Pharmacotherapy of Alzheimer's disease: An overview of systematic reviews. *European Journal of Clinical Pharmacology, 78*(10), 1567–1587. https://doi.org/10.1007/s00228-022-03363-6.
39. Dunn, L. B., Alici, Y., & Roberts, L. W. (2015). Ethical challenges in the treatment of cognitive impairment in aging. *Current Behavioral Neuroscience Reports, 2*(4), 226–233.
40. Stephens, C. (2006). Social injustice and public health. In B. S. Levy & V. W. Sidel (Eds), *International Journal of Epidemiology* (Vol. 35, pp. 1108–1109). Oxford University Press.
41. Yeo-Teh, N. S. L., & Tang, B. L. (2023). A review of scientific ethics issues associated with the recently approved drugs for Alzheimer's disease. *Science and Engineering Ethics*, 29(1), 2.
42. Braun, B., Demling, J., & Loew, T. H. (2022). Alzheimer's disease: History, ethics and medical humanities in the context of assisted suicide. *Philosophy, Ethics, and Humanities in Medicine, 17*(1), 4.
43. Götzelmann, T. G., Strech, D., & Kahrass, H. (2021). The full spectrum of ethical issues in dementia research: Findings of a systematic qualitative review. *BioMed Central Medical Ethics, 22*(1), 32.
44. Mintzer, J. (2018). Perspective from clinical research: Ethical issues in Alzheimer's disease research. *The Journal of Law, Medicine and Ethics, 46*(3), 699–703. https://doi.org/10.1177/1073110518804231.
45. Davis, D. S. (2017). Ethical issues in Alzheimer's disease research involving human subjects. *Journal of Medicine Ethics, 43*(12), 852–856.
46. Keith, S. J. (2001). Evaluating characteristics of patient selection and dropout rates. *Journal of Clinical Psychiatry, 62*(Suppl 9), 11–14.
47. Karlawish, J., Cary, M. S., Rubright, J., & Tenhave, T. (2008). How redesigning AD clinical trials might increase study partner's willingness to participate. *Neurology, 71*(23), 1883–1888.
48. Connell, C. M., Shaw, B. A., Holmes, S. B., & Foster, N. L. (2001). Caregivers' attitudes toward their family members' participation in Alzheimer disease research: Implications for recruitment and retention. *Alzheimer Disease and Associated Disorders, 15*(3), 137–145. https://doi.org/10.1097/00002093-200107000-00005.

49. Harper, B., & Neuer, A. (2009). A strategic formula to enhance subject enrolment and retention. *The Monitor*, *23*, 59–63.
50. Knebl, J. A., & Patki, D. (2010). Recruitment of subjects into clinical trials for Alzheimer disease. *The Journal of the American Osteopathic Association*, *110*(9 Suppl 8), S43–S49.
51. Department of Health, Education, and Welfare; National Commission for the Protection of Human Subjects of Biomedical and Behavioral Research. 2014 (). The Belmont Report. Ethical principles and guidelines for the protection of human subjects of research. *The Journal of the American College of Dentists*, *81*(3), 4–13.
52. Lynch, F. (2020). The right to withdraw from controlled human infection studies: Justifications and avoidance. *Bioethics*, *34*(8), 833–848.
53. Sun, W., Matsuoka, T., & Narumoto, J. (2021). Decision-making support for people with Alzheimer's disease: A narrative review. *Frontiers in Psychology*, *12*, 750803
54. Beattie, B. L. (2007). Consent in Alzheimer's disease research: Risk/benefit factors. The *Canadian Journal of Neurological Sciences*, *34*(S1), S27–S31.
55. Wilkins, J. M., & Forester, B. P. (2020). Informed consent, therapeutic misconception, and clinical trials for Alzheimer's disease. *International Journal of Geriatric Psychiatry*, *35*(5), 430–435.
56. Anderson, R. M., Hadjichrysanthou, C., Evans, S., & Wong, M. M. (2017). Why do so many clinical trials of therapies for Alzheimer's disease fail? *Lancet*, *390*(10110), 2327–2329.
57. Singh, M. M. (2024). The Education System in India-Rahul Education. https://rahuleducation.org/our-scribes/what-is-an-education-system/. (accessed May 2, 2024).
58. Dhara, H. Child Centric Education. Ijcrt.org. https://ijcrt.org/papers/IJCRT2204525.pdf (accessed May 2, 2024).
59. Lopes Alves, I., Heeman, F., Collij, L. E., Salvadó, G., Tolboom, N., Vilor-Tejedor, N., Markiewicz, P., Yaqub, M., Cash, D., Mormino, E. C., Insel, P. S., Boellaard, R., van Berckel, B. N. M., Lammertsma, A. A., Barkhof, F., & Gispert, J. D. (2021). Strategies to reduce sample sizesin Alzheimer's disease primary and secondary prevention trials using longitudinal amyloid PET imaging. *Alzheimer's Research & Therapy*, *13*(1), 82. https://doi.org/10.1186/s13195-021-00819-2.
60. Muthuswamy, V. (2013). Ethical issues in clinical research. *Perspectives in Clinical Research*, *4*(1), 9–13.
61. Rush, C. How to comply with HIPAA and EU GDPR in medical device studies. Greenlight Guru. https://www.greenlight.guru/blog/how-to-comply-with-hipaa-and-eu-gdpr-in-medical-device-studies (accessed May 2, 2024).
62. De Deyn, P. P., & D'Hooge, R. (1996). Placebos in clinical practice and research. *Journal of Medicine Ethics*, *22*(3), 140–146. https://doi.org/10.1136/jme.22.3.140.
63. Grill, J. D., Monsell, S., & Karlawish, J. (2012). Are patients whose study partners are spouses more likely to be eligible for Alzheimer's disease clinical trials. *Dementia and Geriatric Cognitive Disorders*, *33*(5), 334–340. https://doi.org/10.1159/000339361.
64. Aisen, P., Langbaum, J., & Zissimopoulos, J. Overcoming barriers to success in Alzheimer's disease clinical trials. USC Schaeffer. https://healthpolicy.usc.edu/evidence-base/overcoming-barriers-to-success-in-alzheimers-disease-clinical-trials/ (accessed May 2, 2024).

65. Abdelnour, C., Agosta, F., Bozzali, M., Fougère, B., Iwata, A., Nilforooshan, R., Takada, L. T., Viñuela, F., & Traber, M. (2022). Perspectives and challenges in patient stratification in Alzheimer's disease. *Alzheimer's Research and Therapy*, *14*(1), 112.
66. Galvin, J. E., Meuser, T. M., Boise, L., & Connell, C. M. (2009). Predictors of physician referral for patient recruitment to Alzheimer disease clinical trials. *Alzheimer Disease and Associated Disorders*, *23*(4), 352–356.
67. McCarten, J. R. (2013). The case for screening for cognitive impairment in older adults. *Journal of the American Geriatrics Society*, *61*(7), 1203–1205.
68. Carr, S. A., Davis, R., Spencer, D., Smart, M., Hudson, J., Freeman, S., Cooper, G. E., Schmitt, F. A., Markesbery, W. R., Danner, D., & Jicha, G. A. (2010). Comparison of recruitment efforts targeted at primary care physicians versus the community at large for participation in Alzheimer disease clinical trials. *Alzheimer Disease and Associated Disorders*, *24*(2), 165–170.
69. Apostolaros, M., Babaian, D., Corneli, A., Forrest, A., Hamre, G., Hewett, J., Podolsky, L., Popat, V., & Randall, P. (2020). Legal, regulatory, and practical issues to consider when adopting decentralized clinical trials: Recommendations from the clinical trials transformation initiative. *Therapeutic Innovation and Regulatory Science*, *54*(4), 779–787.
70. Nakamura, K., Ozawa, H., Shibata, T., Ushirozawa, N., Hata, T., Okita, N., Fuse, N., Sato, N., Ikeda, K., Hanaoka, H., Maruyama, T., Wada, M., Shimizu, S., Kasai, H., Yamamoto, Y., Sakurai, J., Todaka, K., Tashiro, S., & Yamamoto, H. (2022). Survey results and recommendations from japanese stakeholders for good clinical practice renovation. *Therapeutic Innovation and Regulatory Science*, *56*(2), 220–229.
71. Gold, M., Amatniek, J., Carrillo, M. C., Cedarbaum, J. M., Hendrix, J. A., Miller, B. B., Robillard, J. M., Rice, J. J., Soares, H., Tome, M. B., Tarnanas, I., Vargas, G., Bain, L. J., & Czaja, S. J. (2018). Digital technologies as biomarkers, clinical outcomes assessment, and recruitment tools in Alzheimer's disease clinical trials. *Alzheimer's & Dementia (New York, N. Y.)*, *4*(1), 234–242.
72. Kirste, T., Hoffmeyer, A., Koldrack, P., Bauer, A., Schubert, S., Schröder, S., & Teipel, S. (2014). Detecting the effect of Alzheimer's disease on everyday motion behaviour. *Journal of Alzheimer's Disease*, *38*(1), 121–132. https://doi.org/10.3233/JAD-130272.
73. Boschi, V., Catricala, E., Consonni, M., Chesi, C., Moro, A. & Cappa, S. F. (2017). Connected speech in neurodegenerative language disorders: A review. *Frontiers in Psychology*, *8*, 269.
74. Anguera, J. A., Boccanfuso, J., Rintoul, J. L., Al-Hashimi, O., Faraji, F., Janowich, J., Kong, E., Larraburo, Y., Rolle, C., Johnston, E., & Gazzaley, A. (2013). Video game training enhances cognitive control in older adults. *Nature*, *501*(7465), 97–101.
75. Tarnanas, I., Tsolaki, A., Wiederhold, M., Wiederhold, B., & Tsolaki, M. (2015). Five-year biomarker progression variability for Alzheimer's disease dementia prediction: Can a complex instrumental activities of daily living marker fill in the gaps? *Alzheimer's & Dementia (Amsterdam, Netherlands)*, *1*(4), 521–532.
76. Seo, Y., Jang, H., & Lee, H. (2022). Potential applications of artificial intelligence in clinical trials for Alzheimer's disease. *Life (Basel)*, *12*(2), 275.

CHAPTER 7

International Efforts in Alzheimer's Research and Treatment: A Worldwide Outlook on Alzheimer's Disease

RICHA GUPTA[1], UDIT CHAUBE[2], LALIT PRATAP CHANDRVANSHI[3], SHRUTI PORIYA[1], SANJAY KHATTRI[4], and RAJENDRA K. SHUKLA[5,*]

[1]*Department of Pharmacology, Institute of Pharmacy, Nirma University, Ahmedabad, Gujarat, India*

[2]*Department of Pharmaceutical Chemistry, Institute of Pharmacy, Nirma University, Ahmedabad, Gujarat, India*

[3]*Department of Forensic Science, Sharda University, Greater Noida, Uttar Pradesh, India*

[4]*Derpartment of Pharmacology, King George'e Medical University, Lucknow, Uttar Pradesh, India*

[5]*Department of Biochemistry, MSD Autonomous State Medical College & Maharishi Balark Hospitals, Bahraich, Uttar Pradesh, India*

*Corresponding author

ABSTRACT

Alzheimer's disease (AD) presents a global challenge that affects millions of people across the world, and thus imposes a significant burden on our healthcare system and society. This chapter provides an overview of international efforts in Alzheimer's research and treatment, highlighting collaborative endeavors, innovative approaches, and emerging trends. From North America to Europe, Asia, and beyond,

Cellular and Molecular Mechanisms in Alzheimer's Disease Therapeutics, Volume 2: Molecular Mechanisms. Prashant Tiwari, Sunil Kumar Kadiri, and Kunjbihari Sulakhiya (Eds)

DOI: 10.1201/9781779641816-7

researchers, clinicians, policymakers, and advocacy groups are actively engaged in understanding the complexities of AD and developing effective interventions. This chapter explores key initiatives, such as multinational research consortia, cross-disciplinary partnerships, and global funding initiatives, driving progress in treatment for Alzheimer's and research related to it. It explores the prevalence, impact, challenges, and advancements in research and treatment of AD worldwide. Furthermore, it examines the challenges and opportunities inherent in international collaborations, including cultural differences, resource disparities, and ethical considerations. It highlights the importance of global cooperation in advancing our understanding of the disease, developing innovative treatment strategies, and improving care for individuals affected by Alzheimer's. By examining the worldwide landscape of Alzheimer's research and treatment efforts, this chapter aims to give insight into the multifaceted nature of Alzheimer's and inspire collective action toward enhancing outcome and improving quality of life for individuals suffering with AD.

7.1 ALZHEIMER'S DISEASE: OVERVIEW AND FACTS

According to the World Health Organisation (WHO), dementia is the seventh most prevalent disability disorder globally, affecting the abilities of millions of individuals and significantly impacting their ability to fulfill routine tasks and lead a normal lifestyle. Alzheimer's disease (AD) is a leading cause of dementia worldwide, concerning health professionals and research scientists across the globe. Alzheimer's has become a major critical health crisis, despite extensive efforts to understand the molecular mechanisms and identify the target in the medial temporal lobe. Alzheimer's is a progressive neurodegenerative disorder with gradual decline in cognitive function, behavior changes, and memory impairment being its characteristic features, which affect daily life and impact patients' quality of life [1].

Alzheimer's is caused due to abnormal accumulation of proteins deposition in and around the cells of the brain. These proteins include beta-amyloid, which forms plaques and tau tangles. Due to these pathological changes, the communication between nerve cells is disrupted which leads to degeneration and eventual death of neurons. The primary reason of AD is still unclear; predisposing effects, such as genetics, advanced age, lifestyle, family history, and environmental factors, are thought to be responsible for the increased risk of developing the disease. Around 95% of AD occurs in individuals over the age of 60–65, termed as late onset, while approximately 5% are classified as early onset. Diagnosis of AD is based on clinical and neuropathological evaluation. The appearance of beta-amyloid plaques and neurofibrillary tangles in neurons, containing tau protein and amyloid protein, serves as an indicator of AD.

As Alzheimer's continues to evolve, efforts to identify early biomarkers, develop novel treatments, and implement preventive strategies are essential in addressing this significant public health challenge and improving outcomes for affected individuals and their families. AD, a complex condition, demands an interdisciplinary approach to effectively address its myriad challenges. The lack of understanding of the cause and effective treatment, increased failure rates of clinical trials, hurdles in implementing treatments, and varied nature of AD highlight the importance for enhanced collaboration [70].

7.2 GLOBAL EPIDEMIOLOGICAL LANDSCAPE OF AD

The global dementia patient population has surpassed 50 million individuals, resulting in an economic burden exceeding $1 trillion annually. Among the leading causes of death worldwide, dementia ranks as the fifth and is the fourth leading cause of disability-adjusted life years lost in the age group of 75 years and above [59].

Between 1990 and 2019, the incidence of AD and other forms of dementia witnessed a substantial surge, increasing by 147.95%. This notable rise in incidence rates highlights the escalating burden of the disease on global populations. As more individuals are affected by dementia, there is a growing urgency to address the associated challenges in healthcare, social support, and research to effectively manage and mitigate its impact on societies worldwide. In 2019, the worldwide prevalence of dementia was significantly higher among women when compared to men, with a female to male ratio of 1.69. This gender imbalance in dementia occurrence is a significant aspect to address in public health strategizing and intervention [63]. Various countries exhibit distinct impacts on AD, resulting in different prevalence rates. A study conducted in China from 2010 to 2020 unveiled a dementia prevalence rate of 6% [83].

India has witnessed a remarkable rise in life expectancy, nearly doubling from 36.98 years in 1950 to 69.27 years between 2015 and 2020. In South Asia, the prevalence of dementia was 1.9% in 2005, with an estimated 3.6 million cases in 2020. Projections indicate a significant increase to 7.5 million cases by 2040 [16]. Prevalence of AD in Europe is estimated to be around 5.05%. This prevalence exhibits variations based on factors, such as age and gender, with higher rates typically observed among older age groups and women [55].

Research findings have demonstrated divergent trends across countries and regions regarding the incidence of AD. Japan and China have notably experienced a rapid rise in age-standardized incidence rates, with 35 million people over the age of 65. In the United States, 48 million people have an age-related disorder. Contrasting with countries, such as Luxembourg, Spain, and Belgium, where stable or declining trends in incidence rates have been observed [11, 47].

As the global demographics shift toward an aging population, AD has emerged as a major challenge for public health. It is imperative to monitor the prevalence and impact of dementia systematically. Such data-driven approaches are essential for informing the development of targeted and evidence-based health strategies and policies aimed at mitigating the escalating burden posed by AD [59]. The impact of an aging population on the burden of AD is profound and multifaceted. As global demographics shift toward older age groups, the prevalence of AD increases significantly. This demographic trend is a primary driver of the rising burden of AD worldwide. Aging is the single prominent predisposing risk factor for dementia, and with people living longer, the numbers affected by the disease are growing.

Age is a significant risk factor in development of Alzheimer's. AD incidence increases with age, especially after the age of 65 years. Research, such as the Aging Demographic and Memory Study (ADAMS), has estimated the prevalence of AD and other dementias. Estimates of AD in 2002 from ADAMS were approximately 50% lower when compared to Chicago Health and Aging Project [84].

Regional disparities and trends in AD epidemiology highlight the complexity of this neurodegenerative condition's impact across different geographical areas. Variations in prevalence and incidence rates are influenced by a multitude of factors, including demographics, socioeconomic status, and cultural norms.

7.3 IMPORTANCE OF COLLABORATION IN ALZHEIMER'S RESEARCH AND TREATMENT

International collaboration is vital for middle- and lower-income countries to overcome the significant challenges they face in AD research and treatment, such as limited resources, infrastructure, and funding. The Alzheimer's Association acknowledges the importance of including these countries in global research efforts. International collaboration can provide access to funding, expertise, and technology that may not be available domestically, enabling these countries to build capacity and conduct high-quality research [67].

International societies like ISOAD and the UNESCO executive committee play crucial roles in promoting interdisciplinary collaborative research on AD. They can facilitate this by creating platforms and infrastructure for collaborative research, organization, and discussion. These efforts are essential for advancing our understanding of Alzheimer's and developing effective strategies for prevention and treatment [70].

This chapter aims to provide comprehensive insights into AD, offering succinct yet thorough information on its pathology, epidemiology, and global impact. It review the current state of AD worldwide, elucidating its prevalence and the demographic profile of those affected. Central to the disclosure is the remarkable international collaboration that has emerged in the realm of AD associated research and treatment. This chapter highlights the collaboration efforts that transcend geographical boundaries and combine the collective expertise, resources of researchers, and clinicians from diverse regions. Through this collaborative approach, countries pooled their scientific investigations on the etiology of Alzheimer's and its associated risk factors, shedding light into its pathophysiology and important therapeutic targets.

The collaborated initiative has yielded significant strides in the progress of treatment and intervention for AD. By sharing data, methodology, and clinical trial findings, researchers have collectively advanced our understanding of AD. This collaborative approach has resulted in identification of promising therapeutic candidates and novel treatment modalities.

This chapter emphasizes the pivotal role of international collaboration in AD, emphasizing its transformative impact on research, treatment, and care strategies.

7.4 INTERNATIONAL COLLABORATIVE INITIATIVES

7.4.1 *OVERVIEW OF COLLABORATIVE EFFORTS IN ALZHEIMER'S RESEARCH*

In the broad field of Alzheimer's research, partnerships become essential forces for progress, bringing together researchers, physicians, and institutions from across the world. One such initiative is the AD Neuroimaging Initiative (ADNI), a pioneering endeavor fostering interdisciplinary collaboration to advance biomarker discovery and clinical trial methodologies. Concurrently, the Global Alzheimer's Association Interactive Network (GAAIN) serves as a hub for international data sharing, facilitating groundbreaking insights into the disease. The AD Consortium (EDAC) fosters collaboration among European researchers, bolstering efforts to understand and combat AD. These collaborative endeavors underscore the collective commitment to unraveling the complexities of Alzheimer's, heralding a future of enhanced diagnosis, treatment, and care for those affected by this devastating condition.

The Alzheimer's Disease Cooperative Study (ADCS) was established in 1991 through a partnership between the National Institute on Aging (NIA) and the University of California, San Diego, representing an endeavor within Alzheimer's research. Serving as a comprehensive platform, the ADCS directs its efforts toward advancing various facets of AD management. This encompasses the exploration and validation of interventions and medical devices with respect to the needs of patients suffering with AD [69].

7.4.2 *SUCCESSFUL INTERNATIONAL PARTNERSHIPS AND CONSORTIA*

ADNI is one of the landmark longitudinal multicentric studies that is designed to develop genetic, clinical, imaging, and biochemical biomarkers for the purpose of early detection and tracking of AD. ADNI involves public-private partnerships, strategically engineered to catalyze breakthroughs in both clinical and fundamental scientific investigations concerning AD [47, 74].

GAAIN is a pioneer platform facilitating comprehensive data exploration, access, and analysis of AD and related dementias (ADRD) research [4]. GAAIN is an international data-sharing initiative that aims to accelerate Alzheimer's research by providing researchers access to larger datasets from around the world. By facilitating data sharing and collaboration, GAAIN helps researchers uncover insights into the disease [72].

Collaborative Approach for Asian American and Pacific Islanders Research and Education is a specialized initiative geared toward understanding the distinct challenges faced by Asian Americans and Pacific Islanders (AAPI) in aging, ADRD, and roles related to care giving. It's dedicated to addressing the critical issue of AAPI individuals in clinical research, particularly in the context of ADRD, which is becoming increasingly vital due to the significant demographic growth of the AAPI community in the United States [72].

Alzheimer's Disease International (ADI) serves as a global federation comprising 75 Alzheimer's Associations worldwide, with a primary objective to heighten awareness regarding dementia. Its focus extends to fortifying Alzheimer's Associations, exchanging knowledge, and ultimately enhancing the quality of life of patients with dementia and their families [61].

The expansion of the original ADNI research culminated in the creation of the worldwide Alzheimer's Disease Neuroimaging Initiative (WW-ADNI). Overseen by the Alzheimer's Association, WW-ADNI incorporates data from diverse regions, such as Argentina, China, Taiwan, and Korea. This initiative marks a significant stride toward fostering a comprehensive global understanding of AD, with major focus on enhancing diagnostic and treatment methodologies across various demographic groups [40].

The International Alzheimer's Disease Research Portfolio (IADRP) is a comprehensive database dedicated for monitoring of research funding and assistance for AD and associated dementia, specifically focussing on low and middle-income countries. IADRP encompasses data from both public and private funding sources, aiming to streamline collaboration and coherence among researchers. The IADRP serves as a pivotal instrument for assessing advancements in Alzheimer's research and pinpointing areas necessitating further funding and support [67].

7.4.3 ROLE OF DATA EXCHANGE AND OPEN SCIENCE IN ADVANCING ALZHEIMER'S RESEARCH

Data sharing and open science are pivotal in advancing Alzheimer's research by fostering collaboration, accelerating discoveries, and maximizing scientific impact. By sharing data, researchers access diverse information, enabling analysis of larger datasets for identifying biomarkers, genetic risk factors, and therapeutic targets. The collaborative approach enhances research efficiency and facilitates breakthroughs that are not achievable in isolated studies.

In the field of research related to Alzheimer's, data-sharing initiatives, like ADNI, GAAIN, and support from organizations like NIA play pivotal roles. The ADNI longitudinal study captures clinical, imaging, and genetic data from participants with varying cognitive states. The GAAIN platform facilitates streamlined access and analysis of ADRD research data, promoting collaboration and discovery [30]. Through these collaborative endeavors, researchers worldwide can access diverse datasets, fostering innovation and accelerating progress toward understanding and treating AD [8, 51, 72].

7.5 RESEARCH FUNDING AND INVESTMENT

7.5.1 GOVERNMENTAL AND NON-GOVERNMENTAL FUNDING SOURCES

Governmental funding for Alzheimer's research predominantly comes from agencies, such as the National Institutes of Health (NIH) in the United States, the Medical Research Council (MRC) in the United Kingdom, and the Canadian Institutes of Health Research (CIHR). Non-governmental organizations also play a vital role, with entities like the Alzheimer's Association and the Alzheimer's Drug Discovery Foundation providing grants and support for innovative research projects [3, 5].

7.5.2 FUNDING ALLOCATION AND PRIORITIZATION

The allocation of research funding for AD varies globally, reflecting regional priorities, healthcare systems, and economic considerations. While high-income countries generally allocate more funding to Alzheimer's research, emerging economies are increasingly recognizing the importance of addressing the disease. There is a growing trend toward collaborative international initiatives like the GAAIN, aiming to harmonize data and accelerate research efforts worldwide [30, 59].

7.5.3 IMPACT OF FUNDING ON THE IMPROVEMENT OF AD RESEARCH

Research funding has a profound impact on the progress of AD research, influencing the scope, depth, and pace of scientific advancements. Adequate funding enables the exploration of diverse therapeutic targets, the conduct of large-scale clinical trials, and the development of innovative technologies for diagnosis and treatment. Insufficient funding, however, can impede progress and hinder the translation of promising discoveries into tangible clinical outcomes, underscoring the critical need for sustained investment in Alzheimer's research [38, 43].

7.6 CLINICAL TRIALS AND DRUG DEVELOPMENT

7.6.1 CURRENT STATE OF WORLDWIDE CLINICAL TRIALS OF AD

AD clinical trials are undergoing a transformative phase worldwide, marked by an increased focus on diverse therapeutic strategies targeting various aspects of the disease. With the advent of precision medicine and biomarker-driven approaches, there's a shift toward early intervention trials aimed at preventing or delaying cognitive decline. Large-scale international collaborations, such as the ADNI and the European Prevention of Alzheimer's Dementia (EPAD) project are facilitating the standardization of protocols and the sharing of data, enhancing the efficiency and impact of clinical trials. However, challenges such as participant recruitment, trial design complexities, and the need for innovative outcome measures remain significant hurdles in advancing Alzheimer's therapeutics [19, 22, 75, 76].

7.6.2 PROMISING THERAPEUTIC APPROACHES AND DRUG TARGETS

Recent advances in understanding the pathophysiology of AD have led to the identification of promising therapeutic approaches and drug targets. Targeting amyloid-beta and tau protein aggregation, inflammation, synaptic dysfunction, and neuroprotection are among the key strategies under investigation. Monoclonal antibodies, small molecules, and novel gene therapies are among the innovative interventions being evaluated in preclinical and clinical studies (Figure 7.1). Additionally, repurposing drugs approved for other indications and exploring multimodal interventions show potential in addressing the complex nature of Alzheimer's pathology. While challenges persist, such as ensuring drug delivery to the brain and overcoming treatment resistance, these therapeutic avenues offer hope for effectively modifying the course of AD [21, 34, 35, 50].

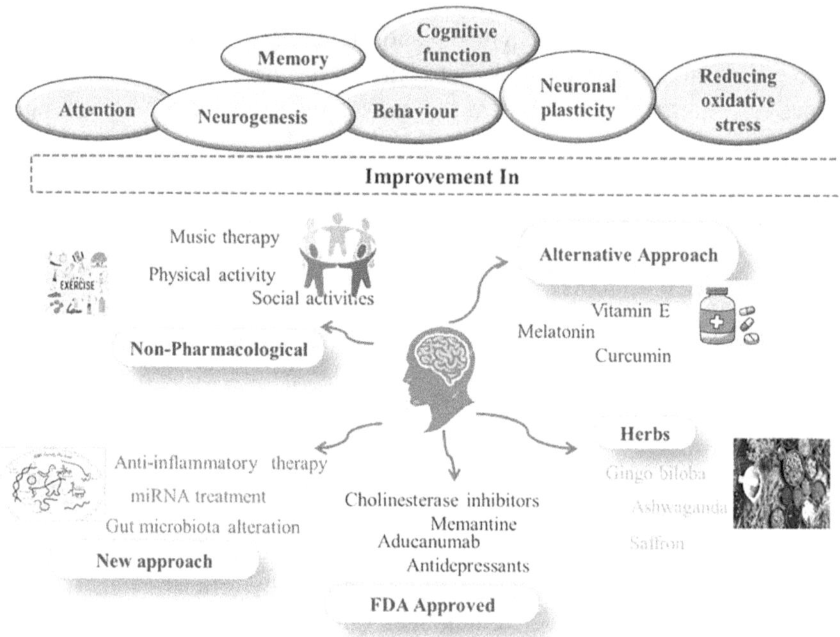

FIGURE 7.1 Therapeutic concepts to treat dementia/AD with advanced and alternate approaches.

7.6.3 CHALLENGES AND OPPORTUNITIES IN TRANSLATING RESEARCH FINDINGS INTO CLINICAL PRACTICE

Translating research findings into clinical practice represents a critical yet complex endeavor in AD management. Despite significant scientific progress, the gap between bench and bedside persists due to various factors, including regulatory hurdles, limited resources, and the heterogeneous nature of the disease. Harnessing real-world data, implementing personalized medicine approaches, and fostering interdisciplinary collaborations between researchers, clinicians, industry stakeholders, and policymakers are essential for bridging this translational gap. Moreover, enhancing public awareness, promoting early detection, and improving healthcare infrastructure are imperative for facilitating the timely adoption of evidence-based interventions and optimizing patient outcomes in clinical practice [22, 75].

7.7 INTERNATIONAL POLICY AND ADVOCACY

7.7.1 ROLE OF INTERNATIONAL ORGANIZATIONS IN SHAPING AD POLICY

International organizations play a pivotal role in shaping AD policy by fostering collaboration, setting global agendas, and advocating for increased funding and awareness. Entities such as the World Health Organization (WHO), the Organization for Economic Co-operation and Development (OECD), and AD International (ADI) provide platforms for sharing best practices, coordinating research efforts, and developing policy frameworks to address the growing impact of dementia worldwide. These organizations work closely with governments, healthcare professionals, and advocacy groups to promote policies that prioritize early detection, improve access to care, and support dementia-friendly communities. By leveraging their expertise and influence, international organizations contribute to the development of comprehensive strategies aimed at reducing the burden of AD on individuals, families, and societies [52, 79].

7.7.2 POLICY RECOMMENDATIONS AND GUIDELINES FOR DEMENTIA CARE AND SUPPORT

Policy recommendations and guidelines for dementia care and support are essential tools in addressing the complex needs of individuals living with AD and their caregivers. Organizations such as the Alzheimer's Association, the National Institute for Health and Care Excellence (NICE), and the International Psycho geriatric Association (IPA) provide evidence-based guidance on dementia diagnosis, treatment, and management. These recommendations encompass a wide range of topics, including care coordination, caregiver support, dementia-friendly environments, and end-of-life care. By promoting person-centered approaches and integrating social, psychological, and medical interventions, these policies aim to enhance quality of life, optimize functional independence, and reduce the societal impact of AD [2, 47, 54].

7.7.3 ADVOCACY EFFORTS TO RAISE AWARENESS AND REDUCE STIGMA SURROUNDING AD

Advocacy efforts play a crucial role in raising awareness and reducing stigma surrounding AD, fostering a supportive environment for affected individuals and their families. Organizations like US Against Alzheimer's, Alzheimer's Society, and Dementia Alliance International engage in public education campaigns, media outreach, and community events to promote understanding and empathy for those living with dementia. By sharing personal stories, dispelling misconceptions, and challenging ageist attitudes, these advocacy groups aim to combat stigma and discrimination associated with AD. Moreover, they advocate for policy changes that prioritize dementia research, improve access to care, and support caregivers, amplifying the voices of those affected by the disease and driving societal change [6, 73].

7.8 CULTURAL AND ETHICAL CONSIDERATIONS IN ALZHEIMER'S RESEARCH AND TREATMENT

7.8.1 CULTURAL PERCEPTIONS OF DEMENTIA AND AGING IN DIFFERENT SOCIETIES

Human beings affected by Alzheimer's are growing at an epidemic pace globally. There are various approaches that can be utilized to observe, including scientific, anthropological, or sociocultural. Yet, dementia is understood differently across the world by various communities or social groups [14, 28, 56, 57, 77]. The use of explanatory models that are specific to societal or cultural environments is necessary for understanding physical or mental aspects [44]. In the present time, everybody now looks more to scientific explanations than spiritual or religious ones, relying on empirical evidence, technological advancements, and rational inquiry to understand natural phenomena, health, and existence, reflecting a shift toward logic-based worldviews and secular reasoning. Evidence from the various scientific studies suggested that minority ethnic communities' people are diagnosed at an advanced stage of Alzheimer's due to the lack of knowledge and awareness about this disease [17, 53]. Studies investigating memory dysfunctions in dementia patients in minority ethnic communities often attributed the symptoms to a normal process of aging. Interestingly, Indian, African, Caribbean, and other low-income countries have a lack of awareness on dementia but predominantly suffer from dementia related symptoms, including memory loss and confusion that have been reported in these communities. Instead of concentrating on the dementia patient, Indian people were especially concerned about the impact of dementia on the family, including the stigma and estrangement experienced by the family. The people from Africa, Caribbean countries, and East and Central European countries reported effects and consequences of dementia on patients. Variations in age perception are linked with the cultural differences in different societies. In this context, people's ability to age successfully, the amount spent on age-related costs, and people's willingness to support age-friendly policy measures, even when those policies interfere with their own interests, are important considerations [32, 44]. It has been suggested that discovered facts on cultural variation in the age attitudes, perceptions, and outcomes for older adults may be a potential way for elucidating the connection of health and well-being disparities across cultures [10, 44]. Studies also revealed that people who lived in more collectivistic societies reported feelings of being more compassionate toward the older adults and also suggested that same cultures may have more negative attitudes toward the elderly. Till date, most of the studies have been focused on older adults and have found that perceptions on aging vary across different societies or communities. However, other recent studies have reported this phenomenon in younger subjects also [65, 66].

7.8.2 ETHICAL ISSUES IN RESEARCH, TREATMENT, AND STRATEGIES FOR CULTURALLY COMPETENT DEMENTIA CARE DELIVERY

Research and clinical practice with persons affected by dementia/AD and their families raise an increasing number of complex ethical problems around dementia/AD research [15]. Conducting research on dementia patients is potentially challenging for the researchers. Poor education and diminished physical activity levels have been observed in AD patients [13]. Further, quality education and lifestyle including physical activity may lower the risk of decline in cognitive functions. Nevertheless, getting approval for and carrying out studies is very puzzling when working with dementia patients because of a number of ethical issues relating to research risks in vulnerable people [13]. Diagnosis and treatment of dementia must be protected by ethical principles that meet the needs of society and individuals. Critical care planning, consistent approach, and standardized assessment tools would all aid in minimizing the ethical obstacles in dementia research.

7.9 FUTURE DIRECTIONS AND EMERGING TRENDS

7.9.1 ADVANCES IN BIOMARKER DISCOVERY AND EARLY DETECTION OF AD

Robust biomarkers for the diagnosis of AD at an early stage of disease will be crucial for better treatment. Early detection biomarkers are mostly derived from the saliva and cerebrospinal fluid and advanced imaging testing may successfully identify and detect dementia at its earliest stages [9, 45]. However, biomarkers from other fluids, including blood, urine, and tear have also been used for improving the diagnosis of dementia [41, 71, 82]. Initial findings of AD have been introduced for pre-symptomatic diagnosis and symptomatic medications [26, 60]. Recently, International Working Groups have stated that people who are cognitively normal yet have biomarkers should be regarded as potentially at-risk for developing AD [26]. Clinical phenotypes are becoming new biomarkers for the clinical diagnosis of AD to help differentiate between various illnesses and distinct AD phenotypes, especially in the early stages and to help identify individuals who may be at risk for developing symptoms of AD. For the critical care, support and treatment, diagnosis is very important at the early stages of AD. Salivary lactoferrin and CSF amyloid beta, and tau have been recommended for use in the diagnosis of AD [11]. Various imaging techniques, like positron emission tomography can be used regularly to estimate the plaque density in AD. But some biomarkers have not reached clinical settings for diagnosis of AD. An advanced and interdisciplinary approach is required to accelerate the research on the identification of new reliable and clinically adaptable biomarkers for AD. Plasma biomarkers, such as Aβ42/40, p-tau181, p-tau231, and p-tau217 are the diagnostics tools for AD pathology [33, 42]. Apart from these blood biomarkers, many studies have identified predicative diagnostic proteins and metabolite biomarkers in plasma [18–25]. Recently, machine learning approaches have been used to extract and identify reliable predictive diagnostic biomarkers at the early stage of AD from neuroimaging, electroencephalography, and speech-based digital data [12, 62, 64]. Artificial intelligence may facilitate the extraction of new biomarkers by multimodal data and multilayer predictive modeling. However, early diagnosis of AD is yet to be explored which could lead to the identification of new therapeutic targets.

7.9.2 PRECISION MEDICINE APPROACHES TO AD TREATMENT AND PREVENTION

Diagnosis of AD based on clinical symptoms is less sensitive and specific than other diagnostic methods due to the overlapping of symptoms with normal aging. Various number of brain disease conditions co-exist with AD [36]. Biomarkers play a critical role in the diagnosis of AD and guiding appropriate treatment and care. AD treatment solely is based on the biomarkers, clinical symptoms, including behavior performance, and brain pathology. Several potential drugs have been developed for AD based on the diagnostic research, but some have failed to produce any improvement on cognition or AD in clinical trials. These failures might be the result of AD patients receiving advanced forms of treatment, so they could clinically benefit from them. Therefore, therapeutic treatment depends upon the highly validated biomarkers, based on correct diagnosis of AD. Donepezil, galantamine, and rivastigmine are cholinesterase inhibitors that have been proved to reduce cognitive decline in AD patients [81, 85]. Galantamine may also help in slowing the progression of severe dementia [81]. Memantine, an N-methyl-D-aspartate (NMDA) receptor antagonist, is a neuroprotective drug that delays the deterioration of cognitive function in both mild and moderate AD [48, 49, 80]. However, side effects, including nausea, loose bowels, fatigue, diarrhea, weight loss, and bowel incontinence occur even after the treatment with these drugs. Two MABs have been approved for treating mild AD and decline amyloid deposits in the brain [19]. Donanemab and aducanumab drugs are capable of dissolving the plaque in 85–95% of AD patients [23, 31]. Some antiviral agents, reducing insulin resistance agents, and immunomodulators, such as amlodipine, metformin, valacyclovir, and sirolim, have been investigated as a possible treatment of AD [20]. Several scientific studies have found neuromodulation therapy, including transcranial magnetic stimulation and transcranial direct current stimulation, to have improved cognition in AD patients [33, 37]. Multiple interventions, such as diet and cognitive and physical activity help to slow the progression of cognition decline [39].

Physical intervention can be good for patients, but it is not necessary that these interventions work for everyone; it has been observed that daily aerobic exercise can prevent the progression of disease and improve cognitive skills [24]. Myokines promote hippocampal neuroprotection and synaptic neuroplasticity by increasing brain-derived neurotrophic factor through exercise [29].

7.10 CASE STUDIES AND SUCCESS STORIES

7.10.1 INTERNATIONAL COLLABORATION IN ALZHEIMER'S RESEARCH AND TREATMENT

International collaboration in Alzheimer's research is based on private-public partnerships via non-profit organizations, which continue to address the complex challenges in the AD/dementia research field. Collaboration in and among registered organizations is with multiple stakeholders, including with government, non-profit organizations, industry, and/or the public sector [69]. This chapter presents analysis to understand and elaborate the technological development in the area of AD diagnosis and treatment, analyzes the collaborative work more comprehensively, and proposes to facilitate additional support for modernization [46]. AD and similar associated problems are a worldwide challenge, with more than 50 million people living with it. Alzheimer's Association, world's major voluntary health group dedicated to AD and other dementia, is committed to its vision of a dementia free world and emphazises the needs, opportunities, and challenges for dementia research in all parts of the world [59, 67].

7.10.2 INNOVATIVE APPROACHES TO DEMENTIA CARE DELIVERY

In recent years, beginning in 2021, the community has focused on continuing approaches, identifying the loop holes in dementia/AD research, treatment, and resources along with enhancing research collaboration and communication among scholars at leading global forums like the Alzheimer's Association International Conference (AAIC) [67]. The Alzheimer's Association has observed that for the last few decades integral components of international research efforts to eradicate dementia/AD have been missing, leaving all countries in a perilous position to address the challenges posed by AD. An impactful number of projects are directed toward research and treatment in the low- and middle-income countries. Many such projects are sponsored in the collaboration with the Global Brain Health Institute (GBHI) and Alzheimer's Society (UK) [58]. Such approaches are not self-determining, and the situation will be the cumulative effect of these tactics that will finally eradicate AD/dementia.

7.10.3 GLOBAL EFFORTS IN AD RESEARCH AND TREATMENT

Presently, AD patients worldwide are reported to be round 24 million. By the year 2050, the number of patients with dementia/AD is expected to increase up to 4 times. Although Alzheimer's is a common public health issue, till date there are only two classes of drugs permitted for treatment of AD, that include cholinesterase inhibitors and antagonists to NMDA neurotransmitters [27, 68]. Meanwhile few clinical trials on Alzheimer's have started in the last few decades, but their consequences were majorly disappointing. An extraordinary collaborative effort by government, industry, and community on a large scale is needed to address this debilitating disease, as well as to develop a research plan toward successfully emerging effective treatment and diagnosis [7].

Recent research showed that 61% of population suffering from dementia belong to low- and middle-income countries. Up to 2019, the economic burden of dementia was appraised at US$ 1313.4 billion. Seventy four percent of this cost has been spent in high income countries. In the present time, everybody now looks more to scientific explanations than spiritual or religious ones, relying on empirical evidence, technological advancements, and rational inquiry to understand natural phenomena, health, and existence, reflecting a shift toward logic-based worldviews and secular reasoning. Approximately more than 200 research projects have been abandoned or have been unsuccessful in the last few decades [78].

7.11 CONCLUSION

7.11.1 SUMMARY OF KEY FINDINGS AND INSIGHTS FROM THIS CHAPTER

Dementia/AD is currently a major healthcare problem in developed and middle-income nations. In this chapter, we have discussed numerous mechanisms that have been investigated to understand and diagnose the pathophysiology of AD in order to explore effective treatments. However, there is no effective treatment available for slowing down disease development. In the recent times, a key focus toward the research on innovative pharmacotherapies has been the amyloidogenic hypothesis of dementia/ Alzheimer's, which suggests that the β amyloid peptide is primarily responsible for learning and memory impairments and neuronal damage.

7.11.2 COLLECTIVE ACTION ON THE GLOBAL AD CRISIS

Recent research focuses on the progress of therapeutic approaches to reduce or slow down disease development. The attention has also shifted toward several novel aspects, including the pathophysiology of the illness, the development of new diagnostic biomarkers, and the accurate diagnosis of patient conditions, as well as the creation of new strategies for clinical trials in hospital setups. Additionally, development of new drugs and investigation for dementia/AD has become more challenging with the inclusion of pre-clinical and prodromic AD patients in current trials. Existing directions given by the FDA for dementia/Alzheimer's clinical trials includes various fluid or neuroradiological specimens. Additionally, we strongly recommend the use of clinical trial trainers, statistics, and variable trial strategies.

7.11.3 ACTION FOR ALZHEIMER'S RESEARCH AND TREATMENT WORLDWIDE

Mediterranean food, logical activity, and awareness of higher teaching all may decrease the development of AD/dementia, and may improve cognitive functions like learning and memory. Numerous studies have demonstrated that lifestyle interventions involving diet, exercise, and cognitive training, along with management of cardiovascular risk factors, can reduce pathophysiology of AD in elder persons. Novel biomarker credentials of AD, symptomatic of prodromic phases of AD, can develop to primary diagnosis and expand predictive conclusions. Every society needs a strategy its growing population of people with dementia. There is a boundless requirement for cost studies, especially in low and middle income countries. We have faith in the discreetly developed nutritional treatments, combined with pharmacotherapies that target the various pathways tangled in dementia/AD.

In conclusion, achieving successful treatment of dementia/AD depends on its initial drug management and follow-up patient intensive care for disease development and using advance biomarkers for diagnosis.

CONFLICT OF INTERESTS

We declare that we have no conflict of interests in this chapter.

CONSENT FOR PUBLICATON

We, the undersigned, give our consent for the publication of identifiable details within this chapter to be published in the Apple Academic Press.

KEYWORDS

- **Alzheimer's Disease Neuroimaging Initiative**
- **Alzheimer's Disease International**
- **Alzheimer's Disease Cooperative Study**
- **Collaborative Approach for Asian American and Pacific Islanders Research and Education**
- **dementia**
- **Global Alzheimer's Association Interactive Network**

REFERENCES

1. Ackerman, L. S., & Chopik, W. J. (2021). Cross-cultural comparisons in implicit and explicit age bias. *Personality and Social Psychology Bulletin, 47*(6), 953–968.
2. Alzheimer's Association. (2020). Alzheimer's Association Guidelines. https://www.alz.org/professionals/health-systems-clinicians/research/clinical-studies/guidelines
3. Alzheimer's Association. (2021). Research Grants. https://www.alz.org/research/alzheimers_grants
4. Alzheimer's disease and related dementias: Update and areas of immediate need. *Alzheimer's & Dementia*. 2019 Feb 1, *15*(2), 292–312.
5. Alzheimer's Drug Discovery Foundation. (2021). Our impact. https://www.alzdiscovery.org/our-impact
6. Alzheimer's Society. (2021). About us. https://www.alzheimers.org.uk/about-us
7. Anderson, R. M., Hadjichrysanthou, C., Evans, S., & Wong, M. M. (2017). Why do so many clinical trials of therapies for Alzheimer's disease fail? *Lancet, 390*, 2327–2329.

8. Ashish, N., Bhatt, P., & Toga, A. W. (2016). Global data sharing in Alzheimer disease research. *Alzheimer Disease & Associated Disorders, 30*(2), 160–168.
9. Ashton, N. J., Ide, M., Zetterberg, H., & Blennow, K. (2019). Salivary biomarkers for Alzheimer's disease and related disorders. *Neurology and Therapy, 8*, 83–94.
10. Ayalon, L., & Cohn-Schwartz, E. (2019). Ageism from a cross-cultural perspective: Results from a national survey of Israelis over the age of 50. *International Psychogeriatrics, 34*(9), 779–787.
11. Babulal, G. M., Quiroz, Y. T., Albensi, B. C., Arenaza-Urquijo, E., Astell, A. J., Babiloni, C., Bahar-Fuchs, A., Bell, J., Bowman, G. L., Brickman, A. M., & Chételat, G. (2019). Perspectives on ethnic and racial disparities in Alzheimer's disease and related dementias: Update and areas of immediate need. *Alzheimer's & Dementia, 15*(2), 292–312.
12. Battista, P., Salvatore, C., Berlingeri, M., Cerasa, A., & Castiglioni, I. (2020). Artificial intelligence and neuropsychological measures: The case of Alzheimer's disease. *Neuroscience & Biobehavioral Reviews, 114*, 211–28.
13. Beydoun, M. A., Beydoun, H. A., Gamaldo, A. A., Teel, A., Zonderman, A. B., & Wang, Y. (2014). Epidemiologic studies of modifiable factors associated with cognition and dementia: Systematic review and meta-analysis. *BMC Public Health, 14*, 1–33.
14. Blay, S. L., & Peluso, É. T. (2010). Public stigma: The community's tolerance of Alzheimer disease. *The American Journal of Geriatric Psychiatry, 18*(2), 163–171.
15. Chandra, M., Harbishettar, V., Sawhney, H., & Amanullah, S. (2021). Ethical issues in dementia research. *Indian Journal of Psychological Medicine, 43*(5_suppl), S25–30.
16. Chaudhary, N. (2023). Dementia in India: An overview. Available at SSRN 4353348.
17. Cooper, C., Mukadam, N., Katona, C., Lyketsos, C. G., Ames, D., Rabins, P., Engedal, K., de Mendonça Lima, C., Blazer, D., Teri, L., & Brodaty, H. (2012). Systematic review of the effectiveness of non-pharmacological interventions to improve quality of life of people with dementia. *International Psychogeriatrics, 24*(6), 856–870.
18. Costa, A. C., Joaquim, H. P., Forlenza, O. V., Gattaz, W. F., & Talib, L. L. (2020). Three plasma metabolites in elderly patients differentiate mild cognitive impairment and Alzheimer's disease: A pilot study. *European archives of Psychiatry and Clinical Neuroscience, 270*, 483–488.
19. Cummings, J., Aisen, P., Lemere, C., Atri, A., Sabbagh, M., & Salloway, S. (2021). Aducanumab produced a clinically meaningful benefit in association with amyloid lowering. *Alzheimer's Research & Therapy, 13*(1), 98.
20. Cummings, J., Lee, G., Nahed, P., Kambar, M. E., Zhong, K., Fonseca, J., & Taghva, K. (2022). Alzheimer's disease drug development pipeline: 2022. *Alzheimer's & Dementia: Translational Research & Clinical Interventions, 8*(1), e12295.
21. Cummings, J., Feldman, H. H., Scheltens, P., Scherrer, B., Ketter, N., Goldwater, R., & van Dyck, C. (2020). The "rights" of precision drug development for Alzheimer's disease. *Alzheimer's & Dementia: Translational Research & Clinical Interventions, 6*(1), e12062.
22. Cummings, J., Lee, G., Ritter, A., & Sabbagh, M. (2019). Alzheimer's disease drug development pipeline: 2019. *Alzheimer's & Dementia: Translational Research & Clinical Interventions, 5*, 272–293.

23. DeTure, M. A., & Dickson, D. W. (2019). The neuropathological diagnosis of Alzheimer's disease. *Molecular Neurodegeneration, 14*(1), 32.
24. Devanand, D. P., Masurkar, A. V., & Wisniewski, T. (2023). Vigorous, regular physical exercise may slow disease progression in Alzheimer's disease. *Alzheimer's & Dementia, 19*(4), 1592–1597.
25. Di Costanzo, A., Paris, D., Melck, D., Angiolillo, A., Corso, G., Maniscalco, M., & Motta, A. (2020). Blood biomarkers indicate that the preclinical stages of Alzheimer's disease present overlapping molecular features. *Scientific Reports, 10*(1), 15612.
26. Dubois, B., Villain, N., Frisoni, G. B., Rabinovici, G. D., Sabbagh, M., Cappa, S., Bejanin, A., Bombois, S., Epelbaum, S., Teichmann, M., & Habert, M. O. (2021). Clinical diagnosis of Alzheimer's disease: Recommendations of the International Working Group. *The Lancet Neurology, 20*(6), 484–496.
27. Eldufani, J., & Blaise, G. (2019). The role of acetylcholinesterase inhibitors such as neostigmine and rivastigmine on chronic pain and cognitive function in aging: A review of recent clinical applications. *Alzheimer's & Dementia: Translational Research & Clinical Interventions, 5*, 175–183.
28. Enjolras, F. (2005). Incidence du pronostic sur la construction des modeles explicatifs de la maladie d'Alzheimer a l'ile de la Reunion. *Sciences sociales et santé, 23*(3), 69–94.
29. Folch, J., Petrov, D., Ettcheto, M., Abad, S., Sánchez-López, E., García, M. L., Olloquequi, J., Beas-Zarate, C., Auladell, C., & Camins, A. (2016). Current research therapeutic strategies for Alzheimer's disease treatment. *Neural Plasticity*.
30. GAAIN. (2021). Global Alzheimer's Association Interactive Network. https://www.gaain.org/
31. Gandy, S., & Ehrlich, M. E. (2023). Moving the needle on Alzheimer's disease with an anti-oligomer antibody. *New England Journal of Medicine, 388*(1), 80–81.
32. Giasson, H. L., & Chopik, W. J. (2020). Geographic patterns of implicit age bias and associations with state-level health outcomes across the United States. *European Journal of Social Psychology, 50*(6), 1173–1190.
33. Gonsalvez, I., Baror, R., Fried, P., Santarnecchi, E., & Pascual-Leone, A. (2017). Therapeutic noninvasive brain stimulation in Alzheimer's disease. Curr Alzheimer Res, *14*(4), 362–376.
34. Gonzalez-Ortiz, F., Kac, P. R., Brum, W. S., Zetterberg, H., Blennow, K., & Karikari, T. K. (2023). Plasma phospho-tau in Alzheimer's disease: Towards diagnostic and therapeutic trial applications. *Molecular Neurodegeneration, 18*(1), 18.
35. Hampel, H., Cummings, J., Blennow, K., Gao, P., Jack Jr, C. R., & Vergallo, A. (2021). Developing the ATX (N) classification for use across the Alzheimer disease continuum. *Nature Reviews Neurology, 17*(9), 580–589.
36. Hampel, H., O'Bryant, S. E., Molinuevo, J. L., Zetterberg, H., Masters, C. L., Lista, S., Kiddle, S. J., Batrla, R., & Blennow, K. (2018). Blood-based biomarkers for Alzheimer disease: Mapping the road to the clinic. *Nature Reviews Neurology, 14*(11), 639–652.
37. Holczer, A., Németh, V. L., Vékony, T., Vécsei, L., Klivényi, P., & Must, A. (2020). Non-invasive brain stimulation in Alzheimer's disease and mild cognitive impairment—A state-of-the-art review on methodological characteristics and stimulation parameters. *Frontiers in Human Neuroscience, 14*, 179.

38. Hurd, M. D., Martorell, P., Delavande, A., Mullen, K. J., & Langa, K. M. (2013). Monetary costs of dementia in the United States. *New England Journal of Medicine, 368*(14), 1326–1334.
39. Isaacson, R. S., Hristov, H., Saif, N., Hackett, K., Hendrix, S., Melendez, J., Safdieh, J., Fink, M., Thambisetty, M., Sadek, G., & Bellara, S. (2019). Individualized clinical management of patients at risk for Alzheimer's dementia. *Alzheimer's & Dementia, 15*(12), 1588–1602.
40. Jones-Davis, D. M., & Buckholtz, N. (2015). The impact of the Alzheimer's Disease Neuroimaging Initiative 2: What role do public-private partnerships have in pushing the boundaries of clinical and basic science research on Alzheimer's disease? *Alzheimer's & Dementia, 11*(7), 860–864.
41. Kalló, G., Emri, M., Varga, Z., Ujhelyi, B., Tőzsér, J., Csutak, A., & Csősz, É. (2016). Changes in the chemical barrier composition of tears in Alzheimer's disease reveal potential tear diagnostic biomarkers. *PLoS One, 11*(6), e0158000.
42. Karikari, T. K., Benedet, A. L., Ashton, N. J., Lantero Rodriguez, J., Snellman, A., Suárez-Calvet, M., Saha-Chaudhuri, P., Lussier, F., Kvartsberg, H., Rial, A. M., & Pascoal, T. A. (2021). Diagnostic performance and prediction of clinical progression of plasma phospho-tau181 in the Alzheimer's Disease Neuroimaging Initiative. *Molecular Psychiatry, 26*(2), 429–442.
43. Khachaturian, Z. S., Barnes, D., Einstein, R., Johnson, S., Lee, V., Roses, A., Sager, M. A., Shankle, W. R., Snyder, P. J., Petersen, R. C., & Schellenberg, G. (2010). Developing a national strategy to prevent dementia: Leon Thal Symposium 2009. *Alzheimer's & Dementia, 6*(2), 89–97.
44. Levy, B. R., Ferrucci, L., Zonderman, A. B., Slade, M. D., Troncoso, J., & Resnick, S. M. (2016). A culture–brain link: Negative age stereotypes predict Alzheimer's disease biomarkers. *Psychology and Aging, 31*(1), 82.
45. Lewczuk, P., Riederer, P., O'bryant, S.E., Verbeek, M.M., Dubois, B., Visser, P.J., Jellinger, K.A., Engelborghs, S., Ramirez, A., Parnetti, L., & Jack Jr, C. R. (2018). Cerebrospinal fluid and blood biomarkers for neurodegenerative dementias: An update of the consensus of the task force on biological markers in psychiatry of the world federation of societies of biological psychiatry. *The World Journal of Biological Psychiatry, 19*(4), 244–328.
46. Li, R., Wang, X., Liu, Y., Zhang, S., & Hanif, O. (2021). Research status and collaboration analysis based on big data mining: An empirical study of Alzheimer's disease. *Technology Analysis & Strategic Management, 33*(4), 379–395.
47. Li, X., Feng, X., Sun, X., Hou, N., Han, F., & Liu, Y. (2022). Global, regional, and national burden of Alzheimer's disease and other dementias, 1990–2019. *Frontiers in Aging Neuroscience, 14*, 937486.
48. Lipton, S. A. (2007). Pathologically-activated therapeutics for neuroprotection: Mechanism of NMDA receptor block by memantine and S-nitrosylation. *Current Drug Targets, 8*(5), 621–32.
49. Livingston, G., Sommerlad, A., Orgeta, V., Costafreda, S. G., Huntley, J., Ames, D., Ballard, C., Banerjee, S., Burns, A., Cohen-Mansfield, J., & Cooper, C. (2017). Dementia prevention, intervention, and care. *The Lancet, 390*(10113), 2673–2734.
50. Long, J. M., & Holtzman, D. M. (2019). Alzheimer disease: An update on pathobiology and treatment strategies. *Cell, 179*(2), 312–339.

51. McNally, J., Bhattacharyya, P., & Lavender, K. (2014). NIA data management plans-help and resources for researchers. Mori, H., Ikeuchi, T. (2014). Alzheimer's disease cooperative study (ADCS). *Brain and Nerve= Shinkei Kenkyu no Shinpo, 66*(6), 691–698.
52. Muir, T. (2015). Addressing dementia: The OECD response. OECD; 2015. https://www.oecd.org/health/addressing-dementia-the-oecd-response.htm
53. Mukadam, N., Cooper, C., & Livingston, G. (2011). A systematic review of ethnicity and pathways to care in dementia. *International Journal of Geriatric Psychiatry, 26*(1), 12–20.
54. National Institute for Health and Clinical Excellence. (2018). Dementia: Assessment, management and support for people living with dementia and their carers. National Institute for Health and Clinical Excellence.
55. Niu, H., Álvarez-Álvarez, I., Guillén-Grima, F., & Aguinaga-Ontoso, I. (2017). Prevalence and incidence of Alzheimer's disease in Europe: A meta-analysis. *Neurología (English Edition), 32*(8), 523–32.
56. Pollitt, P. A. (1996). Dementia in old age: An anthropological perspective. *Psychological Medicine, 26*(5), 1061–1074.
57. Poveda, A. M. (2003). An anthropological perspective of Alzheimer disease. *Geriatric Nursing, 24*(1), 26–31.
58. Prina, A. M., Mayston, R., Wu, Y. T., & Prince, M. (2019). A review of the 10/66 dementia research group. *Social Psychiatry and Psychiatric Epidemiology, 54*, 1–10. DOI: 10.1007/s00127–018–1626–7.
59. Prince, M., Ali, G. C., Guerchet, M., Prina, A. M., Albanese, E., & Wu, Y. T. (2016). Recent global trends in the prevalence and incidence of dementia, and survival with dementia. *Alzheimer's Research & Therapy, 8*, 1–3.
60. Rasmussen, J., & Langerman, H. (2019). Alzheimer's disease–why we need early diagnosis. *Degenerative Neurological and Neuromuscular Disease*, 123–130.
61. Rimmer, E. (2006). Alzheimer's disease international. *International Psychiatry, 3*(4), 22–23.
62. Robin, J., Harrison, J. E., Kaufman, L. D., Rudzicz, F., Simpson, W., & Yancheva, M. (2020). Evaluation of speech-based digital biomarkers: Review and recommendations. *Digital Biomarkers, 4*(3), 99–108.
63. Rocca, W. A., Amaducci, L. A., & Schoenberg, B. S. (1986). Epidemiology of clinically diagnosed Alzheimer's disease. *Annals of Neurology: Official Journal of the American Neurological Association and the Child Neurology Society, 19*(5), 415–424.
64. Rossini, P. M., Miraglia, F., Alù, F., Cotelli, M., Ferreri, F., Di Iorio, R., Iodice, F., & Vecchio, F. (2020). Neurophysiological hallmarks of neurodegenerative cognitive decline: The study of brain connectivity as a biomarker of early dementia. *Journal of Personalized Medicine, 10*(2), 34.
65. Schönstein, A., Ngo, D. T., Stephan, Y., Siè, A., Harling, G., Bärnighausen, T., & Wahl, H. W. (2021). Feeling younger in rural Burkina Faso: Exploring the role of subjective age in the light of previous research from high-income countries. *The Journals of Gerontology: Series, B, 76*(10), 2029–2040.
66. Schönstein, A., Schlomann, A., Wahl, H. W., & Bärnighausen, T. (2023). Awareness of age-related change in very different cultural-political contexts: A cross-cultural

examination of aging in Burkina Faso and Germany. *Frontiers in Psychiatry, 13*, 928564.

67. Sexton, C., Snyder, H. M., Chandrasekaran, L., Worley, S., & Carrillo, M. C. (2021). Expanding representation of low and middle income countries in global dementia research: Commentary from the Alzheimer's association. *Frontiers in Neurology, 12*, 633777.
68. Singh, R., & Sadiq, N. M. (2020). Cholinesterase Inhibitors. https://www.ncbi.nlm.nih.gov/books/NBK544336/ (accessed on 8 Dec 2020).
69. Snyder, H. M., Perlman, W. R., Egge, R., & Carrillo, M. C. (2017). Alzheimer's disease public-private partnerships: Update 2017.
70. Tan, S. Z., Zhao, R. C., Chakrabarti, S., Stambler, I., Jin, K., & Lim, L. W. (2021). Interdisciplinary research in Alzheimer's disease and the roles international societies can play. *Aging and Disease, 12*(1), 36.
71. Teunissen, C. E., Verberk, I. M., Thijssen, E. H., Vermunt, L., Hansson, O., Zetterberg, H., van der Flier, W. M., Mielke, M. M., & Del Campo, M. (2022). Blood-based biomarkers for Alzheimer's disease: Towards clinical implementation. *The Lancet Neurology, 21*(1), 66–77.
72. Toga, A. W., Phatak, M., Pappas, I., Thompson, S., McHugh, C. P., Clement, M. H., Bauermeister, S., Maruyama, T., & Gallacher, J. (2023). The pursuit of approaches to federate data to accelerate Alzheimer's disease and related dementia research: GAAIN, DPUK, and ADDI. *Frontiers in Neuroinformatics, 17*, 1175689.
73. Us Against Alzheimer's. (2021). About us. https://www.usagainstalzheimers.org/about
74. Veitch, D. P., Weiner, M. W., Miller, M., Aisen, P. S., Ashford, M. A., Beckett, L. A., Green, R. C., Harvey, D., Jack Jr, C. R., Jagust, W., & Landau, S. M. (2024). The Alzheimer's Disease Neuroimaging Initiative in the era of Alzheimer's disease treatment: A review of ADNI studies from 2021 to 2022. *Alzheimer's & Dementia, 20*(1), 652–694.
75. Weiner, M. W., Veitch, D. P., Aisen, P. S., Beckett, L. A., Cairns, N. J., Cedarbaum, J., Green, R. C., Harvey, D., Jack, C. R., Jagust, W., & Luthman, J. (2015). 2014 Update of the Alzheimer's Disease Neuroimaging Initiative: A review of papers published since its inception. *Alzheimer's & Dementia, 11*(6), e1–20.
76. Werner, P. (2005). Social distance towards a person with Alzheimer's disease. *International Journal of Geriatric Psychiatry: A Journal of the psychiatry of Late Life and Allied Sciences, 20*(2), 182–188.
77. Whitehouse, P. J., Gaines, A. D., Lindstrom, H., & Graham, J. E. (2005). Anthropological contributions to the understanding of age-related cognitive impairment. *The Lancet Neurology, 4*(5), 320–326.
78. Wimo, A., Seeher, K., Cataldi, R., Cyhlarova, E., Dielemann, J. L., Frisell, O., Guerchet, M., Jönsson, L., Malaha, A. K., Nichols, E., & Pedroza, P. (2023). The worldwide costs of dementia in 2019. *Alzheimer's & Dementia, 19*(7), 2865–2873.
79. World Health Organization. (2017). Global action plan on the public health response to dementia 2017–2025. https://www.who.int/mental_health/neurology/dementia/action_plan_2017_2025/en/
80. Wu, H. M., Tzeng, N. S., Qian, L., Wei, S. J., Hu, X., Chen, S. H., Rawls, S. M., Flood, P., Hong, J. S., & Lu, R. B. (2009). Novel neuroprotective mechanisms of memantine:

Increase in neurotrophic factor release from astroglia and anti-inflammation by preventing microglial activation. *Neuropsychopharmacology, 34*(10), 2344–2357.

81. Xu, H., Garcia-Ptacek, S., Jönsson, L., Wimo, A., Nordström, P., & Eriksdotter, M. (2021). Long-term effects of cholinesterase inhibitors on cognitive decline and mortality. *Neurology, 96*(17), e2220–e2230.
82. Yao, F., Hong, X., Li, S., Zhang, Y., Zhao, Q., Du, W., Wang, Y., & Ni, J. (2018). Urine-based biomarkers for Alzheimer's disease identified through coupling computational and experimental methods. *Journal of Alzheimer's Disease, 65*(2), 421–431.
83. Yin, Y., Tam, H. L., Quint, J., Chen, M., Ding, R., & Zhang, X. (2024). Epidemiology of dementia in China in 2010–2020: A systematic review and meta-analysis. *Healthcare, 12*(3), 334.
84. Zissimopoulos, J. M., Tysinger, B. C., St. Clair, P. A., & Crimmins, E. M. (2018). The impact of changes in population health and mortality on future prevalence of Alzheimer's disease and other dementias in the United States. *The Journals of Gerontology: Series, B, 73*(suppl_1), S38–S47.
85. Zuin, M., Cherubini, A., Volpato, S., Ferrucci, L., & Zuliani, G. (2022). Acetyl-cholinesterase-inhibitors slow cognitive decline and decrease overall mortality in older patients with dementia. *Scientific Reports, 12*(1), 12214.

CHAPTER 8

Role of Neurotrophins in Alzheimer's Disease Pathogenesis to Develop Novel Therapeutics

PRANAY WAL[1,*], HIMANGI VIG[1], BISWAJIT DASH[2], ANKUR AGRAWAL[3], L. KARPAGAVALLI[4], KANGKAN JYOTI SAUD[5], and LAKHYAJIT BORAH[5]

[1]*PSIT-Pranveer Singh Institute of Technology (Pharmacy), Kanpur, Uttar Pradesh, India*

[2]*Amity Institute of Pharmacy, Amity University, Kolkata, West Bengal, India*

[3]*Jai Institute of Pharmaceutical Sciences and Research, Gwalior, Madhya Pradesh, India*

[4]*Saveetha College of Pharmacy, Saveetha Institute of Medical and Technical Sciences, SIMATS, Tamil Nadu, India*

[5]*Arunachal University of studies, Namsai, Arunachal Pradesh, India*

*Corresponding author

ABSTRACT

Alzheimer's disease (AD) is a prevalent neurological condition that is predicted to affect 115 million individuals by 2050. Aberrant extracellular accumulation of amyloid-beta (Aβ) peptide and intraneuronal neurofibrillary knots, which cause neuroinflammation, synapse disorder, neurotransmitter imbalance, neuronal degeneration, and dendritic alterations, are its defining features. There is a theory that neurotrophic factor (NTF) contributes in neurodegenerative illnesses and may be used as

Cellular and Molecular Mechanisms in Alzheimer's Disease Therapeutics, Volume 2: Molecular Mechanisms. Prashant Tiwari, Sunil Kumar Kadiri, and Kunjbihari Sulakhiya (Eds)

DOI: 10.1201/9781779641816-8

a therapeutic agent. One neurotrophin that is extensively expressed in the brain is called brain-derived neurotrophic factor (BDNF). It affects the development and existence of neurons and serves as a neurotransmitter command center. It is also essential for memory and learning. According to available data, BDNF could be a NTF that regulates cognitive processes in both neuropathological and normal circumstances. There has been evidence of a decline in TrkA in the cortical and nucleus basalis, as well as a spike in nerve growth factor and a decrease in the hippocampal BDNF, and some neocortical areas in AD brains. Comparative information on the molecules and receptors of NTF in the brains of research animals and humans, as well as their possible applications to the management of AD, is thus provided in this chapter. Considering that AD is a synaptic illness, BDNF is becoming more and more popular as a possible biomarker for both diagnosis and therapy of AD. While Aβ buildup, tau phosphorylation, neuroinflammation, and neuron death have each been connected to BDNF decline, the precise processes underpinning the impact of compromised BDNF signaling on AD remain to be discovered. This chapter presents an outline of the relationships between BDNF and the variables that control BDNF signaling. It also discusses BDNF's function in AD and the possibility of BDNF-targeting treatments for AD.

8.1 INTRODUCTION

The aging global population is contributing to the rising prevalence of Alzheimer's disease (AD), the most common cause of dementia globally. Approximately 80% of dementia occurrences are caused by AD, making it the most prevalent reason of dementia. The frequency and incidence of dementias, including AD, went up by 160.84 and 147.95%, in that order, during 1990 and 2019 [1]. Approximately, AD is responsible for nearly three quarters of the instances of dementia in people over 65 years of age and beyond, making it the least common forms of dementia. It ranks sixth in the list of causes of death in the United States. Less than 10% of AD patients have a rapid onset which is characterized as starting under the age of 65 [2]. β-amyloid plaque formation and neurofibrillary tangles of hyperphosphorylated tau are the two main reasons that traditionally define this neurodegenerative disease (NDD) progression. This neurodegenerative condition known as AD slowly impairs cognitive and behavioral competencies, including decision-making, reasoning, speaking, retention, and comprehension. It starts slowly and progresses over time. AD cannot be cured, although certain symptoms can be managed using existing medications [3]. Depending on the illness's stage, signs of AD vary. Following short-term memory loss, deficits in judgement, mental agility, problem-solving abilities, motivation, and disorganization can cause issues with multitasking and conceptualization. In the middle to advanced phases, neurological symptoms, such as disinterest, anger, wandering, schizophrenia, and apathy are also frequent. Late in the disease, dyspraxia—trouble performing motor activities that have been taught—olfactory disarray, sleep issues, and extrapyramidal motor symptoms, including dystonia and akathisia manifest. Subprime reflexes, incontinence, and complete reliance on carers come next [4]. However, studies on the speed at which AD is developing have revealed that this rate might be of the utmost importance in assessing the prognosis of a patient, especially those with surged cognitive impairment seem to experience a faster rate of advancement as time passes than those with a less rapid decline [5]. Understanding the underlying processes of AD may aid in the development of novel treatment approaches by identifying characteristics that impact cognitive function in the illness [6]. When provided at the right moment throughout the disease, existing drugs have been demonstrated to improve both the patient's and family's quality of life; nevertheless, medications have no effect on the progression of the disease or the pace of decline.

Clinical investigation is moving in the direction of more efficient remedies for the characteristic pathophysiology of AD in the hope that these measures would slow the illness's ongoing cognitive loss. Numerous research works have assessed the functioning of neurotrophins, including brain-derived neurotrophic factor (BDNF), in the development of AD [7]. A family of growth factors, known as neurotrophins, oversees the development, differentiation, proliferation, and durability of neurons in the peripheral and central nervous systems (CNS).

The process known as neurogenesis is triggered and regulated by these proteins. Among the neurotrophins are BDNF, NT-3, NT-4, and NGF [6]. The most potent neurotrophin of the four is called BDNF. It is widely implicated in psychiatric illnesses, including AD, and is essential to the development and flexibility of the cerebrum [8, 9]. BDNF is primarily produced in the brain's glial and neural cell bodies.

Within the framework of AD, tau phosphorylation, Aβ buildup, neural inflammation, and neuronal degeneration are linked to BDNF deficit [10]. Reducing Aβ-induced neurotoxicity, increasing cell survival, and dramatically upregulating BDNF are all possible outcomes of silencing BDNF antisense RNA.

Additionally, mounting data points to the importance of BDNF signaling in controlling the consequences of Aβ accumulation over a prolonged period in AD. BDNF concentrations in AD patients' brains [35], blood [11], and cerebrospinal fluid (CSF) decrease as the illness advances.

Furthermore, enhanced cognitive efficiency in AD has been linked to elevated blood concentration of BDNF [12]. As a result of these discoveries, BDNF is gaining attention as a possible biomarker for AD diagnosis or treatment. The function of BDNF in AD and the mechanisms by which it slows the development of the illness will be covered in the following chapter, along with the promise of BDNF-targeting treatments for this incurable condition.

8.2 PATHOPHYSIOLOGY

The neurodegenerative disorders known as AD and kindred dementias are typified by a gradual impairment of the brain that appears to proceed in a tiered cell biology sequence consisting of neuronal damage, synaptic failure, and neuronal demise (Figure 8.1). The pathological signs of AD include neuropil threads and neurofibrillary tangles (NFTs). Even if they are just slightly prevalent, they should not be interpreted as signs of aging. Neuropil threads and NFTs are often the first abnormalities to appear. These alterations are believed to have a complex relationship with the onset, progression, and period of the sickness. The last stage of the sickness is characterized by the presence of plaques, either neuritic or amyloid-related. Early neuropil threads and NFTs form at cortical locations. The alterations subsequently disperse to additional areas of the telencephalic cortex in a predictable and non-random way. The slow and steady evolution of the destructive process is taken into consideration in the staging approach, which is based on the successive changes in the lesions' dispersion pattern. Amyloid protein accumulation outside of cells and intraneuronal neurofibrillary alterations are two of the main pathogenic indicators of AD [13].

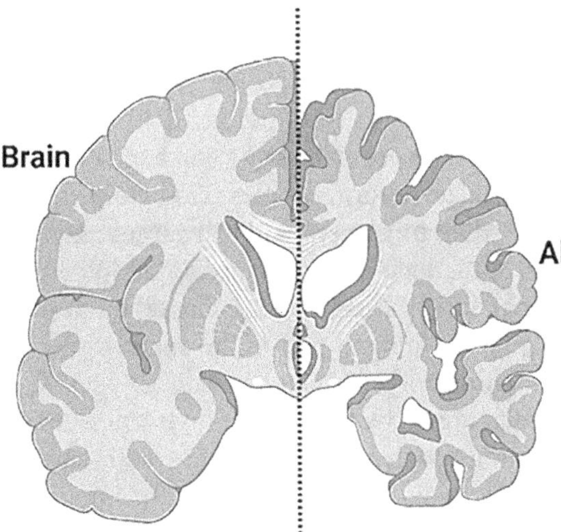

FIGURE 8.1 Alzheimer's brain appearance following neuron degeneration.

Among the clinical aspects of AD is the development of cerebral beta-amyloid (Aβ) plaques, where the predominant peptide component is Aβ42, NFT of coupled helical filaments composed of hyper phosphorylated microtubule-associated protein t (MAPt), and loss of neurons. Portion of the protein known as t, which joins together to create filaments, makes up NFT (Figure 8.2) [14].The degree of dementia is closely correlated with the density of such filaments inside brain neurons. The reason why tangles form is unknown. It is well-established that some gene alleles result in t forms that more likely develop into tangles. To clarify this multifactorial illness, several theories, including the cholinergic, Aβ, tau, and inflammatory hypotheses, have been proposed based on the different causal variables [15]. It was just recently shown that the most widely accepted Aβ theories, which have been in use for the last 20 years, are unable to explain the intricate pathogenesis of this debilitating illness. The involvement of Aβ oligomers in synaptic damage has also been underlined in recent research [16], which implies that they are essentially the only signals amid numerous others that compromise the reliability of brain processes. Furthermore, it appears that the onset of plaque buildup in later life is a relatively late phenomenon [17].

The extracellular accumulations of Aβ that make up senile plaques can have a variety of structural patterns, such as classic, neuritic, diffuse, compact, or dense-cored plaques. APP transmembrane is the basis of Aβ buildup, that is biosynthesized by the breakdown of proteolytic enzymes including β-secretase and γ-secretase [18]. Enzymes like these break down APP into segments of amino acids resulting in Aβ40 and Aβ42. Soluble oligomers that may proliferate across the brain and huge insoluble amyloid particles that may accumulate to form plaques of amyloid are found in this disease. The impact of Aβ42 oligomers in the synapses and mitochondria has been documented to include oxidative harm, tau hyperphosphorylation, and damaging effects [19]. However, it is important to recognize that Aβ42 senile plaques play an integral part in the stimulation of microglia, which releases proinflammatory mediators, such as TNF-α, IL-1β and IFN-y when microglia are activated. Such cytokines then stimulate astrocyte-adjacent neurons to generate more Aβ42 oligomers, which in turn stimulate more Aβ42 output and spreading [20]. Additionally linked to the neuron-astrocyte complex, oligodendroglia (OLGs) is destroyed by Aβ oligomers. In AD brains, aggregations of Aβ oligomers are believed to become the root of the neuron and vasculature deterioration. It causes oxidative damage, which makes OLGs especially vulnerable owing to their decreased glutathione (GSH) level and elevated iron level, leaving them incapable of eliminating oxygen radicals [21].

The main element of NFTs in the cognitive abilities of AD individuals is increased phosphorylation tau protein, and its progression can be seen as reflecting three distinct NFT structural phases: pre-tangle stage, a particular kind of NFT, wherein phosphorylated tau proteins build up in the somatodendritic area lacking the development of paired helical filament (PHF); mature NFTs, that are characterized by filament accumulation of tau protein alongside a relocation of its nucleus to the peripheral part of the soma; and outside the cell tangles, also known as the specter NFTs phase, that is caused by an impairment of neurons due to substantial quantities of filamentous tau protein that is partially resistant to proteolysis [22]. Cognitive decline results from synaptic loss in the limbic and neocortex system, which is often seen in the initial phases of AD. Disorders in axonal transportation, oxidative stress, damaged mitochondria, as well as additional mechanisms that can lead to minute fractions, such as the buildup of tau and A within synaptic sites, are among the causes responsible for synaptic degeneration. Pre-synaptic connections, axonal degeneration, and dendritic spine degeneration are the results of these mechanisms [23].

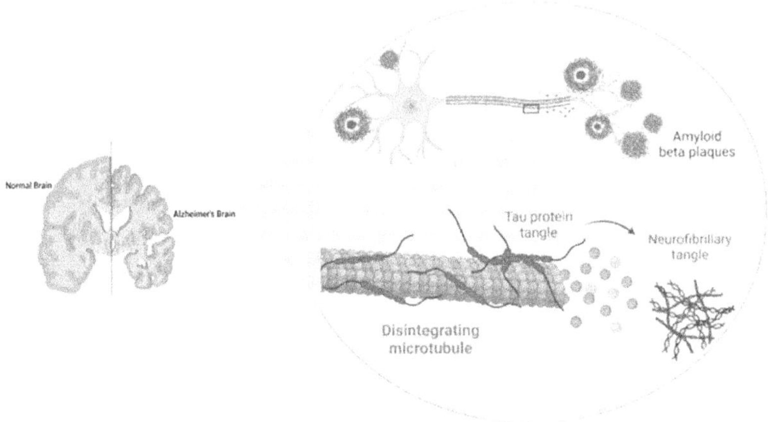

FIGURE 8.2 Pathophysiological features of Alzheimer's disease.

8.3 NEUROTROPHIC FACTORS

Crucial protein secretions known as neurotrophic factors (NTFs) play a variety of roles across both nonneural and neural tissues, influencing the growth, survival, and upkeep of the peripheral and CNS [24]. These pleiotropic particles are crucial for adult development of neurons and neural plasticity. They are also crucial in dendritic growth, synaptic plasticity, axonal growth, and guidance management, which enable the formation of suitable contacts with targeted cells [25].

In general terms, NTFs belong to three families: neurotrophins, glial cell-line derived neurotrophic factor (GDNF) family ligands (GFLs), neuropoeitic factors, and cytokines. Numerous neurotrophic agents have a widely recognized neuroprotective impact through synapse restoration and neuroregeneration [26].

Nerve growth factor (NGF), BDNF, neurotrophin 3 (NT3), and neurotrophin 4 (NT4) are members of the neurotrophin group. Two different kinds of receptors regulate neurotrophin signaling: neurotrophin receptor (p75NTR), tropomyosin receptor kinase (Trk), and p75 [27]. Numerous processes contribute to neurotrophin signaling, but two significant tyrosine kinases are involved: the mitogen-activated protein kinase (MAPK)-extracellular signal-regulated kinase (ERK) pathway and the phosphoinositide 3-kinase (PI3K)-AKT process [28]. Initially produced as proneurotrophins, neurotrophins are contained within secretory vessels.

A C-terminal matured protein and an N-terminal prodomain peptide are released from protoneurotrophins when they undergo proteolytic cleavage [29]. Proneurotrophins may cleave in both external and internal media. Proseurotrophins are not dormant as neurotrophins are. With their association with the p75 neurotrophin receptor (p75NTR), they operate to induce an apoptotic process. Proneurotrophins therefore drive neuronal expansion cone retract and pro-apoptotic effects, primarily amid developmental and diseased situations [30].

Research has shown that individuals with AD have an imbalance in the concentrations of neurotrophin in several brain areas. It has been demonstrated that AD brains contain reduced quantities of BDNF, indicating an inadequacy in trophic assistance in neurodegeneration [31]. In addition to being necessary for the survival and incorporation of newly formed neurons into the hippocampus circuitry, BDNF is also required for the initial stage of adult hippocampal neurogenesis [32].

8.4 BRAIN-DERIVED NEUROTROPHIC FACTOR

One widely recognized A constituent of the family of neurotrophins, BDNF, sometimes called (abrineurin), was first identified in the adult mammal in 1982. Both the glia and neurons in the CNS extensively produce it, with the hippocampal and cerebral cortex exhibiting the maximum levels of expression. Additionally, neurons found in the amygdala, claustrophobic area, and hypothalamus release BDNF. In addition to the hippocampus, the cortex, basal forebrain, kidneys, skeletal muscles, retina, the prostate, and motor neurons all produce BDNF [33]. In human beings, the BDNF gene on chromosomal sequence 11 encodes it, which is synthesized in the endoplasmic reticulum (ER) [34]. The transcription of BDNF is controlled by eight distinct promoters, each of which is associated with a distinct transcript which is spliced to the 3′ encoding exon from any one among the eight untranslated 5′ exons (I–VIII). Strongly induced by calcium ions, promoter IV activity governs the transcription of exon IV, consisting of mRNA sequencing and is mostly governed by a Cre regulating variable, indicating established roles for the transcription factor CREB and the foundation of BDNF's action-reliant processes [35]. It influences both short- and long-term synaptic connections, in addition to the growth and development of the neurons. Pro-BDNF along with mature BDNF are the two main kinds of BDNF; pro-BDNF's terminus region is removed by a separate protein convertase enzyme to produce physiologically functioning mature BDNF [36]. The processes of neuron survival, differentiation, neurotransmitter discharge, dendrite remodeling, and axon development have all been shown to be significantly impacted by BDNF. Furthermore, BDNF could contribute to the origin of cognitive decline as it is necessary for synaptic plasticity, the underlying biological underpinning of memory and learning capacity in the adult brain [37].

8.4.1 BDNF GENETIC EXPRESSION

BDNF is a gene having a complicated morphology that expresses differently in various tissues. Rodent BDNF genes include nine exons and nine distinct operational promoters [38]. The production of BDNF variations that encode the identical BDNF protein is regulated by these promoters. The BDNF genes on the chromosome of humans 11p13 encodes the 247 amino acids that make up complete BDNF protein. BDNF is first produced as pre-pro-BDNF in the ER. This predecessor protein is then transported to the Golgi system, where it cleaves into the pro-BDNF isoform, which has a molecular weight of around 32 kDa. A total of three possible outcomes for pro-BDNF: cleavage by furin or convertases inside the cell, which releases mature BDNF (mBDNF), which is about 14 kDa; pro-BDNF release combined with outside the cell splitting by metalloproteinases 2 (MMP2), MMP9, and plasmin; and pro-BDNF release with no alteration [39]. Tissue plasminogen activator (tPA) regulates pro-BDNF's cleavage transformation [40].

Because of its distinct chromosomal structure, BDNF signaling can be controlled by a variety of variables. Moreover, a unique set of functional consequences can be linked to each BDNF isoform. In particular, the removal of BDNF from promoters IV and VI inhibits GABAergic neural communication, which lowers the activity of the genes within the brain that contribute in peptide and hormonal signaling, such as corticotropin-releasing factor-binding protein (Crhbp), tachykinin (Tac1), cortistatin (Cort), and somatostatin (Sst). The different BDNF transcripts exhibit a specific to the area expression profile in the hypothalamus, prefrontal cortex, and hippocampus, according to quantitative examination of the BDNF protein [41].

The activity-based BDNF transcription is additionally controlled by epigenetic modifications in the organization of chromatin. Exon IV promoter capability—a promoter upstream of BDNF exon IV—and enhanced BDNF synthesis in mice are linked to neural activation. Prior research has indicated that cerebral performance may be impacted by BDNF modulation at the mRNA expression [42].

In the mouse brain, two BDNF transcripts of mRNA were found to support various subcellular localizations. Hippocampal neurons' soma is home to 1 transcript with a short 3′ untranslated section (3′ UTR), whereas the dendrites are home to another transcript with a lengthy 3′ UTR. When mice are genetically engineered to have an alteration in the long 3′ UTR, the transcript is expressed at a shortened length and the BDNF mRNA's ability to recognize dendrites is compromised, causing BDNF production to move from dendrites to the soma. Deficits in dendrite core trimming and specific degradation of long-term strengthening at dendritic terminals are the outcomes of this [43]. BDNF mRNAs exhibit activity-based dendritic positioning *in vitro* in along with transcript class specificity, with transcripts I and IV preferentially influencing transcripts II and VI and dendrites proximities preferentially impacting distal dendrites. The human cerebellum forms dsRNA duplex between BDNF and anti-BDNF transcripts, as demonstrated by the encoding of noncoding endogenous antisense RNAs derived from the anti-BDNF gene into the BDNF gene locus. This indicates that the regulation of BDNF production is mostly dependent on the anti-BDNF transcripts. Anti-BDNF may have a function in controlling splicing of BDNF pre-mRNA and inhibiting either BDNF translation or transcription [44].

8.4.2 BDNF DETECTION METHODS

There have been reports of both conventional methods and cutting-edge strategies for measuring BDNF levels. When studying neurological conditions like AD, wherein concentrations of BDNF are frequently low, measuring BDNF levels might be useful. Every technique has advantages and disadvantages of its own, overall, the best approach will rely on the specific investigation concerns, available resources, and needs for the experiment (Table 8.1) [45].

8.5 BDNF FUNCTIONS IN THE BRAIN

During embryogenesis as well as the postnatal era, BDNF supports the formation and growth of the brain as well as the construction and

TABLE 8.1 List Along with Their Characteristics of Various Common BDNF Detection Techniques.

Sl. No	Detection Method	Characteristic	Advantage	Disadvantage	References
1.	Enzyme-linked immunosorbent assay (ELISA)	The most common approach for identifying BDNF is ELISA because of its well-established procedures, high sensitivity, and ease of use. Sandwich ELISA is another method that may be used to measure the amounts of BDNF proteins in saliva, tissues, blood, and CSF [103]. Commercial ELISA kits for BDNF are accessible in four distinct types [104]. These kits are made to specifically identify pro- or mBDNF; they also include antibodies that target the carboxy-terminal of mBDNF, mBDNF monoclonal, and recombinant monoclonal antibodies.	Prevalent and proven method. Comparatively easy to complete. High levels of sensitivity as well as specificity.	Requires specialized equipment and reagents	[46, 47]
2.	Western blotting	Employs antibodies to identify BDNF and differentiates proteins in the specimen depending on size, providing for the visualization of various isoforms.	Detects many protein isoforms at once. Beneficial for researching changes to proteins.	Taking a long time needs specialized tools and knowledge. Do not accurately measure the levels of BDNF.	[48, 49, 50]
3.	Immunohistochemistry (IHC)	Provides spatial data for the location of the BDNF.	Exposes the BDNF distribution patterns in several brain areas. Can be used in conjunction with staining for additional proteins to look at potential associations or co-localization of BDNF with different molecules.	Less accurate	[51, 52]

TABLE 8.1 *(Continued)*

Sl. No	Detection Method	Characteristic	Advantage	Disadvantage	References
4.	Quantitative PCR (qPCR)	The method is highly sensitive; various cell types may have different regulation systems since they have separate transcriptomes. A lot of interest has been shown in single-cell transcriptome analysis to get more precise data on how individual cells react to inputs or when they develop aberrant phenotype.	Permits the examination of gene expression in certain cell groups.	Restricted to mRNA levels. Does not offer information specific to proteins.	[53, 54, 55]
5.	Biosensors	Make use of chips or electrodes coated with BDNF antibodies. An electrical signal is produced according to the concentration when BDNF binds. Able to identify many biomarkers in therapeutically applicable specimens in real-time.	Large capacity apparatuses. Quick reaction time and simple to use. Inexpensive Assists multiple analyses.	Sample matrix effects may have an impact on sensitivity. Restricted to measuring BDNF levels alone.	[56, 57]

functionality of neural networks. BDNF has developmental and survival effects on a range of neurons, cortical, hippocampal, and dorsal root ganglion cells. The BDNF system functions in several ways in both immature and adult CNS neurons. Specifically, BDNF's control over plasticity of synapses has been thoroughly studied [58].

8.5.1 NEUROGENESIS

Neurogenesis mostly happens in the womb, but it also persists in some form in the adult brain in areas like the olfactory bulb and the hippocampal regions. Neural stem cells (NSCs) divide to produce progenitor cells, which later develop into neurons, during the process of neurogenesis. Moreover, BDNF contributes to the promotion of neurogenesis [59]. Promoting neurogenesis is another function of BDNF. Consider an intraventricular BDNF delivery or adenoviral-induced BDNF action that can be used to raise the supply of neural cells in the brain striatum, thalamus, septum, and olfactory bulb. This effect can be further enhanced by concurrently inhibiting the glial fluctuation in subependymal precursor cells [60]. BDNF helps existing neurons grow and survive. It supports the upkeep and well-being of neurons, which is necessary to establish an atmosphere that is favorable to neurogenesis. The BDNF also has an impact on the phases of adult neurogenesis, which are when neurons are formed from nerve stem cells and progenitor cells. Furthermore, adult neurogenesis has been demonstrated to develop in a number of brain areas, most notably the olfactory circuit and the hippocampus regions, even though this mechanism is most active during the fetal phase and initial growth [61]. NSCs can be influenced by BDNF to differentiate into adult neurons. NSCs are undifferentiated cells that possess the ability to distinguish between several kinds of brain cells. Such cells are guided toward becoming functioning neurons by BDNF [62]. Therefore, BDNF has been established as a strong component regarding controlling neurogenesis in adults.

8.5.2 SYNAPTIC TRANSMISSION

BDNF controls synaptic plasticity, which is the ability of synapses to gradually grow more or less potent in accordance with stimulus. Long-term potentiation (LTP) is linked to the fortification of synaptic connections and essential to learning and memory formation, is supported by BDNF.

By controlling the release of neurotransmitters like glutamate and GABA, BDNF can directly modify the strength of synaptic connections (gamma-aminobutyric acid). It increases the release of neurotransmitters at excitatory synapses, which raises neuronal excitability and synaptic transmission [63]. The development and creation of new synapses is also encouraged by BDNF. It helps create fresh synaptic connections by promoting the growth of dendritic spines, which are tiny projections on dendrites where most excitatory synapses are found. In addition, neurotransmitter receptors that are important for synaptic transmission and plasticity, including NMDA and AMPA receptors, can be affected by BDNF in terms of their expression and trafficking. BDNF can modify receptor location and levels, which in turn can change how sensitive synapses are to neurotransmitter impulses [64].

8.5.3 COGNITIVE FUNCTIONS

8.5.3.1 BDNF-DEPENDENT SIGNALING PATHWAYS

Two distinct transmembrane receptor proteins are bound and activated by BDNF, both pre- and postsynaptically: the lesser-affinity pan neurotrophin receptor p75NTR and the high-affinity tropomyosin related kinase TrkB receptor. Almost all of the actions of BDNF on synapses are now linked to stimulation of TrkB. Nonetheless, research has shown that proBDNF preferentially attaches p75NTR, which has unique physiological implications. Additionally, TrkB splice modifications which do not have the tyrosine kinase domain necessary for downstream transmission can be bound by BDNF. These variations are mostly, but not solely, seen in glial cells. Thus, by making heterodimers that include the full-length TrkB and internalizing BDNF to operate as a clearance receptor, the adhesion of BDNF to such TrkB variants receptor may function as a dominant-negative blocker of BDNF activation. Thus, it is possible to see proBDNF, p75NTR, and shortened TrkB isoforms as antagonistic regulatory processes of the conventional BDNF–TrkB connection, with potential implications for memory and cognition as well as synaptic plasticity [37].

8.5.3.2 LONG-TERM POTENTIATION

The most well-studied kind of synaptic plasticity, prolonged potentiation is thought to be a biological equivalent of memory and learning. It is described as an ongoing rise in synaptic power brought on by an action. Many signaling cascades, including those triggered by BDNF, are activated in response to the generation of LTP. Just in a few milliseconds, small amounts of BDNF trigger depolarization in the membranes of cortical, cerebellar, and hippocampus neurons, which results in the shooting of an action potential [65]. It was a significant discovery since, up until that point it had only been shown that traditional neurotransmitters could affect neurons' potential at the membrane so quickly. Since then, a substantial amount of study has emerged indicating that BDNF is essential for LTP induction, not solely at hippocampal connections in the dentate gyrus and Schaffer collateral—CA1 synapse [66], but additionally in the retinal cortex. For the last stage of LTP (L-LTP), that occurs at least 8 h after tetanization, BDNF is necessary. L-LTP is dependent on CREB and cAMP communication, as well as protein synthesis and gene transcription. Hippocampal L-LTP requires the transformation of proBDNF to mBDNF via protease tPA-activated plasmin [67]. Whenever the synthesis of proteins was blocked, mBDNF was enough to restore L-LTP, indicating that TrkB signaling, which is stimulated by BDNF, is the main regulator for L-LTP generation. In fact, the induction of L-LTP was hindered by the simultaneous inhibition of TrkB transmission at pre- and post-synaptic locations. Beyond the induction phase, BDNF plays a crucial function in L-LTP and is necessary for its maintenance [68]. In BDNF-dependent L-LTP, the stimulation of cAMP transmission may serve as a catalyst for the ejection of BDNF along with the membrane-based integration of the TrkB receptor. L-LTP is also linked to the ERK1/2 kinases' stimulation, relocation to the cell nucleus, and resulting activation of CREB. This might explain the process by which BDNF mRNA production rises in the hippocampal region 2–4 h after tetanic stimulus is applied [37].

8.5.3.3 SYNAPTIC AND CELLULAR MODIFICATIONS

It is widely known that BDNF has permissive functions in the CNS, where it mediates synaptic plasticity [37]. Dendritic spine ultrastructure at excitatory glutamatergic synapses is altered together with the activity-driven and long-lasting rise in synaptic effectiveness that characterizes LTP. New spine development and a rise in glutamatergic AMPA receptors inside dendritic spines are two examples of these alterations. Synapse structure and receptor accessibility are altered as an outcome of a rapid rise in synapse region following LTP-inducing stimulation, which increases neurotransmission [69]. Similarly, comparable synapse reorganization and AMPA receptor trafficking into CA1 synapses are linked to hippocampal-dependent learning [70].

8.6 BDNF'S FUNCTION IN ALZHEIMER'S DISEASE

While several investigations have proven that concentrations of BDNF in AD brains are less compared to those in a healthy human brain, the etiology and pathological mechanisms of AD remain poorly understood. Additional data suggests that BDNF therapy may control cell death, reduce aberrant Aβ synthesis and restore Aβ-induced impairment, improve synaptic damage and cognitive impairment, and even postpone cognitive loss [71]. Spatial memory is compromised when BDNF signaling via TrkB decreases; on the other hand, memory is improved when TrkB is overexpressed. Furthermore, TrkB and BDNF signaling enhanced LTP at hippocampus synapses. Such characteristics of BDNF therefore gave rise to theories on its function in AD [72]. According to some research, AD individuals' autopsy brains have lower amounts of BDNF mRNA as well as protein [73]. Although BDNF mRNA levels in the human post-mortem AD hippocampal tissue have been documented, it remains unclear if the reported modification in the expression of BDNF also transpires at the protein levels. Additionally, it has been discovered that BDNF therapy might lessen Aβ-mediated cellular death, improve synapse degeneration and cognitive impairment, and even postpone dementia in nonhuman primates, as well as APP transgenic mice models. These findings imply that BDNF could be a potential treatment option for AD, as studies have shown a link between BDNF production deficiencies and neurodegenerative etiology [74].

8.6.1 BDNF AND AB IN ALZHEIMER'S DISEASE

The three prominent physiological alterations associated with AD are neuronal degeneration, intracellular NFTs of increased phosphorylation of tau protein, and Aβ accumulation. Aβ and BDNF in AD have been shown in multiple investigations to have opposing impacts on neuronal survival and efficiency. First off, Aβ contributes to the reduction in BDNF in AD. Numerous investigations have revealed that Aβ therapy has a deleterious impact on BDNF levels both *in vitro* and *in vivo* [75]. In animal studies of

AD, the hippocampus and cortex—which are also the favored sites for Aβ deposition—showed a decrease in BDNF. Aβ also significantly reduced the expression of BDNF in AD patient-derived dendritic cells [73].

The fundamental processes of Aβ-downregulating BDNF are as follows: Aβ may affect the PI3K-Akt-mTOR signaling process, the stimulation of the Ras-MAPK/ERK pathway, along with the generation of activity-regulated cytoskeleton-associated protein (Arc) [76]. It can additionally block the cAMP response element-binding protein (CREB) attachment to the promoter domain of BDNF, that in turn results in reduced BDNF transcription in nerve cells [77]. Additionally, Aβ causes structural alterations in microtubules and destroys neuron axons, which provide a transport hurdle for intracellular BDNF [78].

Furthermore, BDNF has the potential to counteract aberrant Aβ buildup and restore the damage caused by Aβ. Following the successive hydrolysis mechanism catalyzed by β- and γ-secretase, Aβ is extracted from the APP. Nevertheless, APP may also be fragmented via a different mechanism inside the Aβ region via α-secretase, producing a discharged version of APP (sAPPα) that exhibits neuronal protection properties [79]. As a result, the prevalence of AD is decreased by either decreasing the amyloidogenic route of APP or increasing its nonamyloidogenic splitting, which prevents the production of Aβ. By affecting filtering protein-related receptor activity associated with the A-type repetitions gene, a neural transport receptor in APP metabolic processes, BDNF is shown to lessen the amyloidogenic pathway of APP and prevent the creation of Aβ [80, 81].

8.6.2 SYNAPTIC REPAIR AND BDNF IN AD

As is well known, the AD brain exhibits continuous synapse loss, particularly in the frontal, entorhinal cortex, hippocampus, and inferior parietal cortex [82]. Significant evidence has demonstrated specific hippocampal LTP damage in AD animals that ends in cognitive deficiency and memory retention blockage (the two main characteristics of AD). As a result, AD was determined to be a synapse failure. More significantly, in AD etiology, Aβ-induced mild synaptic degradation started before neuronal degeneration [82], providing theoretical support for therapies targeted at synaptic repair in AD therapy.

Among the most fascinating developments in the neurotrophin research is the modulation of synapses by BDNF. As already stated, BDNF is essential for triggering and ongoing regulation of persistent rise in transmission of synapses and dendritic spine development in adult brains [83]. It is true that BDNF and LTP can converse in dendrites to affect persistent memory, and although research has only been done in mouse models, the most intriguing part of understanding BDNF's function in AD synaptic regeneration is the beneficial relationship involving BDNF and LTP. Numerous investigations have unequivocally shown that genetic removal of BDNF in rats markedly reduced hippocampal LTP, as well as proportionately compromised hippocampus-reliant memory [84].

Exogenous BDNF injection or BDNF level augmentation can correct Aβ-induced deficits in LTP and synaptic communication both *in vitro* and *in vivo* [85]. Unlike Aβ activity, BDNF promotes NMDA and AMPA receptor autophosphorylation, which in turn increases CaMKII autophosphorylation and functioning to safeguard synaptic activity. Furthermore, it has been noted that Arc protein is crucial for the BDNF-induced induction of LTP [86].

8.6.3 BDNF POLYMORPHISM IN ALZHEIMER'S DISEASE

Studies focusing on the function of BDNF genetic variations, particularly BDNF Val/Met, in various neurodegenerative and psychiatric conditions have recently come to light. Additional data from animal studies revealed that the Met carrier modified synaptic plasticity, particularly the NMDAR-dependent LTP, and decreased the volume of the hippocampus, which led to a loss in hippocampal-related cognition [87]. These have all been verified by studies on aging healthy individuals and individuals with AD to be affected by BDNF Met carriers. The reduction in the brain's structure might be due to a reduction in BDNF discharge and resulting lack of neurotrophic stimulation of dendritic and axonal growth [88]. AD etiology may be linked to BDNF Val66Met. In accordance with this, Met carriers have reduced episodic cognitive and memory impairment in AD clinical settings [89, 90]. BDNF 270C/T is a well-studied BDNF variation that results in a C/T change at site 270 and is situated in the 5′-untranslated region (UTR) [91]. Like Val66Met, the relevance of SNP C270T in AD is likewise unclear; unfavorable findings were reported in American and Italian literature, and all meta-analyses [92–94] have consistently confirmed the insignificance of the findings.

To put it briefly, opinions about the potential contribution of BDNF polymorphism to AD are divided. Even if the findings are debatable, the diversity of research demonstrates the BDNF gene's intricacy in AD. In the study of

AD pathology, genetic variables have drawn increasing interest. A wealth of data suggests that BDNF may be the most probable gene in AD.

8.7 BDNF-TARGETING METHODS MODIFYING AD

With mixed results, several possible therapies for different neurological disorders have been tested to treat AD. As a result, symptomatic remedies for AD are now advised, and further research is being done on prospective disease-modifying therapies. Increasing BDNF levels constitutes one such symptomatic treatment that is often utilized to stop the course of AD in the cerebrums of AD patients [95].

8.7.1 BDNF-BASED PHARMACOTHERAPY

Given that AD patients have been shown to have a BDNF deficiency, it is conceivable that administering exogenous BDNF could serve as a beneficial treatment strategy (Table 8.2). Problems with the specificity of BDNF activity have been difficult to solve to date. This is because some parts of the brain have reduced BDNF expression, and excessive BDNF flooding is bad for the brain [96].

8.7.2 BDNF GENE THERAPY

Recombinant viruses and synthesized vectors are the two basic types of gene-delivery vehicles [104]. Among the polymer-based vectors employed for translocation of the BDNF gene are hydrogels and nanoparticles [105]. One major benefit of liposomes—natural, biodegradable lipid bilayers—is their resemblance to organic cell membranes. The cationic materials' electrostatic forces which dynamically link to the BDNF plasmid serve as the foundation for these nonviral carriers. Several attractive characteristics of polymer-based vectors include minimal immunogenicity, high safety and stability, simplicity of manufacture, and straightforward target ligand incorporation procedures. Regretfully, compared to recombinant viruses, polymers as gene delivery vectors have a transfection efficacy that is several orders of magnitude lower. Therefore, the main method for delivering the BDNF gene remains using a recombinant virus [106].

In AD, BDNF gene delivery provides protection against diseases associated to tau and Aβ. On the other hand, tau hyperphosphorylation and Aβ accumulation are unaffected by this therapy. After 5 months of treatment

TABLE 8.2 List of Several Pharmacological Therapies Targeting BDNF in AD.

Sl. No	Pharmacological Therapy	Example	Mechanism	References
1.	Cholinergic drugs	Donepezil Galantamine Rivastigmine	Due to their reduced capacity to trigger intracellular signaling cascades, a process that results in more stable and long-lasting modifications to synaptic functioning, they may have erratic and temporary effects.	[97]
2.	Glutamatergic drugs	Memantine	Boost the BDNF receptor's expression and ability to raise indigenous BDNF synthesis, which will have an impact on the whole system that controls the transmission of BDNF signals.	[98]
3.	GABA-B antagonists	SGS742	Increase BDNF expression in the frontal lobe and hippocampal regions considerably (clinical trial phase II).	[99]
4.	Phosphodiesterase inhibitors	Rolipram	Enhance the phosphorylation of the cAMP, CREB, and BDNF expression.	[100]
5.	Cannabinoids	D9–tetrahydrocannabinol	Raise the production of BDNF in several brain regions, indicating an immediate impact on cell viability as well as a long-term impact on factors influencing synaptic plasticity and the adaptive mechanisms that underlie cognition.	[101]
6.	Estrogen		Enhanced levels of BDNF production in the forebrain throughout development and adulthood, which may improve and preserve the neurons' proper functional condition.	[102]
6.	Lithium		Raises BDNF blood levels and lessens cognitive decline.	[103]

with Lenti-BDNF delivery of genes, memory and cognition impairments, synaptic decline, and atrophy were all minimized in J20 APP transgenic mice [107]. Yet, the density of amyloid plaque remained unchanged after this BDNF therapy. Like this, tau transgenic mice with P301L mutations which received mammalian BDNF genetic recombinant employing an AAV8 vector (AAV-BDNF) exhibited increased brain BDNF levels and reduced memory impairments; however, tau protein, GSK3β, and phosphatase PP2A were not directly affected by the AAV-BDNF gene delivery [108]. It has been possible to deliver BDNF precisely and reliably into the entorhinal cortex of nonhuman primates using MR-guided infusion of AAV2-BDNF [109]. Furthermore, intracranial structures have been precisely targeted and the vector's dispersion has been tracked in real time throughout injection thanks to real-time magnetic resonance imaging (MRI) of AAV in monkey brains [110]. AAV mutations have also been thoroughly researched. When BDNF was delivered utilizing the tyrosine triple mutant AAV (tm-scAAV2), it was shown that BDNF's RNA expression was over 300 times greater than the AAV subgroup and that it also generated noticeably more proteins [111].

The application of cell-based vectors, including NSCs, Schwann cells, mesenchymal stem cells (MSCs), CD4 T cells, and fibroblasts, provides an additional strategy for the prolonged transmission of BDNF [112]. BBB blockage can be circumvented by direct BDNF gene transfer utilizing MSC [113]. In earlier studies, irradiated female SJL/J mice received an intravenous infusion of BDNF-transduced bone marrow stromal cells (BMSCs) for 8 weeks. This significantly postponed the development of experimental autoimmune encephalomyelitis and decreased its overall severity [114]. However, these BDNF-producing cells permit BDNF to be delivered over an extended period. Unfortunately, since the supplied BDNF doses rely on cell viability and transfection strength, this approach is challenging to manage precisely. A further apprehension is that cells produced from bone marrow may move and settle in different nonhematopoietic organs, leading to unfavorable outcomes. To ensure consistent and local discharge, it has been suggested to encapsulate these BDNF-producing cells. Survival of cells and the production of BDNF were enhanced when BDNF gene recombination MSCs were transfected using a scaffold modified with the adhesive peptides PPFLMLLKGSTR [115]. Additionally, NSCs- and BMSCs-BDNF have been transported using alginate-based compositions, which have been shown to sustain proliferation of cells as well as longevity alongside the regulated discharge of BDNF [116]. Nevertheless, after a 1.5-month therapy period, no protective effect toward neuronal death was seen following the administration of the BDNF gene to APP transgenic mice "disease onset" [117]. This implies that introduction of the BDNF gene may not be a good treatment approach for AD at all during the illness. Consequently, early and long-term therapies can be needed.

8.7.3 NERVE REGENERATION

Enhancing concentrations of BDNF may be an achievable goal in the management of cognitive loss in AD patients. Ongoing research over the function of BDNF revealed suggesting that BDNF exerts physiological impacts on treatments for cognitive impairment and the decrease of neuronal apoptosis. By promoting the expulsion of neurotransmitters, boosting vesicle docking, and strengthening the glutamate-induced postsynaptic reflex [118], BDNF influences plasticity of neurons. When BDNF attaches itself to the high-affinity receptor for TrkB, it triggers the autophosphorylation and dimerization of Trk, which in turn triggers the intracellular signal transduction pathways of phosphatidylinositol 3 kinase/protein kinase B (PI3K/Akt). After engaging with PI3K, Akt becomes phosphorylated and adheres to the internal layer of the plasma membrane [119].

By explicitly impeding the apoptosis process or by suppressing the activation of the tumor suppressor gene *p53*, phosphorylated Akt prevents cell death and promotes the continuing existence of neurons [120]. Cytochrome c (cyt-c) is found among the flaws within the interior as well as exterior mitochondrial membrane underneath ordinary circumstances. It is released onto the cytoplasm upon the activation of apoptotic signaling that originate in mitochondrial spaces. Phosphorylated (p)-Akt regulates the discharge of cyt-c via mitochondria toward cytoplasm and modulates the expression of Bcl2 family components prior to the production of cyt-c. Following the discharge of cyt-c, pAkt may additionally alter the apoptosome's constituents and prevent the apoptosome from forming, which blocks the mitochondria-mediated apoptotic mechanism [121]. By attaching to the membrane receptor TrkB, BDNF and TrkB jointly regulate the expression of the PI3K/Akt network. This regulates survival of cells and synaptic activity and makes the BDNF/TrkB/PI3K/Akt signaling network a viable target for therapy for Alzheimer's. Previous research has shown a favorable correlation between rats', nonhuman primates', and humans' cognitive levels and the production amount of BDNF. By inducing the BDNF/TrkB/PI3K/Akt signaling process, BDNF primarily controls nerve cell viability and synaptic plasticity, which in turn enhances cognitive function. Furthermore, like BDNF, additional TrkB agonists have a physiological effect on neural protection and cognitive performance [122].

8.7.4 MICROBIOTA

A close relationship exists between the gut microbiota's makeup and the BDNF level. In the amygdala, hippocampus, and cortex, germ-free animals exhibit reduced BDNF mRNA and protein concentrations in comparison to mice with a healthy gut microbiota [123]. Following the shift of faecal microbiota, recipient mice exhibit comparable levels of inflammatory mediators, cognitive behavior, microglia activity, and BDNF compared to donor mice [122]. This process is linked to the AKT-GSK3β/β-catenin mechanism being activated. Following findings imply that the lack of gut microbiota might cause major disruptions to CNS BDNF concentrations, which can then be corrected by microbiota transplantation. Additionally, probiotic supplements help to increase levels of BDNF. VSL#3 is a probiotic blend comprising eight strains of Gram-positive bacteria. Treatment with VSL#3 raises the abundance of Actinobacteria and Bacteroidetes in aged male rats (20–22 months), minimizes microglial activation, and raises BDNF concentrations. Gut microbiota can synthesize and recognize certain neurochemicals, like short-chain fatty acids, butyrate, neurotransmitters, and secondary bile acids [124]. So, by adjusting the functioning of these neurochemicals, gut microbiota may affect how well BDNF functions in the CNS.

8.7.5 EXOGENOUS BDNF TREATMENT

BDNF's little plasma half-life and low BBB permeability restrict its intravenous infusion [125]. Evaluating the regional dissemination and activity of BDNF in specific brain areas is difficult. Preclinical research has demonstrated that administering BDNF to particular brain regions is advantageous for eliciting sustained enhancement of synaptic communication, stimulating neurogenesis and ectopic granule cells, and boosting the expression of BDNF receptors (Table 8.3) [126]. Exogenous BDNF transfer is challenging to implement in clinical stages, as the majority of BDNF direct delivery techniques involve significant invasiveness and lack clarity regarding therapy periods and dosage schedules. Several issues arise when exogenous BDNF is supplemented. Since BDNF is not easily able to move across the BBB, the issue with exogenous BDNF administration relies on how much of the protein reaches the damaged cell. A dosage this little might not be enough to provide the desired results, while a dosage this large could be harmful. In addition, BDNF is easily broken down in a biological media due to its instability. A substantial body of research thus conders the clinical efficiency of employing BDNF delivered intranasally because: the nasal mucosa provides more surface area for drug absorption; intranasal administration surpasses the BBB; the invasion free and simple self-administration enhances patient comfort; it allows for both rapid as well as straight CNS administration of BDNF with more bioavailability by surpassing first-pass metabolism; it results in less systemic exposure; a little dose may be utilized without causing adverse reaction; and no drug modification is necessary.

TABLE 8.3 Various Examples of Exogenous BDNF Delivery Routes [127].

Sl. No.	Targeted Brain Region	Delivery Route	BDNF Delivery Vehicle	Inference	References
1.	Hippocampus	Intrahippocampal infusion	BDNF-comprising PBS	Enhances the dentate gyrus's long-lasting potential of synaptic activity.	[128]
2.	Ventral hippocampus	Intrahippocampal infusion	BDNF mixed in PBS	Enhances infralimbic excitability and assists extinction memories.	[129]
3.	Dorsomedial prefrontal cortex	Intracortical infusion	BDNF in saline (0.9%)	Prevents the paradoxical inversion of learning. Improvement brought on due to stress or injury to the prefrontal cortex.	[130]
4.	Nucleus accumbens	Intra-nucleus accumbens injection	BDNF in saline	Reduced dopamine discharge and dopamine-generated activities caused by methamphetamine.	[131]
5.	Nasal cavity	Intranasal delivery	BDNF mixed in sterile PBS	Relieves localized inflammation of the brain brought on by ischemia and reperfusion.	[132]
6.	Cochlear	Intracochlear infusion	BDNF-comprising saline and (1%) BSA	Increased of spiral ganglion cells viability.	[133]

8.8 CONCLUSION AND FUTURE PROSPECTIVE

AD is a terrible neurodegenerative disease that is complicated and diverse. The absence of a viable treatment makes it a major global health care load. Cognitive impairment is mostly caused by neurological and synapse failure, which is a common hallmark of AD regardless of the many etio-pathogenic pathways involved. Because of their proven proneurogenic properties and the possibility of synaptic repair, neurotrophic variables provide a compelling case for their usage. Various stages of AD are associated with impairments in BDNF production and secretion, which can cause aberrant buildup of Aβ and synaptic disruption, ultimately leading to cognitive loss. Furthermore, it has been demonstrated that BDNF gene polymorphism may contribute to the pathophysiology of AD, and Met carrier status is associated with reduced levels of BDNF, specific shrinkage of the brain's structure, and decreased cognitive function. It is indisputable that increased BDNF levels alleviate the cognitive impairment associated with AD, even if certain findings contradict the claim. Considering the debatable circumstances, further thorough investigations will be needed in the future.

While there are rare instances in which the BDNF level stays constant, changes in the brains of AD patients are linked with mild advanced levels of senile dementia. A hallmark of early-stage AD is TrkB-mediated impairment of BDNF signaling. Cognition and psychological processes are controlled by the hippocampus and amygdala areas of the brain, where postmortem tissue from the brain has been found to display higher levels of BDNF. Language, memory, learning, higher order thinking, and nerve cell growth and survival are all regulated by BDNF. As a result, it has been suggested that memory and BDNF are correlated.

There are additional methods documented to bolster the notion that exogenous BDNF therapy can be a different approach to enhance cognitive performance in AD. BDNF's unclear significance in AD presently precludes its use as a biomarker or targeted treatment. Deeper understanding of this field would be attainable with greater investment in clinical studies on BDNF, involving the utilization of imaging to clarify the temporal link of BDNF in correlation with cognitive scores and brain structural, functional, and biochemical alterations. Therapy for AD pathology must address several elements of the pathophysiology pathway, since the disease is linked to several pathways. Medication should target aberrant buildup of Aβ, hyperphosphorylation of tau, and Aβ processing in conjunction with a neuroprotective agent like that of NTF.

KEYWORDS

- **Alzheimer's disease**
- **amyloid β protein**
- **brain-derived neurotrophic factor**
- **long term potentiation**
- **neuroinflammation**
- **neurotrophic factor**
- **neurotrophins**

REFERENCES

1. Kumar, A., Sidhu, J., Goyal, A., & Tsao, J. W. (2020). Alzheimer disease [Updated 2020 Mar 9]. StatPearls. Treasure Island: StatPearls Publishing.
2. Weller, J., & Budson, A. (2018). Current understanding of Alzheimer's disease diagnosis and treatment. *F1000Research, 7.*
3. Tang, Y., Lutz, M. W., & Xing, Y. (2019). A systems-based model of Alzheimer's disease. *Alzheimers & Dementia, 15*(1), 168–171.
4. Mossello, E., & Ballini, E. (2012). Management of patients with Alzheimer's disease: pharmacological treatment and quality of life. *Therapeutic Advances in Chronic Disease, 3*(4), 183–193.
5. Soto, M. E., Andrieu, S., Cantet, C., Reynish, E., Ousset, P. J., Arbus, C., Gillette-Guyonnet, S., Nourhashémi, F., & Vellas, B. (2008). Predictive value of rapid decline in Mini Mental State Examination in clinical practice for prognosis in Alzheimer's disease. *Dementia and Geriatric Cognitive Disorders, 26*(2), 109–116.
6. Wang, Z. H., Xiang, J., Liu, X., Yu, S. P., Manfredsson, F. P., Sandoval, I. M., Wu, S., Wang, J. Z., & Ye, K. (2019). Deficiency in BDNF/TrkB neurotrophic activity stimulates δ-secretase by upregulating C/EBPβ in Alzheimer's disease. *Cell Report, 28*(3), 655–669.
7. Autry, A. E., & Monteggia, L. M. (2012) Brain-derived neurotrophic factor and neuropsychiatric disorders. *Pharmacological Review, 64*(2), 238–258.
8. Ng, T. K. S., Ho, C. S. H., Tam, W. W. S., Kua, E. H., & Ho, R. C. M. (Decreased serum brain derive neurotrophic factor (BDNF) levels in patients with Alzheimer's disease (AD): A systematic review and meta-analysis. *International Journal of Molecular Sciences, 20*(2), 257.

9. Forlenza, O. V., Diniz, B. S., Teixeira, A. L., Radanovic, M., Talib, L. L., Rocha, N. P., & Gattaz, W. F. (2015). Lower cerebrospinal fluid concentration of brain-derived neurotrophic factor predicts progression from mild cognitive impairment to Alzheimer's disease. *NeuroMolecular Medicine, 17*(3), 326–332.
10. Weinstein, G., Beiser, A. S., Choi, S. H., Preis, S. R., Chen, T. C., Vorgas, D., Au, R., Pikula, A., Wolf, P. A., DeStefano, A. L., & Vasan, R. S. (2014). Serum brain-derived neurotrophic factor and the risk for dementia: The Framingham Heart Study. *JAMA Neurology, 71*(1), 55–61.
11. Silvestrelli, G., Lanari, A., Parnetti, L., Tomassoni, D., & Amenta, F. (2006). Treatment of Alzheimer's disease: From pharmacology to a better understanding of disease pathophysiology. *Mechanisms of Ageing and Development, 127*(2), 148–157.
12. Hardy, J., & Selkoe, D. J. (2002). The amyloid hypothesis of Alzheimer's disease: Progress and problems on the road to therapeutics. *Science, 297*, 353–356.
13. Kumar, A., & Dogra, S. (2008). Neuropathology and therapeutic management of Alzheimer's disease—An update. *Drugs Future, 33*(5), 433–446.
14. Hardy, J. (2009). The amyloid hypothesis for Alzheimer's disease: A critical reappraisal. *Journal of Neurochemistry, 110*(4), 1129–1134.
15. Dal Prà, I., Chiarini, A., Gui, L., Chakravarthy, B., Pacchiana, R., Gardenal, E., Whitfield, J. F., & Armato, U. (2015). Do astrocytes collaborate with neurons in spreading the "infectious" Aβ and Tau drivers of Alzheimer's disease? *Neuroscientist, 21*(1), 9–29.
16. Armstrong, R. A. (2009). The molecular biology of senile plaques and neurofibrillary tangles in Alzheimer's disease. *Folia Neuropathology, 47*, 289–299.
17. Rosenmann, H. (2013). Immunotherapy for targeting tau pathology in Alzheimer's disease and tauopathies. *Current Alzheimer Research, 10*(3), 217–228.
18. Galimberti, D., Ghezzi, L., & Scarpini, E. (2013). Immunotherapy against amyloid pathology in Alzheimer's disease. *Journal of the Neurological Sciences, 333*(1), 50–54.
19. Metaxas, A., & Kempf, S. J. (2016). Neurofibrillary tangles in Alzheimer's disease: Elucidation of the molecular mechanism by immunohistochemistry and tau protein phospho-proteomics. *Neural Regeneration Research, 11*, 1579–1581.
20. Overk, C. R., & Masliah, E. (2014). Pathogenesis of synaptic degeneration in Alzheimer's disease and Lewy body disease. *Biochemical Pharmacology, 88*, 508–516.
21. Sampaio, T. B., Savall, A. S., Gutierrez, M. E., & Pinton, S. (2017). Neurotrophic factors in Alzheimer's and Parkinson's diseases: implications for pathogenesis and therapy. *Neural Regeneration Research, 12*(4), 549–57.
22. Ledda, F., & Paratcha, G. (2016). Assembly of neuronal connectivity by neurotrophic factors and leucine-rich repeat proteins. *Frontiers in Cellular Neuroscience, 10*, 199.
23. Lu, B., Nagappan, G., Guan, X., Nathan, P. J., & Wren, P. (2013). BDNF-based synaptic repair as a disease-modifying strategy for neurodegenerative diseases. *Nature Reviews Neuroscience, 14*, 401–416.
24. Chao, M. V. (2003). Neurotrophins and their receptors: A convergence point for many signalling pathways. *Nature Reviews Neuroscience, 4*, 299–309.

25. Kazim, S. F., & Iqbal, K. (2016). Neurotrophic factor small-molecule mimetics mediated neuroregeneration and synaptic repair: Emerging therapeutic modality for Alzheimer's disease. *Molecular Neurodegeneration, 11*, 1–6.
26. Bothwell, M. (2014). NGF, BDNF, NT3, and NT4. *Handbook of Experimental Pharmacology, 220*, 3–15.
27. Hempstead, B. L. (2014). Deciphering proneurotrophin actions. *Handbook of Experimental Pharmacology, 220*, 17–32.
28. Lee, J., Fukumoto, H., Orne, J., Klucken, J., Raju, S., Vanderburg, C. R., Irizarry, M. C., Hyman, B. T., & Ingelsson, M. (2005). Decreased levels of BDNF protein in Alzheimer temporal cortex are independent of BDNF polymorphisms. *Experimental Neurology, 194*, 91–96.
29. Vilar, M., & Mira, H. (2016). Regulation of neurogenesis by neurotrophins during adulthood: expected and unexpected roles. *Frontiers in Neuroscience, 10*, 26.
30. Song, J. H., & Yu, J. T., & Tan, L. (2015). Brain-derived neurotrophic factor in Alzheimer's disease: Risk, mechanisms, and therapy. *Molecular Neurobiology, 52*, 1477–1493.
31. Qiu, L. L., Pan, W., Luo, D., Zhang, G. F., Zhou, Z. Q., Sun, X. Y., Yang, J. J., & Ji, M. H. (2020). Dysregulation of BDNF/TrkB signaling mediated by NMDAR/Ca 2+/calpain might contribute to postoperative cognitive dysfunction in aging mice. *Journal of Neuroinflammation, 17*(1), 1–5.
32. Zheng, F., & Wang, H. (2019). NMDA-mediated and self-induced BDNF exon IV transcriptions are differentially regulated in cultured cortical neurons. *Neurochemistry International, 54*, 385–392.
33. Girotra. P., Behl, T., Sehgal, A., Singh, S., & Bungau, S. (2022). Investigation of the molecular role of brain-derived neurotrophic factor in Alzheimer's disease. *Journal of Molecular Neuroscience, 72*(2), 173–86.
34. Sandhya, V. K., Raju, R., Verma, R., Advani, J., Sharma, R., Radhakrishnan, A., Nanjappa, V., Narayana, J., Somani, B. L., Mukherjee, K. K., Pandey, A., Christopher, R., & Prasad, T. S. (2013). A network map of BDNF/TRKB and BDNF/p75NTR signaling system. *Journal of Cell Communication and Signaling, 7*(4), 301–307.
35. Karpova, N. N. (2014). Role of BDNF epigenetics in activity-dependent neuronal plasticity. *Neuropharmacology, 76*, 709–718.
36. Vafadari, B., Salamian, A., & Kaczmarek, L. (2016). MMP-9 in translation: From molecule to brain physiology, pathology, and therapy. *Journal of Neurochemistry, 139*, 91–114.
37. Pang, P. T., Teng, H. K., Zaitsev, E., Woo, N. T., Sakata, K., Zhen, S., Teng, K. K., Yung, W. H., Hempstead, B. L., & Lu, B. (2004). Cleavage of proBDNF by tPA/plasmin is essential for long-term hippocampal plasticity. *Science, 306*, 487.
38. Maynard, K. R., Hill, J. L., Calcaterra, N. E., Palko, M. E., Kardian, A., Paredes, D., Sukumar, M., Adler, B. D., Jimenez, D. V., Schloesser, R. J., & Tessarollo, L. (2015). Functional role of BDNF production from unique promoters in aggression and serotonin signaling. *Neuropsychopharmacology, 41*, 1943.
39. Martinowich, K., Hattori, D., Wu, H., Fouse, S., He, F., Hu, Y., Fan, G., & Sun, Y. E. (2003). DNA methylation- related chromatin remodeling in activity-dependent BDNF gene regulation. *Science, 302*, 890–893.

40. Li, H., Shang, J., Zhang, C., Lu, R., Chen, J., & Zhou, X. (2020). Repetitive transcranial magnetic stimulation alleviates neurological deficits after cerebral ischemia through interaction between RACK1 and BDNF exon IV by the phosphorylation-dependent factor MeCP2. *Neurotherapeutics, 17*(2), 651–663.
41. Pruunsild, P., Kazantseva, A., Aid, T., Palm, K., & Timmusk, T. (2007). Dissecting the human BDNF locus: bidirectional transcription, complex splicing, and multiple promoters. *Genomics, 90*(3), 397–406.
42. Donegà, S., & Tongiorgi, E. (2019). Detecting BDNF protein forms by ELISA, western blot, and immunofluorescence. *Brain-Derived Neurotrophic Factor (BDNF),* 89–103.
43. Yoshida, T., Ishikawa, M., Iyo, M., & Hashimoto, K. (2012). Serum levels of mature brain-derived neurotrophic factor (BDNF) and its precursor pro BDNF in healthy subjects. *Open Clinical Biochemistry Journal, 5*(1), 7–12.
44. Elfving, B., Plougmann, P. H., & Wegener, G. (2010). Detection of brain-derived neurotrophic factor (BDNF) in rat blood and brain preparations using ELISA: pitfalls and solutions. *Journal of Neuroscience Methods, 187*(1), 73–77.
45. Singer, W., Manthey, M., Panford-Walsh, R., Matt, L., Geisler, H. S., Passeri, E., Baj, G., Tongiorgi, E., Leal, G., Duarte, C. B., & Salazar, I. L. (2018). BDNF-live-exon-visualization (BLEV) allows differential detection of BDNF transcripts in vitro and in vivo. *Frontiers in Molecular Neuroscience, 11*, 325.
46. Le Blanc, J., Fleury, S., Boukhatem, I., Bélanger, J. C., Welman, M., & Lordkipanidzé, M. (2020). Platelets selectively regulate the release of BDNF, but not that of its precursor protein, proBDNF. *Frontiers in Immunology, 11*, 575607.
47. Calabrese, F., Molteni, R., Maj, P. F., Cattaneo, A., Gennarelli, M., Racagni, G., & Riva, M. A. (2007). Chronic duloxetine treatment induces specific changes in the expression of BDNF transcripts and in the subcellular localization of the neurotrophin protein. *Neuropsychopharmacology, 32*(11), 2351–2359.
48. Bharani, K. L., Ledreux, A., Gilmore, A., Carroll, S. L., & Granholm, A. C. (2020). Serum pro-BDNF levels correlate with phospho-tau staining in Alzheimer's disease. *Neurobiology of Aging, 87*, 49–59.
49. Kopec, B. M., Zhao, L., Rosa-Molinar, E., & Siahaan, T. J. (2020). Non-invasive brain delivery and efficacy of BDNF in APP/PS1 transgenic mice as a model of Alzheimer's disease. *Medical Research Archives, 8(2).*
50. Mathys, H., Davila-Velderrain, J., Peng, Z., Gao, F., Mohammadi, S., Young, J. Z., Menon, M., He, L., Abdurrob, F., Jiang, X., & Martorell, A. J. (2019). Single-cell transcriptomic analysis of Alzheimer's disease. *Nature, 570*, 332–337.
51. de Assis, G. G., Hoffman, J. R., & Gasanov, E. V. (2020). BDNF Val66Met polymorphism, the allele-specific analysis by qRT-PCR-a novel protocol. *International Journal of Medical Sciences, 17*(18), 3058.
52. Sahay, A. S., Jadhav, A. T., Sundrani, D. P., Wagh, G. N., & Joshi, S. R. (2020). Differential expression of nerve growth factor (NGF) and brain derived neurotrophic factor (BDNF) in different regions of normal and preeclampsia placentae. *Clinical and Experimental Hypertension, 42*(4), 360–364.
53. Mobed, A., Charsouei, S., Yazdani, Y., Gargari, M. K., Ahmadalipour, A., Sadremousavi, S. R., Farrahizadeh, M., Shahbazi, A., & Haghani M. (2023).

Biosensors, recent advances in determination of BDNF and NfL. *Cell Molecular Neurobiology, 43*(8), 3801–3814.

54. Karaboğa, M. N., & Sezgintürk, M. K. (2022). Biosensor approaches on the diagnosis of neurodegenerative diseases: Sensing the past to the future. *Journal of Pharmaceutical and Biomedical Analysis, 209*, 114479.
55. Park, H., & Poo, M.-M. (2013). Neurotrophin regulation of neural circuit development and function. *Nature Reviews Neuroscience, 14*, 7–23.
56. Obernier, K., & Alvarez-Buylla, A. (2019). Neural stem cells: Origin, heterogeneity and regulation in the adult mammalian brain. *Development, 146*(4), dev156059.
57. Silakarma, D., & Sudewi, A. A. (2019). The role of brain-derived neurotrophic factor (BDNF) in cognitive functions. *Bali Medical Journal, 8*(2), 518–525.
58. Han, Y., Yuan, M., Guo, Y. S., Shen, X. Y., Gao, Z. K., & Bi, X. (2022). The role of enriched environment in neural development and repair. *Frontiers in Cellular Neuroscience, 16*, 890666.
59. Navarro Quiroz, E., Navarro Quiroz, R., Ahmad, M., Gomez Escorcia, L., Villarreal, J. L., Fernandez Ponce, C., & Aroca Martinez, G. (2018). Cell signaling in neuronal stem cells. *Cells, 7*(7), 75.
60. Porcher, C., Medina, I., & Gaiarsa, J. L. (2018). Mechanism of BDNF modulation in GABAergic synaptic transmission in healthy and disease brains. *Frontiers in Cellular Neuroscience, 12*, 273.
61. Srivastava, D, P., Woolfrey, K. M., & Evans, P. D. (2013). Mechanisms underlying the interactions between rapid estrogenic and BDNF control of synaptic connectivity. *Neuroscience, 239*, 17–33.
62. Cunha, C., Brambilla, R., & Thomas, K. L. (2010). A simple role for BDNF in learning and memory? *Frontiers in Molecular Neuroscience, 3*, 865.
63. Kafitz, K. W., Rose, C. R., Thoenen, H., & Konnerth, A. (1999). Neurotrophin-evoked rapid excitation through TrkB receptors. *Nature, 401*, 918–921.
64. Messaoudi, E., Ying, S. W., Kanhema, T., Croll, S. D., & Bramham, C. R. (2002). Brain-derived neurotrophic factor triggers transcription-dependent, late phase long-term potentiation in vivo. *Journal of Neuroscience, 22*, 7453–7461.
65. Barco, A., Pittenger, C., & Kandel, E. R. (2003). CREB, memory enhancement and the treatment of memory disorders: Promises, pitfalls and prospects. *Expert Opinion on Therapeutic Targets, 7*, 101–114.
66. Chen, L., Tracy, T., & Nam, C. I. (2007). Dynamics of postsynaptic glutamate receptor targeting. *Current Opinion in Neurobiology, 17*, 53–58
67. Hu, H., Real, E., Takamiya, K., Kang, M. G., Ledoux, J., Huganir, R. L., & Malinow, R. Emotion enhances learning via norepinephrine regulation of AMPA-receptor trafficking. *Cell, 131*, 160–173.
68. Lübke, J., Idoon, F., Mohasel-Roodi, M., Alipour, F., Hami, J., Ehteshampour, A., Mostafaee, H., & Sadeghi, A. (2021). Neurotrophic factors in Alzheimer's disease: Pathogenesis and therapy. *Acta Neurobiologiae Experimentalis, 81*(4), 314–27.
69. Wan, G., Gómez-Casati, M. E., Gigliello, A. R., Liberman, M. C., & Corfas, G. (2014). Neurotrophin-3 regulates ribbon synapse density in the cochlea and induces synapse regeneration after acoustic trauma. *Elife, 3*, e03564.

70. Meng, Ch., He, Zh., & Xing, D. (2013). Low-level laser therapy rescues dendrite atrophy via upregulating BDNF expression: implications for Alzheimer's disease. *Journal of Neuroscience, 33*, 13505–13517.
71. Coelho, F. G., Vital, T. M., Stein, A. M., Arantes, F. J., Rueda, A. V., Teodorov, E., & Santos-Galduroz, R. F. (2013) Acute aerobic exercise increases brain derived neurotrophic factor levels in elderly with Alzheimer's disease. *Journal of Alzheimer's Disease, 39*(2), 401–408.
72. Ciaramella, A., Salani, F., Bizzoni, F., Orfei, M. D., Langella, R., Angelucci, F., Spalletta, G., Taddei, A. R., Caltagirone, C., & Bossù, P. (2013). The stimulation of dendritic cells by amyloid beta 1–42 reduces BDNF production in Alzheimer's disease patients. *Brain, Behavior, and Immunity, 32*, 29–32.
73. Moghbelinejad, S., Nassiri-Asl, M., Farivar, T. N., Abbasi, E., Sheikhi, M., Taghiloo, M., Farsad, F., Samimi, A., & Hajiali, F. (2014). Rutin activates the MAPK pathway and BDNF gene expression on beta-amyloid induced neurotoxicity in rats. *Toxicology Letters, 224*(1), 108–113.
74. Liu, R., Zhang, T. T., Zhou, D., Bai, X. Y., Zhou, W. L., Huang, C., Song, J. K., Meng, F. R., Wu, C. X., Li, L., & Du, G. H. (2013). Quercetin protects against the Abeta (25–35)-induced amnesic injury: involvement of inactivation of rage-mediated pathway and conservation of the NVU. *Neuropharmacology, 67*, 419–431.
75. Hoppe, J. B., Coradini, K., Frozza, R. L., Oliveira, C. M., Meneghetti, A. B., Bernardi, A., Pires, E. S., Beck, R. C., & Salbego, C. G. (2013). Free and nanoencapsulated curcumin suppress beta-amyloid-induced cognitive impairments in rats: Involvement of BDNF and Akt/GSK-3beta signaling pathway. *Neurobiology of Learning and Memory, 106*, 134–144.
76. Jimenez, S., Torres, M., Vizuete, M., Sanchez-Varo, R., Sanchez-Mejias, E., Trujillo-Estrada, L., Carmona-Cuenca, I., Caballero, C., Ruano, D., & Gutierrez, A. (2011). Age-dependent accumulation of soluble amyloid β (Aβ) oligomers reverses the neuroprotective effect of soluble amyloid precursor protein-α (sAPPα) by modulating phosphatidylinositol 3-kinase (PI3K)/Akt-GSK-3β pathway in Alzheimer mouse model. *Journal of Biological Chemistry, 286*, 18414–18425.
77. Ye, X., Tai, W., & Zhang, D. (2012). The early events of Alzheimer's disease pathology: From mitochondrial dysfunction to BDNF axonal transport deficits. *Neurobiology of Aging, 33*(6), 1122.
78. Rohe, M., Synowitz, M., Glass, R., Paul, S. M., Nykjaer, A., & Willnow, T. E. (2009). Brain-derived neurotrophic factor reduces amyloidogenic processing through control of SORLA gene expression. *Journal of Neuroscience: Official Journal of Society Neuroscience, 29*(49), 15472–15478.
79. Shankar, G. M., & Walsh, D. M. (2009). Alzheimer's disease: Synaptic dysfunction and Abeta. *Molecular Neurodegeneration, 4*, 48.
80. Selkoe, D. J. (2002). Alzheimer's disease is a synaptic failure. *Science, 298*, 789–791.
81. Hiester, B. G., Galati, D. F., Salinas, P. C., & Jones, K. R. (2013). Neurotrophin and Wnt signaling cooperatively regulate dendritic spine formation. *Molecular and Cellular Neuroscience, 56*, 115–127.
82. Lee, P. R., Cohen, J. E., Becker, K. G., & Fields, R. D. (2005). Gene expression in the conversion of early-phase to late-phase long-term potentiation. *Annals of the New York Academy of Sciences, 1048*, 259–271.

83. Gomez-Palacio-Schjetnan, A., & Escobar, M. L. (2013). Neurotrophins and synaptic plasticity. *Current Topics in Behavioral Neurosciences, 15*, 117–136.
84. Ninan, I., Bath, K. G., Dagar, K., Perez-Castro, R., Plummer, M. R., Lee, F. S., & Chao, M. V. (2010). The BDNF Val66Met polymorphism impairs NMDA receptor-dependent synaptic plasticity in the hippocampus. *Journal of Neuroscience: Official Journal of Society Neuroscience, 30*(26), 8866–8870.
85. Yang, X., Liu, P., Sun, J., Wang, G., Zeng, F., Yuan, K., Liu, J., Dong, M., von, Deneen, K. M., Qin, W., & Tian, J. (2012). Impact of brain-derived neurotrophic factor Val66Met polymorphism on cortical thickness and voxel-based morphometry in healthy Chinese young adults. *PLoS One, 7*(6), e37777.
86. Harris, S. E., Fox, H., Wright, A. F., Hayward, C., Starr, J. M., Whalley, L. J., & Deary, I. J. (2006). The brain-derived neurotrophic factor Val66Met polymorphism is associated with age-related change in reasoning skills. *Molecular Psychiatry, 11*(5), 505–513.
87. Szeszko, P. R., Lipsky, R., Mentschel, C., Robinson, D., Gunduz-Bruce, H., Sevy, S., Ashtari, M., Napolitano, B., Bilder, R. M., Kane, J. M., & Goldman, D. (2005). Brain-derived neurotrophic factor val66met polymorphism and volume of the hippocampal formation. *Molecular Psychiatry, 10*(7), 631–636.
88. Kunugi, H., Ueki, A., Otsuka, M., Isse, K., Hirasawa, H., Kato, N., Nabika, T., Kobayashi, S., & Nanko, S. (2001). A novel polymorphism of the brain-derived neurotrophic factor (BDNF) gene associated with late-onset Alzheimer's disease. *Molecular Psychiatry, 6*(1), 83–86.
89. Fukumoto, N., Fujii, T., Combarros, O., Kamboh, M. I., Tsai, S. J., Matsushita, S., Nacmias, B., Comings, D. E., Arboleda, H., Ingelsson, M., Hyman, B. T., Akatsu, H., Grupe, A., Nishimura, A. L., Zatz, M., Mattila, K. M., Rinne, J., Goto, Y., Asada, T., Nakamura, S., & Kunugi, H. (2010). Sexually dimorphic effect of theVal66Met polymorphism of BDNF on susceptibility to Alzheimer's disease: New data and meta-analysis. *American Journal of Medical Genetics Part B: Neuropsychiatric Genetics, 153*, 235–242.
90. Lin, Y., Cheng, S., Xie, Z., & Zhang, D. (2014). Association of rs6265 and rs2030324 polymorphisms in brain-derived neurotrophic factor gene with Alzheimer's disease: A meta-analysis. *PLoS One, 9*(4), e94961.
91. Chen, J., Liang, X., Li, B., Jiang, X., & Xu, Z. (2013). Gender-related association of brain-derived neurotrophic factor gene 196A/G polymorphism with Alzheimer's disease—A meta-analysis including 6854 cases and 6868 controls. *International Journal of Neurosciences, 124*(10), 724–733.
92. Hsiao, Y. H., Hung, H. C., Chen, S. H., & Gean, P. W. (2014). Social interaction rescues memory deficit in an animal model of Alzheimer's disease by increasing BDNF-dependent hippocampal neurogenesis. *Journal of Neuroscience, 34*(49), 16207–16219.
93. Fumagalli, F., Racagni, G., & Riva, M. A. (2006). The expanding role of BDNF: A therapeutic target for Alzheimer's disease? *Pharmacogenomics Journal, 6*(1), 8–15.
94. Seo, H., Isacson, O. Abnormal APP, cholinergic and cognitive function in Ts65Dn Down's model mice. *Experimental Neurology, 193*, 469–480.
95. Marvanova, M., Lakso, M., Pirhonen, J., Nawa, H., Wong, G., & Castren, E. (2001). The neuroprotective agent memantine induces brain-derived neurotrophic factor

and trkB receptor expression in rat brain. *Molecular and Cellular Neuroscience, 18*, 247–258.
96. Gibbs, R. B. (1998). Levels of trkA and BDNF mRNA, but not NGF mRNA, fluctuate across the estrous cycle and increase in response to acute hormone replacement. *Brain Research, 23*, 787–268.
97. Nibuya, M., Nestler, E. J., & Duman, R. S. (1996). Chronic antidepressant administration increases the expression of cAMP response element binding protein (CREB) in rat hippocampus. *Journal of Neuroscience, 16*, 2365–2372.
98. Butovsky, E., Juknat, A., Goncharov, I., Elbaz, J., Eilam, R., & Zangen, A. (2005). In vivo up-regulation of brain-derived neurotrophic factor in specific brain areas by chronic exposure to Delta-tetrahydrocannabinol. *Journal of Neurochemistry, 93*, 802–811.
99. Jezierski, M. K., & Sohrabji, F. (2000). Region- and peptide-specific regulation of the neurotrophins by estrogen. *Brain Research: Molecular Brain Research, 85*, 77–84.
100. Leyhe, T., Eschweiler, G. W., Stransky, E., Gasser, T., Annas, P., & Basun, H. (2009). Increase of BDNF serum concentration in lithium treated patients with early Alzheimer's disease. *Journal of Alzheimer's Disease, 16*(3), 649–56.
101. Moon, J. J., Suh, H., Bershteyn, A., Stephan, M. T., Liu, H., Huang, B., Sohail, M., Luo, S., Ho Um, S., Khant, H., & Goodwin, J. T. (2011). Interbilayer crosslinked multilamellar vesicles as synthetic vaccines for potent humoral and cellular immune responses. *Nature Materials, 10*(3), 243–251.
102. Lopes, C. D. F., Gonçalves, N. P., Gomes, C. P., Saraiva, M. J., & Pêgo, A. P. (2017). BDNF gene delivery mediated by neuron-targeted nanoparticles is neuroprotective in peripheral nerve injury. *Biomaterials, 121*, 83–96.
103. Kells, A. P., Fong, D. M., Dragunow, M., During, M. J., Young, D., & Connor, B. (2004). AAV mediated gene delivery of BDNF or GDNF is neuroprotective in a model of Huntington disease. *Molecular Therapy, 9*(5), 682–688.
104. Nagahara, A. H., Mateling, M., Kovacs, I., Wang, L., Eggert, S., Rockenstein, E., Koo, E. H., Masliah, E., & Tuszynski, M. H. (2013). Early BDNF treatment ameliorates cell loss in the entorhinal cortex of APP transgenic mice. *Journal of Neuroscience, 33*, 15596–602.
105. Jiao, S. S., Shen, L. L., Zhu, C., Bu, X. L., Liu, Y. H., Liu, C. H., Yao, X. Q., Zhang, L. L., Zhou, H. D., Walker, D. G., & Tan, J. (2016). Brain-derived neurotrophic factor protects against tau-related neurodegeneration of Alzheimer's disease. *Translational Psychiatry, 6*(10), e907.
106. Nagahara, A. H., Wilson, B. R., Ivasyk, I., Kovacs, I., Rawalji, S., Bringas, J. R., Pivirotto, P. J., Sebastian, W. S., Samaranch, L., Bankiewicz, K. S., & Tuszynski, M. H. (2018). MR-guided delivery of AAV2–BDNF into the entorhinal cortex of nonhuman primates. *Gene Therapy, 25*(2), 104–114.
107. Fiandaca, M. S., Varenika, V., Eberling, J., McKnight, T., Bringas, J., Pivirotto, P., Beyer, J., Hadaczek, P., Bowers, W., Park, J., & Federoff, H. (2009). Real-time MR imaging of adeno-associated viral vector delivery to the primate brain. *Neuroimage, 47*, T27–T35.
108. Shiozawa, A. L., Igarashi, T., Kobayashi, M., Nakamoto, K., Kameya, S., Fujishita, S., Takahashi, H., & Okada, T. (2020). Tyrosine triple mutated AAV2-BDNF gene

therapy in an inner retinal injury model induced by intravitreal injection of N-methyl-Daspartate (NMDA). *Molecular Vision, 26*, 409–422.

109. Gunther, M. I., Weidner, N., Muller, R., & Blesch, A. (2015). Cell-seeded alginate hydrogel scaffolds promote directed linear axonal regeneration in the injured rat spinal cord. *Acta Biomaterial, 27*, 140–150.
110. Gao, M., Lu, P., Lynam, D., Bednark, B., Campana, W. M., Sakamoto, J., & Tuszynski, M. (2016). BDNF gene delivery within and beyond templated agarose multi-channel guidance scaffolds enhances peripheral nerve regeneration. *Journal of Neural Engineering, 13*(6), 066011.
111. Makar, T. K., Bever, C. T., Singh, I. S., Royal, W., Sahu, S. N., Sura, T. P., Sultana, S., Sura, K. T., Patel, N., Dhib-Jalbut, S., & Trisler, D. (2009). Brainderived neurotrophic factor gene delivery in an animal model of multiple sclerosis using bone marrow stem cells as a vehicle. *Journal of Neuroimmunology, 210*(1), 40–51.
112. Li, L. M., Huang, L. L., Jiang, X. C., Chen, J. C., OuYang, H. W., & Gao, J. Q. (2018). Transplantation of BDNF gene recombinant mesenchymal stem cells and adhesive peptide-modified hydrogel scaffold for spinal cord repair. *Current Gene Therapy, 18*(1), 29–39.
113. Nagahara, A. H., Merrill, D. A., Coppola, G., Tsukada, S., Schroeder, B. E., Shaked, G. M., Wang, L., Blesch, A., Kim, A., Conner, J. M., & Rockenstein, E. (2009). Neuroprotective effects of brain-derived neurotrophic factor in rodent and primate models of Alzheimer's disease. *Nature Medicine, 15*(3), 331–337.
114. Paraskevopoulou, F., Herman, M. A., & Rosenmund, C. (2019). Glutamatergic innervation onto striatal neurons potentiates GABAergic synaptic output. *Journal of Neuroscience, 39*, 4448-4460.
115. Li, Y., Xiang, L., Wang, C., Song, Y., Miao, J., & Miao, M. (2021). Protection against acute cerebral ischemia/reperfusion injury by leonuriherba total alkali via modulation of BDNF-TrKB-PI3K/Akt signaling pathway in rats. *Biomedicine & Pharmacotherapy, 133*, 111021.
116. Chimenti, M. S., Sunzini, F., Fiorucci, L., Botti, E., Fonti, G. L., Conigliaro, P., Triggianese, P., Costa, L., Caso, F., Giunta, A., & Esposito, M. (2018). Potential role of cytochrome c and tryptase in psoriasis and psoriatic arthritis pathogenesis: Focus on resistance to apoptosis and oxidative stress. *Frontiers in Immunology, 9*, 2363.
117. Zhou, L. J., Mo, Y. B., Bu, X., Wang, J. J., Bai, J., Zhang, J. W., Cheng, A. B., Ma, J. H., Wang, Y. W., & Xie, Y. X. (2018). Erinacine facilitates the opening of the mitochondrial permeability transition pore through the inhibition of the PI3K/Akt/GSK-3β signaling pathway in human hepatocellular carcinoma. *Cellular Physiology and Biochemistry, 50*, 851-867.
118. Jiang, T., Wang, X. Q., Ding, C., & Du, X. L. (2017). Genistein attenuates isoflurane-induced neurotoxicity and improves impaired spatial learning and memory by regulating cAMP/CREB and BDNF-TrkB-PI3K/Akt signaling. *Korean Journal of Physiology & Pharmacology, 21*, 579-589.
119. Gao, J., & Liping, L. (2023). Enhancement of neural regeneration as a therapeutic strategy for Alzheimer's disease. *Experimental and Therapeutic Medicine, 26*(3), 1–14.
120. Clarke, G., Grenham, S., Scully, P., Fitzgerald, P., Moloney, R. D., Shanahan, F., Dinan, T. G., & Cryan, J. (2013). The microbiome-gut-brain axis during early life

regulates the hippocampal serotonergic system in a sex-dependent manner. *Molecular Psychiatry, 18*(6), 666–73.
121. Jiang, Y., Liu, Y., Gao, M., Xue, M., Wang, Z., & Liang, H. (2020). Nicotinamide riboside alleviates alcohol-induced depression-like behaviours in C57BL/6J mice by altering the intestinal microbiota associated with microglial activation and BDNF expression. *Food Function, 11*(1), 378–391.
122. Distrutti, E., O'Reilly, J. A., McDonald, C., Cipriani, S., Renga, B., Lynch, M. A., & Fiorucci, S. (2014). Modulation of intestinal microbiota by the probiotic VSL# 3 resets brain gene expression and ameliorates the age-related deficit in LTP. *PLoS ONE, 9*(9), e106503.
123. Maqsood, R., & Stone, T. W. (2016). The gut-brain axis, BDNF, NMDA and CNS disorders *Neurochemical Research, 41*(11), 2819–2835.
124. Sayin, S. I., Wahlström, A., Felin, J., Jäntti, S., Marschall, H. U., Bamberg, K., Angelin, B., Hyötyläinen, T., Orešič, M., & Bäckhed, F. (2013). Gut microbiota regulates bile acid metabolism by reducing the levels of tauro-beta-muricholic acid, a naturally occurring FXR antagonist. *Cell Metabolism, 17*(2), 225–235.
125. Poduslo, J. F., & Curran, G. L. (1996). Permeability at the blood-brain and blood-nerve barriers of the neurotrophic factors: NGF, CNTF, NT-3, BDNF. *Molecular Brain Research, 36*(2), 280–286.
126. Scharfman, H., Goodman, J., Macleod, A., Phani, S., Antonelli, C., & Croll, S. (2005). Increased neurogenesis and the ectopic granule cells after intrahippocampal BDNF infusion in adult rats. *Experimental Neurology, 192*(2), 348–356.
127. Gao, L., Zhang, Y., Sterling, K., & Song, W. (2022). Brain-derived neurotrophic factor in Alzheimer's disease and its pharmaceutical potential. *Translational Neurodegeneration, 11*(1), 1–34.
128. Messaoudi, E., Bårdsen, K., Srebro, B., & Bramham, C. R. (1998). Acute intrahippocampal infusion of BDNF induces lasting potentiation of synaptic transmission in the rat dentate gyrus. *Journal of Neurophysiology, 79*(1), 496–499.
129. Rosas-Vidal, L. E., Do-Monte, F. H., Sotres-Bayon, F., & Quirk, G. J. (2014). Hippocampal–prefrontal BDNF and memory for fear extinction. *Neuropsychopharmacology, 39*(9), 2161–2169.
130. Graybeal, C., Feyder, M., Schulman, E., Saksida, L. M., Bussey, T. J., Brigman, J. L., & Holmes, A. (2011). Paradoxical reversal learning enhancement by stress or prefrontal cortical damage: rescue with BDNF. *Nature Neuroscience, 14*(12), 1507–1509.
131. Narita, M., Aoki, K., Takagi, M., Yajima, Y., & Suzuki, T. (2003). Implication of brain-derived neurotrophic factor in the release of dopamine and dopamine-related behaviors induced by methamphetamine. *Neuroscience, 119*(3), 767–775.
132. Jiang, Y., Wei, N., Lu, T., Zhu, J., Xu, G., & Liu, X. (2011). Intranasal brain-derived neurotrophic factor protects brain from ischemic insult via modulating local inflammation in rats. *Neuroscience, 172*, 398–405.
133. Agterberg, M. J., Versnel, H., van Dijk, L. M., de Groot, J. C., & Klis, S. F. (2009). Enhanced survival of spiral ganglion cells after cessation of treatment with brain-derived neurotrophic factor in deafened guinea pigs. *Journal of the Association for Research in Otolaryngology, 10*, 355–367.

CHAPTER 9

Behavioral Intervention and Cognitive Training in Disease Management in Alzheimer's Disease

SHAHBAZ KHAN[1,*], DIKSHA DIWAKAR[2], RAJIB SINGH[3], ARSHI[4], BRIJ RAJ SINGH[5], and ALKA LOHANI[6]

[1]*Pharmacy Academy, Faculty of Pharmacy, IFTM University, Moradabad, Uttar Pradesh, India*

[2]*Sahu Onkar Saran School of Pharmacy, Faculty of Pharmacy, IFTM University, Moradabad, Uttar Pradesh, India*

[3]*Shakti College of Pharmacy, Balrampur, Uttar Pradesh, India*

[4]*S.D. College of Pharmacy, Patel Nagar, Muzaffarnagar, Uttar Pradesh, India*

[5]*Malti Memorial Trust CSM Group of Institutions, Faculty of B. Pharmacy, Prayagraj, Uttar Pradesh, India*

[6]*Amity Institute of Pharmacy, Amity University, Noida, Uttar Pradesh, India*

*Corresponding author

ABSTRACT

One of the most common chronic illnesses affecting the elder personality is Alzheimer's disease (AD). However, the efficacy of licensed antidementia treatments is restricted; they only offer a moderate level of clinical symptom alleviation. One noninvasive therapy that may help in AD treatment and prevention is cognitive intervention. Research indicates that people with AD may benefit therapeutically from specially created cognitive interventions,

Cellular and Molecular Mechanisms in Alzheimer's Disease Therapeutics, Volume 2: Molecular Mechanisms. Prashant Tiwari, Sunil Kumar Kadiri, and Kunjbihari Sulakhiya (Eds)

DOI: 10.1201/9781779641816-9

which are linked to significant biological alterations in the brain. Furthermore, data suggests that a mix of nonpharmacological and pharmaceutical therapies may be able to reduce clinical symptoms more effectively than either one used alone. We now know more about the underlying neurobiological mechanisms of aging and neurodegeneration, thanks to studies on functional and structural magnetic resonance imaging (MRI), but there is still much to learn about using neuroimaging to study how cognitive intervention affects the brain. An overview of cognitive intervention's use to both healthy older people and AD patients is given in this review, along with a summary of recent research showing the strategy's efficacy. We address the possible consequences of this therapy on disease modification and offer suggestions for future study on the application of cognitive therapies in AD.

9.1 INTRODUCTION

Alzheimer's disease (AD) is a long-term, severely debilitating neurological condition that is characterized by a decline in memory and cognition, a growing impairment in daily living activities, a variety of neuropsychiatric symptoms and behavioral abnormalities. There were 26.6 million cases of AD globally in 2006; by 2050, this figure is expected to triple. With 15 million cases worldwide, mild cases account for the majority of instances of AD. It is a progressive neurodegenerative disorder characterized by cognitive decline, memory loss, and changes in behavior. It affects millions of individuals worldwide, posing significant challenges for patients, caregivers, and healthcare systems [1]. While there is currently no cure for Alzheimer's, various behavioral interventions have shown promise in managing symptoms, improving quality of life, and enhancing the well-being of affected individuals. This chapter provides an overview of behavioral interventions commonly employed in the care of individuals with AD. Drawing upon research and clinical evidence, it explores the effectiveness of these interventions and their impact on cognitive function, behavior management, and overall quality of life. By understanding and implementing these strategies, caregivers and healthcare professionals can better support individuals with Alzheimer's and promote their independence and dignity. With 15 million cases worldwide, mild cases account for the majority of instances of AD but in recent times, there has been a growing investigation into the efficacy of behavioral therapies, leading to a shift in viewpoint. Understanding that memory is not a unitary phenomenon, realizing that some types of memory are comparatively spared in AD, realizing that conceptual knowledge is largely intact, and gaining a better understanding of how people learn and remember are some of the factors that are leading to this shifting perspective. Individuals with AD can now learn new material and continue to function better throughout the course of the disease thanks to intervention strategies that leverage their preserved memory systems and learning and memory principles. Nostalgia treatment and perception orientation are the cognitive-behavioral therapies that have been studied the most for treating AD patients. These interventions aim to preserve or restore temporal and spatial orientation and chronologic memory through frequent presentation of orientation and personal information. While these interventions have shown a slight improvement in orientation and personal knowledge, the effects are transient [2].

The American Psychiatric Association came to the conclusion in 1997 that "the slight improvements observed with some of these treatments have not lasted beyond the treatment sessions and thus do not appear to warrant the risk of adverse effects" after reviewing the literature on the benefits of cognitive interventions. Numerous protective factors against AD have been found, including lifestyle characteristics pertaining to social, physical, and cognitive activities. There are also substantial epidemiological connections between engaging in mentally challenging hobbies and good cognitive health [3].

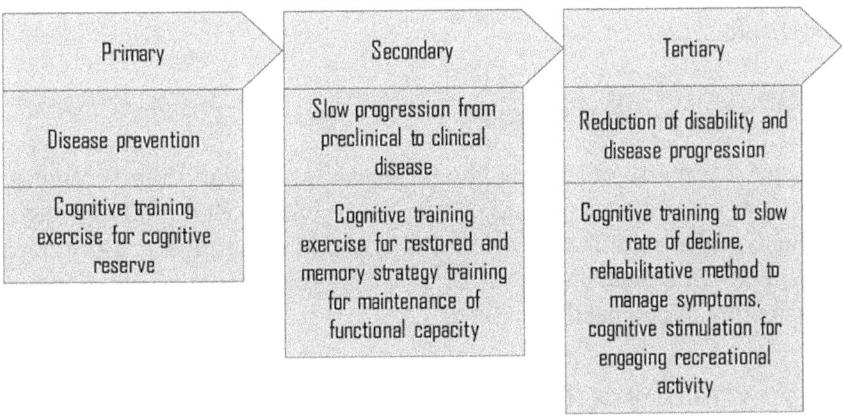

FIGURE 9.1 Stages of dementia prevention integrated with cognitive training interventions.

9.2 ETIOLOGY OF ALZHEIMER'S DISEASE

AD, named after the German physician Alois Alzheimer who first described it in 1906, is a neurodegenerative disorder characterized by progressive cognitive decline, memory loss, and behavioral changes. While the exact cause of AD is not fully understood, several factors are believed to contribute to its etiology [4].

9.2.1 GENETICS

There are both early-onset familial forms and late-onset sporadic forms of AD. Mutations in certain genes, such as the *amyloid precursor protein* (*APP*), *presenilin* 1 (*PSEN*1), and *presenilin* 2 (*PSEN*2) genes, have been identified in familial cases. However, these mutations are relatively rare and account for only a small percentage of Alzheimer's cases. The *apolipoprotein E* (*APOE*) gene, particularly the *APOE ε4* allele, is a significant genetic risk factor for late-onset AD.

9.2.2 AMYLOID PLAQUES

One of the hallmarks of AD is the accumulation of beta-amyloid plaques in the brain. These abnormal protein fragments clump together and interfere with neuronal function, leading to cell death and cognitive decline. The exact role of amyloid plaques in the development of AD is still under investigation, but it is believed to be a key contributor to the neurodegenerative process.

9.2.3 TAU PROTEIN TANGLES

Another hallmark of AD is the presence of neurofibrillary tangles, which are twisted fibers of tau protein found within neurons. Tau protein normally helps stabilize microtubules in neurons, but in AD, it becomes hyper phosphorylated and forms tangles, disrupting neuronal function and contributing to cell death.

9.2.4 NEUROINFLAMMATION

Chronic inflammation in the brain, mediated by microglia and astrocytes, may play a role in the progression of AD. Inflammation can exacerbate neuronal damage and contribute to the formation of amyloid plaques and tau tangles.

9.2.5 OXIDATIVE STRESS

Imbalance between the production of reactive oxygen species (ROS) and the body's ability to neutralize them can lead to oxidative stress, damaging neurons and contributing to neurodegeneration.

9.2.6 MITOCHONDRIAL DYSFUNCTION

Mitochondria, the cellular organelles responsible for energy production, play a crucial role in neuronal function. Dysfunction of mitochondria can lead to energy deficits in neurons and contribute to their degeneration.

9.2.7 ENVIRONMENTAL FACTORS

Some environmental factors, such as head trauma, exposure to certain toxins, and cardiovascular risk factors (e.g., hypertension, diabetes, obesity), have been implicated in the development of AD. These factors may interact with genetic predispositions and other biological mechanisms to increase the risk of developing the disease.

9.2.8 AGE

Advancing age is the greatest known risk factor for AD. While Alzheimer's can occur in younger individuals, the risk increases significantly with age, especially after 65.

9.3 BEHAVIORAL INTERVENTION IN ALZHEIMER'S DISEASE

Behavioral interventions in AD encompass a range of strategies aimed at addressing cognitive, emotional, and behavioral symptoms. These interventions are tailored to the individual needs and preferences of patients and may involve various approaches, including structured routines, cognitive stimulation, environmental modifications, and caregiver support. Behavioral interventions are integral to the care and management of AD, offering personalized approaches to address cognitive, emotional, and behavioral symptoms. By implementing these interventions, caregivers and healthcare professionals can improve the quality of life for individuals with Alzheimer's and enhance their overall well-being. Below are key behavioral interventions supported by research and clinical practice:

9.3.1 STRUCTURED ROUTINE

Establishing a consistent daily routine can provide stability and predictability for individuals with Alzheimer's. Regular activities such as meals, exercise, and social interactions help maintain cognitive function and reduce agitation [5].

9.3.2 COGNITIVE STIMULATION

Engaging in mentally stimulating activities such as puzzles, games, and reminiscence therapy can help preserve cognitive function and memory skills. These activities also promote social interaction and emotional well-being [6].

9.3.3 ENVIRONMENTAL MODIFICATIONS

Simplifying the environment by reducing clutter, organizing belongings, and minimizing distractions can enhance the safety and comfort of individuals with Alzheimer's. Environmental modifications help reduce confusion and anxiety [7].

9.3.4 VALIDATION THERAPY

Validation therapy involves acknowledging and validating the emotions and experiences of individuals with Alzheimer's, even if they may not align with reality. Empathetic listening and reassurance help reduce distress and improve communication [8].

9.3.5 BEHAVIORAL MANAGEMENT TECHNIQUES

Strategies such as redirection, distraction, and positive reinforcement are used to manage challenging behaviors like aggression, agitation, and wandering. Identifying triggers and addressing underlying needs is crucial for effective management [9].

9.3.6 PHYSICAL EXERCISE

Regular physical activity, tailored to individual abilities, promotes mood, sleep quality, and overall well-being in individuals with Alzheimer's. Activities like walking, dancing, and gentle exercises have beneficial effects [10].

9.3.7 MUSIC AND ART THERAPY

Music therapy and art activities offer opportunities for emotional expression, reminiscence, and enjoyment. Listening to familiar music and engaging in creative activities can reduce stress and enhance quality of life [11].

9.3.8 SOCIAL ENGAGEMENT

Maintaining social connections through group outings, volunteer work, and support groups is essential for individuals with Alzheimer's. Social engagement fosters a sense of belonging and support [12].

9.3.9 CAREGIVER SUPPORT AND EDUCATION

Providing education and support for caregivers is vital for effective care management. Access to caregiver support groups, respite care services, and educational programs improves caregiver well-being and enhances patient care [13].

9.3.10 MEDICATION MANAGEMENT

In some cases, medications may be prescribed to manage symptoms such as agitation, anxiety, and depression in individuals with Alzheimer's. Regular review and adjustment of medication regimens are necessary to optimize treatment outcomes [14].

9.4 COGNITIVE TRAINING IN ALZHEIMER'S DISEASE

Cognitive training interventions are increasingly recognized as valuable components in the management of AD, aiming to preserve cognitive function and enhance quality of life for affected individuals. These interventions encompass a variety of approaches tailored to target different cognitive domains affected by the disease. Memory training exercises, problem-solving activities, attentional drills, language tasks, and executive function training are among the common strategies employed in cognitive training programs for AD patients [15]. Additionally, advancements in technology have facilitated the development of computerized cognitive training programs and smartphone applications, providing accessible and engaging platforms for cognitive stimulation [16]. Group-based interventions also play a significant role, offering social interaction and peer support alongside cognitive activities. Although the effectiveness of cognitive training in AD varies depending on factors such as disease severity and individual characteristics, evidence suggests potential benefits in preserving cognitive function and improving aspects of daily living [17]. Moreover, combining cognitive training with other therapeutic approaches, such as physical exercise and pharmacological interventions, may yield synergistic effects in promoting cognitive health [18]. As part of a comprehensive treatment plan, cognitive training interventions complement pharmacotherapy, lifestyle modifications, and caregiver support, contributing to a holistic approach to AD management. Cognitive training has emerged as a promising intervention in AD management, aiming to enhance cognitive function, slow down cognitive decline, and improve overall quality of life for individuals affected by the disease.

9.4.1 TYPES OF COGNITIVE TRAINING

Cognitive training interventions vary widely and can include activities such as memory exercises, problem-solving tasks, attention training, and language exercises. These interventions are typically tailored to the individual's cognitive strengths and weaknesses [19].

9.4.2 BENEFITS OF COGNITIVE TRAINING

Cognitive Function Improvement: Some studies suggest that cognitive training can lead to improvements in specific cognitive domains, such as memory, attention, and executive function.

Functional Independence: By improving cognitive function, cognitive training may help individuals with AD maintain their independence in daily activities for a longer period.

Quality of Life: Engaging in cognitive training programs can enhance overall quality of life by providing mental stimulation and a sense of accomplishment.

Neuroplasticity and Cognitive Training: Cognitive training takes advantage of the brain's ability to reorganize and form new neural connections, known as neuroplasticity. By repeatedly engaging in cognitive tasks, individuals with AD may strengthen existing neural pathways or establish new ones, potentially slowing cognitive decline.

9.4.3 EVIDENCE SUPPORTING COGNITIVE TRAINING

Meta-Analyses and Systematic Reviews: Several meta-analyses and systematic reviews have reported positive effects of cognitive training on cognitive function and everyday functioning in individuals with AD. For example, a meta-analysis by Bahar-Fuchs et al. (2019) found small to moderate effects of cognitive training on various cognitive domains [20].

Randomized Controlled Trials (RCTs): Numerous RCTs have investigated the efficacy of cognitive training interventions in AD. While results vary across studies, many have demonstrated beneficial effects on cognitive function and related outcomes.

Longitudinal Studies: Longitudinal studies provide valuable insights into the long-term effects of cognitive training. Some longitudinal studies suggest that the benefits of cognitive training may persist over time, even in individuals with progressive neurodegenerative diseases like AD.

9.4.4 CONSIDERATIONS AND CHALLENGES

Individualized Approach: Cognitive training interventions should be tailored to the individual's cognitive abilities, preferences, and goals.

Sustainability: Long-term engagement in cognitive training may be challenging for individuals with AD due to factors such as disease progression and cognitive decline.

Comprehensive Approach: Cognitive training is often more effective when combined with other interventions, such as physical exercise, social engagement, and pharmacotherapy.

9.5 FUTURE DIRECTIONS

Ongoing research aims to further refine cognitive training interventions, identify optimal timing and duration of interventions, and explore novel approaches, such as technology-based interventions and multidomain interventions targeting multiple aspects of cognition simultaneously.

9.6 CONCLUSION

Behavioral intervention and cognitive training represent pivotal components in the comprehensive management of AD. These approaches, designed to alleviate symptoms and enhance cognitive function, offer promise in improving the quality of life for individuals afflicted by AD and easing the burden on caregivers. Behavioral interventions, encompassing strategies to manage behavioral and psychological symptoms, aim to create supportive environments and routines tailored to the unique needs of each individual. Meanwhile, cognitive training programs focus on stimulating cognitive abilities through targeted exercises and activities. Despite facing challenges such as implementation consistency and disease progression, numerous studies have demonstrated their efficacy in mitigating symptoms and fostering positive outcomes. However, personalized and early interventions are key, as they enable tailored support and potentially delay cognitive decline. Moving forward, ongoing research is imperative to refine these approaches, optimize delivery methods, and establish long-term effectiveness. By embracing a multifaceted approach and prioritizing individualized care, behavioral intervention and cognitive training can continue to serve as integral pillars in the holistic management of AD, offering hope and support to those affected by this debilitating condition.

KEYWORDS

- **aging**
- **cognition**
- **dementia**
- **management**
- **memory**
- **rehabilitation**

REFERENCES

1. Brookmeyer, R., Johnson, E., Ziegler-Graham, K., & Arrighi, H. M. (2007). Forecasting the global burden of Alzheimer's disease. *Alzheimer's & Dementia, 3*(3), 186–191.

2. Taulbee, L. R., & Folsom, J. C. (1966). Reality orientation for geriatric patients. *Psychiatric Services*, *7*(5), 133–135.
3. Wilson, R. S., Segawa, E., Boyle, P. A., & Bennett, D. A. (2012). Influence of late-life cognitive activity on cognitive health. *Neurology*, *78*(15), 1123–1129.
4. Aalten, P., De Vugt, M. E., Jaspers, N., Jolles, J., & Verhey, F. R. (2005). The course of neuropsychiatric symptoms in dementia. Part I: Findings from the two-year longitudinal Maasbed study. *International Journal of Geriatric Psychiatry*, *20*(6), 523–530.
5. Brodaty, H., Green, A., & Koschera, A. (2003). Meta-analysis of psychosocial interventions for caregivers of people with dementia. *Journal of the American Geriatrics Society*, *51*(5), 657–664.
6. Gitlin, L. N., Corcoran, M., Winter, L., Boyce, A., & Hauck, W. W. (2001). A randomized, controlled trial of a home environmental intervention: Effect on efficacy and upset in caregivers and on daily function of persons with dementia. *The Gerontologist*, *41*(1), 4–14.
7. Feil, N. (1993). *The validation breakthrough: Simple techniques for communicating with people with "Alzheimer's-type dementia"*. Health Professions Press.
8. Cohen-Mansfield. J., Thein, K., Marx, M. S., & Dakheel-Ali, M. (2012). What are the barriers to performing nonpharmacological interventions for behavioral symptoms in the nursing home? *Journal of the American Medical Directors Association*, *13*(4), 400–405.
9. Forbes, D., Forbes, S. C., Blake, C. M., Thiessen, E. J., & Forbes, S. (2015). Exercise programs for people with dementia. *Cochrane Database of Systematic Reviews*, (4).
10. Sarkamo, T., Tervaniemi, M., Laitinen, S., Forsblom, A., Soinila, S., Mikkonen, M., Autti, T., Silvennoinen, H. M., Erkkilä, J., Laine, M., & Peretz, I. (2008). Music listening enhances cognitive recovery and mood after middle cerebral artery stroke. *Brain*, *131*(3), 866–876.
11. Fratiglioni, L., Paillard-Borg, S., & Winblad, B. (2004). An active and socially integrated lifestyle in late life might protect against dementia. *The Lancet Neurology*, *3*(6), 343–353.
12. Olazarán, J., Reisberg, B., Clare, L., Cruz, I., Peña-Casanova, J., Del Ser, T., Woods, B., Beck, C., Auer, S., Lai, C., & Spector, A. (2010). Nonpharmacological therapies in Alzheimer's disease: a systematic review of efficacy. *Dementia and Geriatric Cognitive Disorders*, *30*(2), 161–178.
13. Sink, K. M., Holden, K. F., & Yaffe, K. (2005). Pharmacological treatment of neuropsychiatric symptoms of dementia: A review of the evidence. *JAMA*, *293*(5), 596–608.
14. Clare, L., Woods, R. T., Moniz Cook, E. D., Orrell, M., & Spector, A. (2003). Cognitive rehabilitation and cognitive training for early-stage Alzheimer's disease and vascular dementia. *Cochrane Database of Systematic Reviews*, *4*, CD003260.
15. Bahar-Fuchs, A., Martyr, A., Goh, A. M., Sabates, J., & Clare, L. (2019) Cognitive training for people with mild to moderate dementia. *Cochrane Database of Systematic Reviews*, (3).
16. Gates, N. J., Sachdev, P. S., Fiatarone Singh, M. A., & Valenzuela, M. (2011). Cognitive and memory training in adults at risk of dementia: A systematic review. *BMC Geriatrics*, *11*, 1–4.

17. Ngandu, T., Lehtisalo, J., Solomon, A., Levälahti, E., Ahtiluoto, S., Antikainen, R., Bäckman L, Hanninen, T., Jula, A., Laatikainen, T., & Lindström, J. (2015). A 2 year multidomain intervention of diet, exercise, cognitive training, and vascular risk monitoring versus control to prevent cognitive decline in at-risk elderly people (FINGER): a randomised controlled trial. *The Lancet, 385*(9984), 2255–2263.
18. Bahar-Fuchs, A., Clare, L., & Woods, B. (2013). Cognitive training and cognitive rehabilitation for mild to moderate Alzheimer's disease and vascular dementia. *Cochrane Database of Systematic Reviews*, (6).
19. Gauthier, S., Reisberg, B., Zaudig, M., Petersen, R. C., Ritchie, K., Broich, K., Belleville, S., Brodaty, H., Bennett, D., Chertkow, H., & Cummings, J. L. (2006). Mild cognitive impairment. *The Lancet*, *367*(9518), 1262–1270.
20. Scheltens, P., Blennow, K., Breteler, M. M., De Strooper, B., Frisoni, G. B., Salloway, S., & Van der Flier, W. M. (2016). Alzheimer's disease. *The Lancet*, *388*(10043), 505–517.

CHAPTER 10

Therapeutic Potential of NMDA Receptor in the Alzheimer's Disease

VIRENDRA KUSHWAHA, SHRIRAM SISODIA, MOHD. ASAD FAROOQI, and KANTROL KUMAR SAHU*

Institute of Pharmaceutical Research, GLA University, Mathura, Uttar Pradesh, India

*Corresponding author

ABSTRACT

Alzheimer's disease (AD) is identified as the degeneration of brain cells, which primarily causes dementia, marked by a decline in cognitive abilities and the capability to perform routine activities independently. N-methyl-D-aspartate receptors (NMDARs), which are a type of glutamate-gated ion channels, are important for both normal and pathological brain activities. These receptors have a critical function in synaptic transmission as well as synaptic plasticity, which essentially affects learning and memory. NMDARs facilitate a gradual part of excitatory synaptic transmission and are broadly spread throughout the central nervous system, where they control synaptic plasticity. These receptors are thoroughly investigated for their pivotal involvement in the synaptic plasticity which is vital for memory, learning, as well as their contribution to neurotoxicity observed in acute brain injuries and neurodegenerative disorders. In AD, the initial disturbance of synapses is linked with higher levels of oligomeric amyloid-beta peptide. This leads to synaptic depression, which is dependent on NMDA receptors and the removal of spines. While synaptic NMDARs offer neuroprotection, excessive activation of NMDARs outside the synapse results in decline of mitochondrial membrane potential, as well

Cellular and Molecular Mechanisms in Alzheimer's Disease Therapeutics, Volume 2: Molecular Mechanisms. Prashant Tiwari, Sunil Kumar Kadiri, and Kunjbihari Sulakhiya (Eds)

DOI: 10.1201/9781779641816-10

as subsequent cell death. The potential of NMDA receptor antagonists to inhibit glutamate-mediated excitotoxicity has been reported, making them promising therapeutic targets for numerous neurodegenerative disorders, including AD. This chapter discusses about the importance of NMDA receptor and its antagonist for management of AD.

10.1 INTRODUCTION

The main prevalent reason of irreversible Alzheimer's disease (AD) is brought on by a confluence of age, hereditary, and environmental factors [1]. According to estimates, the disease affects over 35 million people globally and by 2050, the number of patients is expected to triple. As a result, AD is a serious public health issue that has drawn the attention of numerous studies. Nevertheless, significant advances in fundamental science have not yet produced medications that are clinically beneficial. As of right now, the two approved medications for AD are memantine, an N-methyl-D-aspartate receptor (NMDAR) antagonist that reduces glutamate excitotoxicity and acetylcholinesterase inhibitors (AChEIs), which are effective in reversing cholinergic deficiency. These are symptomatic treatments that provide patients with very little and transient relief [2].

Over-activation of NMDAR, a kind of glutamate receptor, may result in abnormal handling of synaptic calcium, potentially resulting in synaptic dysfunction [3]. Glutamate, the brain's major neurotransmitter (excitatory), communicates with ionotropic and metabotropic glutamate receptors. Ionotropic glutamate receptors (iGluRs) are classified into three subfamilies: α-amino-3-hydroxy-5-methyl-4-isoxazole propionic acid (AMPA) receptors, kainate receptors, and NMDARs. These receptors facilitate rapid brain transmission at excitatory synapse [4]. Nevertheless, excitotoxicity is the outcome of overstimulating glutamatergic signaling [5]. Various research studies demonstrate that glutamate excitotoxicity takes part in late, gradually progressive neurodegeneration in addition to its acute effects [6, 7]. Since NMDARs showed a significantly more calcium ion permeation than other iGluRs [8, 9], excessive Ca2C entry is the primary mechanism causing the toxicity [10, 11]. The NMDARs have vital roles in both neurotoxicity and nervous system development. Additionally, it is vital for synaptic transmission and synaptic plasticity, processes believed to be fundamental for remembrance and learning. Recently, it's confirmed that a connection has been established between NMDAR activity and synaptic impairment in AD [12, 13].

Persistent loss of synaptic activity and eventual neuronal cell death caused by the aberrant amount of Ca2C signaling are clinically correlated with the development of disordered brain architecture and the progressive impairment in cognition/memory observed in AD patients. Thus, clinical studies of memantine as an NMDAR antagonist for alleviating AD symptoms and providing neuroprotection are justified [9, 14, 15]. Memantine, a noncompetitive NMDA receptor antagonist, is permitted for treatment of temperate as well as severe AD. Despite the fact that it did not alleviate excessive agitation or hippocampus shrinkage or total brain shrinkage, it has been extensively recommended to treat AD symptoms and enhance standard of living [16, 17].

However, the brain regions mostly damaged by AD are largely constituted of NMDARs with GluN2A and GluN2B subunits. Additionally, targeting particular GluN2B subunit antagonists may be an intriguing method to prevent synaptic disturbance in AD. This is particularly convincing because extra synaptic NMDARs carrying GluN2B have been linked to overactivity in AD [18].

10.2 PATHOPHYSIOLOGY OF NMDA RECEPTOR IN AD

AD is marked by the defeat of remembrance and intellectual decline, marking it as a neurodegenerative disorder. Currently, about 50 million individuals worldwide are affected with AD, a number likely to surge to 150 million by 2050 [19].

AD occurs in two familial forms: Early-onset, primarily influenced by genetics and affecting individuals in middle age, and late-onset or sporadic, typically occurring in those over the age of 65. The exact cause of late-onset AD remains elusive, believed to involve a more complex interplay of factors than previously thought of. Alongside genetic predisposition, other factors should include aging, cardiovascular issues, female gender, depression, head injuries, diabetes mellitus, insomnia, and deficiencies in vitamin B [20].

AD is illustrated by two abnormal protein buildups in the brain region: extracellular plaques made of β-amyloid (Aβ) protein and intracellular neurofibrillary tangles [21]. Amyloid-β monomers are generated through the cleavage of amyloid precursor protein (APP) within neural membranes by β-secretase and γ-secretase enzymes [22]. Aβ 1–42 is a poisonous peptide that inhibits long-term hippocampal potentiation (LTP), produces reactive oxygen species (ROS), and causes apoptosis with disruption of synaptic function [23].

Activation of extra synaptic NMDA GluN2B receptors results in LTD, reduction in spine size, and loss of synapses [24], contributing to AD through excitotoxicity caused by glutamate. These receptors, located close to mitochondria, could facilitate the migration of Ca2+ ions toward mitochondrial membranes [25], elevating ROS [26] and nitric oxide (NO) production. Subsequently, these molecules gather around mitochondria, working together with death-associated protein kinase 1 and ultimately leading to death of cells [27]. Notably, Aβ's show effects on neurons and glial cells result in increased extracellular glutamate levels. As the expression of NMDA GluN2A receptors declines, additional insight into molecular mechanisms is presented in this section. Consequently, the glutamate system, especially NMDA GluN2B receptors, plays a considerable role in AD pathology [28].

NMDA receptors play a crucial role as ligand-gated ion channels vital for memory formation. They comprise seven GluN1 subunits, four distinct GluN2 subunits (GluN2A-D), and two GluN3 subunits (GluN3A-B). These subunits assemble in diverse combinations, forming various subtypes with distinct physiological roles [29].

The activation of NMDA receptors is essential for glutamate binding to the GluA2 subunit of AMPA receptors, and depolarization removes the Mg2+ obstruction of NMDA receptors. Postsynaptic NMDA receptors aid in the influx of Ca2+ ions and the transmission of the action potential [30].

Recent research studies showed that the regulation of NMDA GluN2A receptors supports cellular survival and NMDA GluN2B receptors, predominantly situated in extra synaptic locations, have adverse effects [31]. Furthermore, GluN2C and GluN2D subunits have been recognized for their beneficial contributions to transmission of synaptic nerve and working memory [32]. Another category of NMDA receptors functions as metabotropic receptors, comprising eight forms of G-protein coupled receptors known as mGluR1-8, initiate milder synaptic responses [4]. Significantly, NMDA receptors have the capacity to transmit signals metabotropically without requiring calcium entry with the channel [30]. Activation of ionotropic NMDA receptors triggers LTP, while the stimulation of metabotropic receptors initiates long-term depression (LTD), setting off p38 mitogen-activated protein kinases (p38MAPK) signaling cascades and synaptic shortfall like those seen in AD [33].

10.2.1 SIGNALING PATHWAYS LINKED WITH NMDARS IN AD

Numerous researchers have discovered the important cellular communication concerned in the progression of AD. Among that pathway, one of the key players is CREB, a transcription factor that is regulated by signals and has a critical function in maintaining LTP [34]. CREB-mediated gene expression is linked to synaptic plasticity, as well as the action of knowledge and recall within the hippocampus [35]. CREB at serine residue 133 gets phosphorylated and has a substantial part in its transcriptional activity, which is reduced in AD [36]. CREB phosphorylation in neurons is initiated by diverse signaling pathways. One pathway involves elevated intracellular calcium levels, which can arise from the activation of voltage- or ligand-gated channels such as NMDARs. Another pathway involves the starting of receptor tyrosine kinase by growth factors. Calcium influx into the neuron via synaptic NMDARs facilitates CREB phosphorylation and the regulation of brain-derived neurotrophic factor (BDNF). BDNF is pivotal in synaptic plasticity and the establishment of fresh connections between neurons. Conversely, the stimulation of extra synaptic NMDARs has opposite action on the CREB pathway. This leads to the inhibition of CREB activity, suppressing the genes activity connected to plasticity of synaptic and remembrance formation. Amyloid-beta (Aβ), a peptide associated to AD, interferes with CREB phosphorylation by inactivating protein kinase A (PKA). This inhibition of PKA decreases CREB phosphorylation, which in turn impairs the generation of LTP, a cellular activity underlying cognition and remembrance. Overall, these findings highlight the importance of CREB-mediated gene activity and its directive in synaptic plasticity, cognition, and remembrance processes, as well as the role of Aβ in disrupting this pathway in AD [37]. This suggests that NMDAR activation triggers a pathway that leads to the movement of Jacob to the nucleus. Understanding the role of Jacob and its interaction with caldendrin in the nucleus is important for unraveling the mechanisms behind synaptic plasticity and the processes engaged in cognition and remembrance. Researchers studying the Jacob regulatory and its effects on CREB phosphorylation and synaptic contacts can provide valuable insights into these fundamental brain functions [38]. The phosphorylation status of Jacob is an essential determinant of its capacity to enhance cell survival and synaptic plasticity, or to trigger cell death. Following phosphorylation, Jacob is translocated to the nucleus upon synaptic NMDARs activation.

In this state, phosphorylated Jacob is associated with neuroprotection, offering a defense mechanism for the cells. Conversely, in a non-phosphorylated form, Jacob is transported to the nucleus following extra synaptic NMDARs activation. Non-phosphorylated Jacob is linked to reduced CREB activity, resulting in diminished neuronal plasticity and negatively impacting dendritic complexity and synaptic density [39].

The forkhead transcription factor (FoxO3a) is linked to AD neuropathology and becomes triggered and relocated to the nucleus in reaction to increased extra synaptic NMDAR activity [40]. Synaptic activation of NMDAR hinders the activity of FOXO through the facilitation of phosphorylation by Akt. Conversely, extra synaptic NMDAR activation encourages the transfer of FOXO in nucleus, ultimately leading to excitotoxic cell death [41]. Activation of extra synaptic NMDARs triggers the start of calpain, causing the cleavage of STEP61 protein into STEP33. This is associated with a rise in the level of phosphorylated p38 protein, which plays a pivotal role in cellular stress responses, neuroinflammation, and apoptotic signaling pathways, contributing to synaptic dysfunction and progression of neurodegenerative conditions like Alzheimer's disease. In contrast, when synaptic NMDARs are activated, STEP61 undergoes ubiquitination and subsequent degradation. This disturbance of STEP61 correlates with the phosphorylation of ERK1/2 protein [42]. Synaptic NMDARs play a role in maintaining the starting of the ERK1/2 pathway, which is known to activate CREB and promote neuroprotection. In contrast, when NMDARs are globally activated by bath application of NMDA, initial activation of ERK is observed, subsequently its deactivation. Importantly, the localization of NMDARs determines their influence on ERK signaling. Synaptic NMDARs trigger the activation of ERK, while extra synaptic NMDARs induce its deactivation [43]. Therefore, synaptic and extra synaptic NMDARs exhibit mutual antagonism in ERK signaling.

10.2.2 AB AND NMDA RECEPTOR

The connection between NMDAR activation and the accumulation of Aβ is complex. The activation of extra synaptic NMDARs has a pivotal function in the production of Aβ. Conversely, the presence of Aβ in the brain stimulates the activation of extra synaptic NMDA receptors [44]. Some specific neuronal activities that are controlled by NMDA receptors lead to an increase in the β-secretase activity, compared to α-secretase. This preferential enhancement in β-secretase activity leads to a higher production of Aβ, which is an indication of the early stages of AD [45]. Furthermore, the activation of NMDA receptors that are located outside of the synapses leads to an enhancement in the production and phosphorylation of tau proteins. This process contributes to the formation of neurofibrillary tangles, which are a hallmark pathological feature of AD [46]. While the initial stage of AD development is identified by the deposition of Aβ in the medial parietal cortex, it is worth noting that tau deposition in the medial temporal lobe appears before the accumulation of Aβ in cognitively healthy older age individuals [47]. NMDAR antagonists completely block Aβ uptake, indicating the involvement of this receptor in the peptide's reuptake [48]. Aβ oligomerization begins intracellularly rather than extracellularly [49]. Increasing evidence indicates that Aβ oligomers are more effective than Aβ fibrils in causing impairments in synaptic function. Specifically, Aβ1-42 oligomers have been identified as particularly harmful and are responsible for the initial cognitive impairments seen in AD [50].

The deposition of intracellular Aβ within the mitochondria is associated with the various mitochondrial abnormalities observed in AD [51]. Aβ is moved into mitochondria through the translocase of the outer membrane machinery and is localized within the mitochondrial cristae [52]. In postmortem brain samples from individuals with AD, as well as in cellular and animal AD models, Aβ has been shown to initiate mitochondrial dysfunction. The dysfunction induced by Aβ is associated with its interaction with a range of mitochondrial proteins located across the outer mitochondrial membrane, intermembrane space, inner mitochondrial membrane, and matrix. This interaction disrupts oxidative phosphorylation and mitochondrial dynamics, while increasing the generation of ROS [53].

10.2.2.1 AB AND NMDA SUBUNITS

The expression patterns of NMDAR subunits differ across various brain regions and undergo significant modifications during development. The GluN1 subunit, which binds glycine/D-serine and is crucial for all NMDA receptor subtypes, is present throughout the CNS, from the embryonic stage E14 to adulthood. This widespread expression reflects the distribution of NMDARs across various regions of the CNS [54]. The patterns of expression of the four different GluN2 subunits (GluN2A-D), which bind glutamate and regulate NMDA receptor properties, vary during development and across different regions of the brain. In embryonic brain development, predominant expression of GluN2B and GluN2D subunits is observed, with GluN2D being particularly localized in caudal regions. As the brain matures during the first 2 weeks after birth, there is a gradual increase in GluN2A expression. By adulthood, GluN2A is abundantly expressed in virtually every region of the CNS. These variations in GluN2 subunit expression contribute to the diverse functions and responses of NMDA receptors across various brain regions [12]. In the mature CNS, especially in advanced brain regions, such as the hippocampus and cortex, GluN2A and GluN2B subunits become the primary subunits, highlighting their crucial roles in synaptic function and plasticity [55].

10.2.3 TAU AND NMDA RECEPTOR

Tau is an important component of neurofibrillary tangles (NFTs), which are distinctive characteristics of AD. In a normal brain, tau is present mainly in axons and helps in the organization and stability of microtubules. However, in individuals with AD, tau undergoes excessive phosphorylation and forms abnormal fibrils. These fibrils are seen as neuropil threads in dendrites and as NFTs in various parts of the cell, including the soma, dendrites, and axons. Evidence suggests that in familial autosomal dominant AD, the deposition of cerebral amyloid (Aβ) occurs before the development of tau pathology. On the other hand, in sporadic AD, the presence of NFTs precedes the accumulation of Aβ in the most affected regions [56]. Lowering tau levels has been found to have an impact on the levels and aggregation of Aβ, thereby preventing the progression of Aβ-related pathology. Tau serves a dual purpose in the brain by maintaining the structure of axons through microtubule stability and regulating synaptic function. It contributes in the activation of NMDAR mediated by Fyn in the postsynaptic density (PSD). Moreover, tau is necessary for inducing LTD and facilitating morphological plasticity that relies on BDNF.

Fyn phosphorylates the NMDAR subunit GluN2B at Y1472, enhancing the interaction between NMDARs and PSD-95 in the PSD, consequently boosting the activity of GluN2B-containing NMDARs [57, 58]. While tau is typically highly concentrated in axons compared to dendrites [59], exposure to Aβ prompts extensive redistribution of tau into the somato-dendritic compartment [60]. In individuals with AD, there is an imbalance of Fyn and tau proteins in the dendrites of neurons. This imbalance causes an increase in the activity of NMDA receptors, which leads to an excessive entry of calcium into the dendrites. This influx of calcium is harmful and can result in excitotoxicity, a condition where excessive stimulation and calcium levels damage neuronal structures. As a consequence, the postsynaptic sites, which are critical for neuronal communication, are impaired, ultimately leading to the death of neurons [61].

10.2.3.1 TAU AND NMDA SUBUNITS

Similar to amyloid beta and tau, glutamate excitotoxicity has been linked to synaptic dysfunction, contributing to neuronal damage, impaired neurotransmission, and cognitive decline in Alzheimer's disease through overstimulation of NMDA receptors and calcium-mediated neurodegenerative signaling cascades. When tau is phosphorylated, it moves from axons to dendritic spines, leading to reduced expression of AMPA receptors, deficits in LTP, and subsequent synaptic dysfunction [62]. Increasing the manifestation of phosphorylated tau promotes the transmission of NMDA receptors, which in turn facilitates the induction of LTD. The phosphorylation of tau can be triggered by the activation of NMDA receptors and exposure to Aβ. However, it is crucial to note that the phosphorylation of tau induced by acute activation of NMDA receptors can be reversed [63].

When human tau is overexpressed in primary neuronal cultures, it leads to the death of these cells. However, this cell death can be prevented by the use of ifenprodil, a specific antagonist that targets GluN2B-containing extrasynaptic NMDARs. This indicates that tau-induced neurotoxicity is related to the activation of extrasynaptic NMDARs that contain the GluN2B subunit.

The presence of these activated NMDARs triggers the activation of calpain, which is an enzyme that is activated by the influx of calcium through the NMDARs. Once active, calpain degrades the tau protein, breaking it down into N-terminal peptides. These toxic peptides ultimately contribute to the neuronal cell death [64]. In the rat hippocampal slice model, the presence of GluN2A-containing NMDA receptors has an impact on the level of tau phosphorylation. When these receptors are activated, there is a decrease in tau phosphorylation. This decrease occurs through the involvement of a pathway that includes protein kinase C (PKC) and glycogen synthase kinase 3 (GSK3). The activation of GluN2A-containing NMDA receptors leads to the activation of PKC, which in turn inhibits GSK3 activity. This inhibition of GSK3 ultimately results in a reduction in tau phosphorylation [65]. Tau plays a crucial role in neuronal cell death triggered by Aβ and can independently lead to synaptic dysfunction [66]. Additional research is required to explore both the mechanisms influenced by tau itself and those independent of tau, as well as a more extensive understanding of the varied pathological consequences of NMDA receptors.

10.3 SOME UNCONVENTIONAL NMDAR SIGNALING IN AD

10.3.1 PRESYNAPTIC NMDAR

NMDARs had always been assumed to be found in the postsynaptic membrane, but new morphological and physiological data suggest that they could also be present at presynaptic terminals. Presynaptic glutamate release can be controlled by presynaptic NMDARs (preNMDARs), which can also modify synaptic plasticity and transmission [67]. Therefore, the preNMDAR subunit composition plays an important role in determining the location and manner in which glutamate release is regulated by preNMDARs. There is notable regional heterogeneity in the makeup of the PreNMDAR subunits; GluN3A and all four GluN2 subunits (Glu2A-D) can be included, depending on the brain area. There are several brain areas where the presynaptic GluN2B subunit is present, including the entorhinal cortex, somatosensory and visual cortex, the cerebellum, and the hippocampus [68]. The GluN2A subunit can also be integrated into preNMDAR sites, although appearing and peaking later in the developmental process. At cerebellar parallel fiber-Purkinje cell synapses, preNMDARs predominantly consist of diheteromeric GluN1/GluN2A subunits. Additionally, GluN2B and GluN3A subunits play crucial roles as preNMDARs in the developing of L2/3 visual cortex. They most likely combine to form the triheteromeric GluN1/GluN2B/GluN3A. PreNMDARs have been suggested to control both evoked and spontaneous release. While it was often believed that both types of release used the same machinery, new research suggests that preNMDARs regulate both spontaneous and evoked release through different pathways. L5 pyramidal cells' preNMDARs control both spontaneous and induced release through pathways that depend on JNK2 and RIM1αβ, respectively. This suggests that while preNMDARs modulate evoked release in a frequency-dependent manner through the conventional Mg^{2+}-dependent pathway, they might regulate spontaneous release regardless of Mg^{2+} and Ca^{2+} presence. Moreover, preNMDAR activation is crucial for the induction of LTD [69]. However, it's important to remember that preNMDARs' responsibilities change as people grow. When LTD is induced in young mice's (before postnatal 20 days) visual cortical pyramidal cells, presynaptic NMDARs must be activated, whereas in older animals, postsynaptic NMDARs must be activated. However, research indicates that in the somatosensory cortex of 2–3-week-old rats and mice, LTD induction necessitates postsynaptic, rather than preNMDARs. This contradictory finding may be the result of different brain regions being observed. In addition to being necessary for LTD induction, preNMDARs are also involved in LTP induction [70]. PreNMDARs are demonstrated to be crucial for triggering LTP at corticostriatal synapses in mice. Activation of preNMDARs elevates Ca^{2+} signals in the axonal terminals, leading to the production of BDNF [71], suggesting that preNMDARs are just as important for LTP induction as postsynaptic NMDARs. The glutamatergic presynaptic bouton density of people with moderate cognitive impairment (MCI) was discovered to exhibit a counterintuitive rise, which subsequently depletes and lowers with the advancement of the condition. These findings demonstrated that in the latter stages of AD, dystrophic neurite production and decreased presynaptic bouton densities had a deleterious effect on neurotransmission and cognitive performance. Despite considerable progress, more work has to be done to elucidate the precise roles and molecular processes of pre-NMDAR [72].

10.3.2 GLIAL NMDAR

NMDARs have been extensively researched in neurons, but they are additionally expressed in astrocytes and other non-neuronal cells. Astrocytic NMDA receptors are less well characterized than neuronal receptors. Recent data indicates that astrocytic NMDARs exhibit unique structural and functional characteristics, such as reduced Ca^{2+} permeability and a limited sensitivity to Mg^{2+} blocking [73]. Both mouse neocortex and cultured astrocytes have shown NMDAR expression and function in astrocytes [74]. All seven of the known subunits of NMDAR (GluN1, GluN2A-D, and GluN3AB) were detected in human primary astrocyte material. Increasing research suggests that astrocytes generate triheteromeric

NMDARs, composed of GluN1, either GluN2C or GluN2D, along with a GluN3 subunit [75]. Astrocytic NMDARs can be activated by QUIN and glutamate, which raises Ca2+ influx and starts a signaling cascade [76]. Functional NMDARs in astrocytes respond to neuronal glutamatergic input, leading to a dynamic rise in intracellular calcium levels, which in turn triggers gliotransmitter-mediated modulation of synaptic activity [77]. To put it in simpler terms, astrocyte NMDARs in the brain may be involved in neuroinflammation and changes in the structure of astrocytes, which are cells that support and protect neurons. These receptors can promote to the development of AD due to their involvement in a process called glutamate excitotoxicity, which can damage neurons. Our research found that blocking specific types of astrocytic NMDARs can make early damage to synapses caused by Aβ protein, which is a hallmark of AD. Additionally, a protein called nerve growth factor (β-NGF) may have a role in protecting synapses when astrocytic NMDARs are activated.

Before exposing astrocytes to Aβ1–40 in a co-culture system, they were treated with either 1 μM or 10 μM NMDA to activate GluN2A or GluN2B [78]. This activation prevented the introduction of PSD-95 by Aβ and decreased the levels of a protein called synaptophysin. When GluN2A or GluN2B in astrocytes were blocked with TCN-201 or ifenprodil respectively, the harmful effects of Aβ on synapses were worsened. Additionally, NMDARs in oligodendrocyte lineage cells play a role in increasing intracellular calcium levels, which leads to a reduction in the survival of oligodendrocytes and injury to white matter.

Currents induced by NMDA in mature oligodendrocytes are primarily driven by the release of elevated extracellular potassium ions (K+) that occurs when astrocytic or neuronal NMDA receptors are activated [79]. Activation of these NMDARs results in elevated extracellular K+ levels, thereby affecting NMDA receptor-mediated currents in oligodendrocytes. Moreover, the transient receptor potential cation channel (TRP) A1 contributes to regulating the influx of calcium ions (Ca2+) into oligodendrocytes. When TRPA1 is activated, it facilitates the increase in intracellular calcium levels in oligodendrocytes.

NMDARs originating from oligodendrocyte precursor cells (OPCs) might also play a role in the process of myelination [80]. NMDAR activation boosted migration, myelin basic protein production, and differentiation in OPC cells [81]. Currently, it was discovered that microglia, a type of immune cell in the brain, also express NMDA receptors. This finding is noteworthy because NMDARs are typically associated with neurons and contribute in synaptic transmission. When exposed to NMDAR co-agonists, microglia exhibited inward currents and an increase in intracellular calcium levels, both of which are characteristic responses to NMDAR activation. Interestingly, these effects were inhibited by MK801, a non-competitive blocker of the NMDAR channel.

In addition to these cellular responses, the activation of NMDARs in microglia also triggered significant phosphorylation of the ERK1/2 signaling pathway. This pathway plays a role in cellular proliferation, survival, and inflammatory reactions. In neurons, the interaction between NMDARs and the mitogen-activated protein kinase (MAPK) signaling pathway is dependent on both calmodulin (CaM) and neuronal calcium sensor-1 (NCS1). However, in microglia, NMDAR signaling appears to be solely dependent on CaM. Furthermore, microglia taken from transgenic mice models of AD with increased amyloid plaque formation exhibited enhanced NMDAR function. This suggests that the connection between NCS1 and NMDARs is particularly important for receptor function in microglia in the context of AD. These findings shed new light on the involvement of microglia in NMDAR-mediated signaling and its potential relevance to neuroinflammatory processes in AD [45].

10.3.3 METABOTROPIC NMDAR

It was previously believed that Ca^{2+} influx alone governed NMDAR-dependent synaptic plasticity, with the postsynaptic neuron's elevated cytoplasmic Ca^{2+} serving as a secondary messenger [30]. Without calcium conductance, new and intriguing research suggests that NMDARs can cause long-term alterations in synaptic function upon binding glutamate [33]. Put differently, NMDAR is capable of functioning as a metabotropic receptor and facilitating metabotropic signaling without requiring Ca^{2+} input via the channel [82]. According to a widely held perspective, metabotropic receptors arise as a result of synaptic depression induced by low-frequency stimulation (LFS), a phenomenon recognized as NMDAR-dependent LTD, whereas NMDARs initiate LTP by a high degree of Ca^{2+} influx [83]. Recent findings indicate that binding of glutamate can trigger conformational alterations in NMDARs, which in turn cause p38MAPK signaling cascades and LTD. Remarkably, an increase in Ca^{2+} can cause LTP. Conversely, activation of metabotropic NMDARs leads to synaptic depression and needs baseline intracellular Ca^{2+} levels, triggering downstream signaling cascades that modulate gene expression, reduce synaptic strength, and contribute to long-term depression mechanisms implicated in neurodegenerative and cognitive disorders. Researchers have discovered that preNMDARs can regulate spontaneous release without relying on Mg^{2+} or Ca^{2+} requirements, while evoked release remains susceptible to Mg^{2+}, aligning with this metabotropic signaling mechanism. Partially through PKC signaling, preNMDARs facilitate transmitter release. These findings bolster the growing theory that evoked and spontaneous release occur through different mechanisms and imply that preNMDARs may signal through a metabotropic mechanism. While there is growing evidence in favor of an ion flux-independent mechanism for NMDAR-dependent LTD, several recent studies have cast doubt on it. It is more likely that there are two different types of NMDAR-dependent LTD at work; one kind needs ion flux, while the other does not. Furthermore, certain research findings indicate that astrocytic NMDARs might potentially function via metabotropic signaling pathways. These routes could entail phospholipase C-induced elevation of Ca^{2+} in the endoplasmic reticulum and activation of PKC. However, further investigation is necessary to elucidate this intricate process. Increased levels of Aβ oligomers are associated with initial synaptic dysfunction in AD, leading to rapid synaptic depression and spine loss that is dependent on NMDAR activity. However, other research studies have revealed that AP-5, but not MK-801, may inhibit Aβ-induced NMDAR-dependent synaptic depression, indicating that NMDARs' metabotropic action plays a role in Aβ-induced synaptic dysfunction. There could be a similar mechanism behind both Aβ-induced synaptic depression and metabotropic NMDAR-dependent LTD [45].

10.4 TREATMENT ASPECTS OF NMDARS FOR AD

AD etiology includes both anatomical and operational issues. AD causes architectural lesion in the brain, including senile plaques composed of Aβ, NFTs having phosphorylated tau, and profound synaptic depletion. AD is characterized by severe oxidative stresses and mitochondrial abnormalities [84]. There is also evidence of severe synapse damage and neuronal death.

There was no cure for Alzheimer's until recently. There are a few treatments, though. The National Institute on Aging's AD treatments fact sheet divides the currently approved FDA prescription treatments for those with the disease into two groups. AChEIs are a therapy option for AD. The other medication is employed to treat moderate, severe AD; it includes an antagonist NMDAR, which is activated by glutamate.

This chapter aims to examine the function of glutamatergic neurotransmission in AD. It focuses primarily on the significance of NMDAR signaling in AD [85].

10.4.1 GLUTAMATE AND THE SIGNALING RELATED TO GLUTAMATE ACTIVITY

Glutamate, an abundant excitatory neurotransmitter in the human CNS, is predominantly found within cells. It is widely distributed throughout the CNS and can be synthesized through various metabolic pathways. Glutamate concentrations in moist tissue vary between 5 and 15 μ mol/g, while the resting in the synaptic cleft is nearly 0.6 Mm. Glutamate concentrations can reach 10 μM during synaptic transmission in some extracellular regions. The glutamate uptake/transporter system removes any leftover glutamate. The amount of accessible extracellular glutamate is strictly regulated to provide optimum signaling levels. Glutamate and its associated receptors, specifically iGluRs, have a crucial role to play in the majority of excitatory neurotransmission in the human CNS. These receptors are particularly important for synaptic plasticity, which is the chemical process underlying memory and learning. Due to their crucial functions in excitatory neurotransmission, iGluRs are advantageous goals for therapeutic treatment since a variety of neuropathological illnesses and diseases, like Alzheimer's, Parkinson's, Huntington's, multiple sclerosis, epilepsy, and stroke are associated with an impairment of normal signaling [85].

10.4.2 GLUTAMATERGIC SIGNALING MEDIATED BY NMDAR IN SYNAPTIC PLASTICITY AND NEURONAL SURVIVAL

An agonist called NMDA preferentially gates a subgroup of iGluRs; this subgroup is known as the NMDA receptor (NMDAR). Differentiating from other iGluRs, NMDAR stands out due to its unique ability to be activated voltage-dependently through the removal of Mg^{2+} blockage, its high permeability to Ca^{2+}, and its relatively slow kinetics when gated by ligands.

These distinctive features make NMDAR crucial for ensuring the proper functioning and adaptability of synapses. At about –70 mV, the resting membrane potential of NMDAR is blocked by Mg^{2+}. Conversely, when LTP is induced, powerful and sustained glutamate discharge from the presynaptic terminal activates AMPARs. This depolarization then eliminates the Mg^{2+} obstruction of the NMDAR channel and lets Ca^{2+} ions to enter. Increased synaptic strength is the ultimate result of a signaling cascade driven by protein kinase II (CaMKII) and Ca^{2+}/calmodulin-dependent NMDAR activation. In difference, LTD is conducted by phosphatases and is brought on by a slight activation of NMDARs that outcome in a moderate rise in postsynaptic Ca^{2+} [86]. Beyond their role in plasticity and synaptic transmission, NMDARs seem essential for neuron survival by activating pathways crucial for neuronal viability. This is corroborated by research showing that neuronal death and degeneration result from blocking NMDAR function. Pro-survival transcription factors are triggered and apoptosis is repressed as the main NMDAR-dependent neuroprotection. Activation of NMDAR at synapses enhances the genes activity that support cell life through the activation of Ca^{2+}–dependent transcription factors like CREB. Furthermore, it inhibits caspases and apoptotic pathways, including PUMA, a gene that promotes cell death. The transcription factor FOXO (Forkhead box protein O) and the activated signaling cascade induced the production of pro-death genes [87].

10.4.3 GLUTAMATE EXCITOTOXICITY WITH ATYPICAL NMDAR ACTIVATION DURING AD

Insufficient synaptic NMDAR signaling can damage the survival of neuronal cells, even as extreme glutamatergic signaling can lead to brain trauma, such as stroke or excitotoxicity, causing damage or death of nerve cells. There is increasing data that glutamate excitotoxicity is linked in the gradual neurodegeneration seen in delayed processes, in addition to its acute effects. The toxicity primarily occurs due to excessive entry of calcium ions, predominantly through NMDARs, which exhibit a higher passage to Ca^{2+} in contrast to additional iGluRs [88]. This impairment in glutamate uptake and recycling leads to increased extracellular glutamate levels, which in turn further activates NMDARs and exacerbates calcium influx. Additionally, alterations in the function of NMDAR channels themselves can also contribute to abnormal NMDAR signaling in AD [89].

The clinical research of memantine as an NMDAR antagonist in AD is based on the rationale that reducing excessive NMDAR activation and calcium influx can help alleviate the neurodegenerative processes and symptoms associated with the disease. Memantine acts by binding to the NMDAR and blocking its excessive activation, while still allowing for physiological levels of NMDAR signaling necessary for cell survival and neuronal function.

By modulating NMDAR signaling, memantine aims to restore a healthier balance in calcium homeostasis and mitigate the downstream effects of excessive glutamate excitotoxicity. This can potentially slow down cognitive decline and delay neuronal cell death, thereby providing symptomatological and neuroprotective benefits in AD. Overall, the therapeutic approach of targeting NMDAR signaling with memantine in AD is supported by the consideration of the pathological action involving excessive calcium influx and glutamate excitotoxicity. By restoring a more balanced NMDAR signaling, memantine offers a potential strategy for managing the neurodegenerative processes and improving clinical outcomes in AD patients [18].

Research has shown that AD disrupts the proper functioning of "excitatory amino acid transporter 2" (EAAT2), mainly situated within presynaptic astrocytes. Experimental studies involving different forms of Aβ peptides in neuronal cell cultures have provided evidence that these harmful peptides can potentially enhance the levels of glutamate by interfering with its uptake and recycling mechanisms. This increased glutamate availability is likely a contributing factor to the excitotoxicity and neurotoxicity associated with AD.

The health of the presynaptic neurotransmitter-releasing system can also influence glutamate availability. Aβ has been shown to substantially inhibit the production of presynaptic proteins involved in neurotransmitter release, like syntaxin, synaptophysin, and synaptotagmin. Additionally, studies have shown that the modulation of action-dependent synaptic vesicle release is significantly affected by endogenous Aβ. These shortcomings in the machinery responsible for releasing presynaptic vesicles can hinder the availability of glutamate, reducing the likelihood of excitotoxicity [90].

Moreover, AD might also affect signaling of NMDAR by modifying receptors themselves. Various research studies have demonstrated that Aβ influences the electrophysiological activity of NMDARs. NMDAR antagonists, including MK-801 and D-APV, have been investigated, can attenuate or inhibit the increased synaptic currents and collateral toxicity mediated by NMDARs caused by Aβ. These NMDAR antagonists can also counteract the action of Aβ, such as synaptic defeat, similar to the effects of memantine. Additionally, Aβ may physically interact with NMDARs with the help of synaptic proteins like PSD95, as suggested by various researchers. AD may also influence the levels of NMDAR coagonists, specifically D-serine or glycine, which are required for the full activation of NMDARs by glutamate and have a significant modulatory effect on NMDAR function. Reports have indicated the involvement of D-serine and the activity of serine racemase, the enzyme accountable for D-serine production, are raised in the hippocampus of AD patients or Aβ-treated cultured microglia. Furthermore, deletion of serine racemase effectively lowers the concentration of D-serine in the forebrain and lessened NMDA or Aβ-induced neurotoxicity [91].

AD is commonly linked with varying glutamate availability and dysfunction in NMDAR channels, which are influenced by Aβ. These alterations are supposed to have an important part in the neurotoxicity and degeneration that occur in this disease.

10.4.4 THE SIGNALING OF NMDARS AND ITS RELATIONSHIP TO AD

Cell fate appears to be mostly determined by the Ca^{2+} level through NMDAR signaling; excessive signaling results in excitotoxicity and neurodegeneration, while insufficient signaling fails to sustain cell viability. Emerging data, nevertheless, suggests that this is not the whole picture. NMDARs' membrane position and regionalized signaling may be crucial to the pathogenesis of AD. The two population groupings of NMDARs on membrane—synaptic and extrasynaptic—can be broadly distinguished. Extra synaptic NMDARs have been the subject of fewer investigations than synaptic NMDARs, and until recently, little was known about their function. Recent research suggests that glutamate excitotoxicity and cell death may be related to or caused by extrasynaptic NMDARs. The commonly accepted model, which is still subject to conflicting reports, holds that extrasynaptic NMDAR activation causes cell death while synaptic NMDAR activation enhances cell survival. The imbalance between synaptic and extrasynaptic (SE) NMDAR action is believed to contribute to neuronal disbalance, including both acute cell trauma and chronic neurodegenerative diseases. Additionally, studies have demonstrated that D-serine and glycine, two separate endogenous coagonists, activate both SE-NMDARs, respectively. It is interesting to note that memantine, an NMDAR antagonist, has been shown to selectively target extrasynaptic NMDAR. Responses elicited by extrasynaptic NMDAR appear to be closely linked to the physiological alterations that occur in AD. Recent studies showed that memantine countered the deleterious action of Aβ, which caused synaptic loss, and that Aβ particularly activated extrasynaptic NMDARs. Aβ-induced reduction in NMDAR expression can also be addressed with this regionalized NMDAR signaling model. The ratio of synaptic to extrasynaptic NMDARs may be upset by the reduction in NMDARs. It appears to be well-understood what happens to the synaptic component of NMDARs in AD.

It has been demonstrated in numerous investigations that soluble Aβ inhibits synaptic plasticity and reduces synaptic glutamatergic transmission. As one study found, NMDAR-mediated currents are lowered and synaptic NMDARs internalize when Aβ1-42 is applied to cultured cortical neurons. One common characteristic of AD pathogenesis is generally AD-associated synapse degradation or weakened synaptic function. The extrasynaptic NMDAR-mediated signaling pathway in AD inhibits the cell survival route controlled by synaptic NMDARs by oxidative stress and pro-death signaling. Additionally, it is involved in the inactivation of CREB and FOXO transcription factors. Moreover, various processes can further compromise the proper functioning of neuronal NMDARs. Consequently, the delicate equilibrium of NMDAR signaling is disrupted, causing an imbalance in SE regions. This imbalance promotes the beginning of downstream signaling pathways, ultimately resulting in the harmful outcome of damage of neuronal cells in AD [90].

10.4.5 THE ROLE OF MEMANTINE: AN NMDAR ANTAGONIST IN AD

Presence of increased concentration of oligomeric Aβ peptide in AD has been linked with early synaptic dysfunction, resulting in synaptic depression and spine elimination that is dependent on NMDAR activity. Memantine, a low-affinity antagonist that blocks NMDAR channel, has been utilized as a treatment for moderate to severe AD. The mechanisms and connection between AD and NMDAR impairment lack conclusive data. This chapter highlights unique aspects of NMDAR function and variations in the expression of different NMDAR subunits [93].

10.5 WHAT IS MEMANTINE?

Memantine is a drug used to treat AD; it is also marketed under the Namenda brand belonging to the class of NMDAR antagonist drugs. Memantine is approved to assist patients with moderate to severe AD in managing their memory, attention, and cognitive functions. AD illness is a neurological disorder that progressively deteriorates over time, causing cognitive impairment, behavioral abnormalities, and memory loss. Although the precise etiology of AD is unidentified, it is considered to be related to change in the action of several neurotransmitters, such as glutamate [94].

Memantine functions by preventing the brain's excessive glutamate from having any negative effects. A neurotransmitter called glutamate is involved in learning and memory, but too much of it can cause damage to the brain by overexciting brain cells. Memantine helps control glutamate activity by inhibiting NMDA receptors, avoiding excessive activation, and shielding brain cells from harm.

While there is presently no known treatment for AD, memantine may be able to assist some people manage their symptoms and slow down the illness's progression. Usually, the prescription is written as a part of a thorough treatment plan that generally includes lifestyle changes, supportive care, and other medications.

Memantine should be taken under a doctor's supervision because, like any medicine, it may have negative effects. A doctor will decide on a patient's unique needs and circumstances to determine the precise dosage and length of treatment [95].

10.6 MECHANISM OF ACTION OF MEMANTINE

The brain's NMDA receptor's mechanisms by which memantine works involve uncompetitive antagonism, selectively blocking excessive glutamate activity to prevent excitotoxicity, while preserving normal synaptic transmission, thereby protecting neurons and improving cognitive function in moderate-to-severe Alzheimer's disease. One kind of glutamate receptor is the NMDA receptor. Glutamate has a vital position as a neurotransmitter, contributing to purpose, such as remembrance formation, learning processes, and effective communication between nerve cells.

An excess of NMDA receptor activation results from a glutamate imbalance in the brain through AD. An increase in calcium entering the neurons due to this overactivation may result in excitotoxicity and brain tissue destruction [85].

Memantine interacts with NMDA receptors differently than glutamate does because it is a collegial antagonist of these receptors. Memantine binds to the receptor to prevent an excessive amount of calcium from entering the neurons. By controlling the receptor's activity rather than blocking normal receptor function, it lessens the negative effects of excessive glutamate stimulation.

Memantine helps to restore the proper glutamate balance in the brain and shields neurons from excitotoxicity and cell damage by modifying the activation of NMDARs. The clinical benefits of memantine in AD, with enhanced behavior, memory, and cognitive function are believed to be a result of this mechanism of action.

It is noteworthy that the precise mechanisms behind the onset and advancement of AD remain incompletely known, and that memantine's effects on brain function in this particular context are multifaceted. On the other hand, one important component of its therapeutic impact is thought to be its capacity to modify NMDA receptor activation.

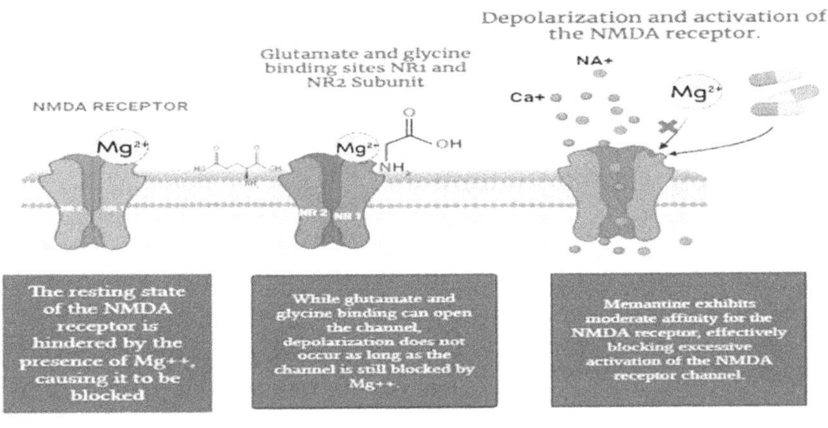

FIGURE 10.1 Mechanism of action of memantine.

By selectively inhibiting the NMDA receptor with low affinity, memantine acts on the glutamatergic system to regulate and prevent excessive activity of glutamate:

- Compared to Mg+ ions, memantine binds to the NMDAR more firmly. It preserves the transient physiological initiation of the channels brought on by greater planes of glutamate released synoptically but faces the voltage-dependent inhibition of NMDARs by Mg+ ions and the persistent influx of Ca+ ions through the NMDA channel (induced by high glutamate levels).

    1. Neurons rely on tight regulation of intracellular $Ca^{2}+$ concentration to ensure their survival and maintain proper physiological function.
    2. Memantine, being voltage-dependent, undergoes dissociation in response to an average physiological signal, facilitating the continuation of normal transmission.
    3. Memantine doesn't interfere with glutamate's physiological effect, which is essential for healthy brain function because it is non-competitive.
    4. Memantine inhibits extra synaptic NMDARs more potently, which decreases excitotoxicity while maintaining unvarying synaptic transmission. Prolonged activation of NMDARs leads to neurotoxicity.
    5. Memantine's unique and beneficial off-rate from the NMDA receptor channel determines both its therapeutic efficacy and tolerability. A slower off-rate would permit the drug to build up and obstruct regular neurotransmission, whereas a rapid off-rate would be ineffective.

10.7 COMBINATION THERAPY

Research is now being done on the combination of NMDA receptor-targeting medications with other therapeutic approaches. For instance, AChEIs, a different class of medications used to treat AD, and NMDA receptor antagonists together have demonstrated some promise in enhancing cognitive functions in AD patient.

TABLE 10.1 Research on the Usage of Memantine and AChEIs in Combined Therapy for AD.

Combination Therapy	Effects	References
AChEIs + memantine	There was no discernible difference in the two research periods' rates of change in total brain volume. Memantine therapy has also been associated with improved executive functioning and a decrease in right hippocampus atrophy.	[96]
Rivastigmine + memantine	There were no noticeable differences in the effectiveness of the treatments. The tolerability and safety of the treatment groups showed no discernible variations.	[91]
Memantine + rivastigmine	The tolerability among the treatment groups was similar, with no noticeable variations. There were also no noticeable disparities in cognition or overall functioning between the groups.	[97]
AChEI + memantine	Prolonging the admission period to a care home significantly.	[95]
Rivastigmine + memantine	Improvement in executive function and attention as secondary memory develops.	[94]
AChEI + memantine	The treatment demonstrated notable enhancements in data from the Severe Impairment Battery (SIB) and the Clinician's Interview-Based Impression of Change Plus (CIBIC+), suggesting its safety and tolerability.	[93]

10.7.1 THERAPEUTIC POTENTIAL OF NMDA RECEPTORS OTHER THAN AD

NMDAR is a kind of glutamate receptor which is chief for memory, learning, and synaptic plasticity in the CNS. Though AD has been linked to NMDAR malfunction, there are other possible therapeutic uses for this receptor. Targeting the NMDA receptor has the following additional therapeutic applications.

1. **The involvement of the NMDA receptor in major depressive disorder (MDD):**

   Studies indicate that disturbance in the glutamatergic system, particularly involving NMDA receptors, is associated with major depressive disorder (MDD). Alterations in the activity of NMDAR subtypes with disturbances in signaling pathways are generally noticed in a person with MDD. NMDA receptor inhibitors, like ketamine, have demonstrated quick-acting antidepressant action in cases of treatment resistant depression. But, the precise neurobiological action behind these effects is not completely stated. It is supposed that signaling pathways linked to synaptic plasticity have a central role in the antidepressant action of ketamine and other NMDA receptor antagonists. Understanding these mechanisms may aid in the development of new antidepressants that replicate the benefits of ketamine without its drawbacks. This chapter explores the involvement of NMDA receptors in the neurobiology of MDD, including changes in receptor subtypes and signaling pathways observed in postmortem studies. The impression of NMDA receptor antagonists on neuroplasticity in mood-related brain circuits is discussed, providing insights into the underlying mechanisms of their rapid antidepressant effects [98].

2. **The participation of the NMDAR in antidepressant activity:**

   Animal study has established that antidepressants have the capability to modify the countenance of NMDAR in specific sections of the brain. This can lead to alterations in the composition of the receptor subunits. These alterations in NMDA receptor subunits may serve as a common pathway through which antidepressants exert their effects and might even have the potential to predict the clinical effectiveness of these medications. The adaptive changes in NMDA receptors may be influenced by region-specific variations in the levels of excitant amino acids. Taken together, the modifications in the subunit composition of NMDAR could be a crucial factor in determining the therapeutic effects of different antidepressant drugs [99].

3. **The participation of the NMDA receptor in post-traumatic stress disorder (PTSD):**

   The study measures the impact of lanicemine on behavioral sensitization and BNST reactivity, which are believed to be influenced by NMDARs. Lanicemine has the potential to address the interaction between behavioral sensitization and acute PTSD symptoms by targeting NMDARs. If lanicemine proves to be safe and effective in this challenging population, it could have translational potential for treating not only PTSD but also other illnesses like bipolar disorder or substance use illnesses that comprise behavioral sensitization. However, there are limitations to this study. The results are inadequate to entities with high APS, and certain individuals with comorbidities or taking certain medications for PTSD are excluded. Additionally, research through lanicemine in persons with treatment-resistant depression has revealed a stronger placebo response with rarer depressive symptoms and without antipsychotic medication. The choice to enroll entities with symptoms of PTSD rather than meeting full criteria may further impact the placebo and landmine arms differently, but the findings on placebo effects may not apply to PTSD symptoms or biomarkers of useful mechanisms or target appointment. Overall, this study tracks the impact of lanicemine on a useful mechanism applicable to multiple illnesses, aligning with recent cooperative studies and "Fast-Fail" trials, and can provide a pattern for testing psychiatric drugs [100].

4. **The participation of the NMDA receptor in Parkinson's disease:**

Glutamate receptors play a vital role involving movement control in the basal ganglia and are considered attractive marks for rising novel therapies for Parkinson's disease. These receptors are a type of ion channels made up of many subunits with distinct expression patterns. They have multifaceted controlling properties, including agonist and co-agonist binding spots, as well as regulation through phosphorylation and protein interactions. While they originate in various basal ganglia constructions, the subunit composition differs between them. NMDAR in the corpus striatum is chiefly important for dopamine glutamate connections. The profusion, anatomical structure, and operation of these receptors are affected by dopamine reduction in PD and are additional adapted by pharmacological actions. Animal studies have established that NMDA receptor inhibitors can effectively alleviate PD symptoms and reduce complications associated with chronic dopamine therapy, such as tiring off and dyskinesias. However, the usage of these agents in individuals has been restricted due to side effects resulting from non-selective blocker of NMDA receptor function. The expansion of added potent and choosy drugs offers the potential for an innovative therapeutic method to PD [101].

5. **The connection of the NMDA receptor in acute and chronic pain:**

   While NMDA receptor inhibitors show promise in managing pain, the current literature has limitations, such as minor and faulty studies with short continuations. These limitations contribute to mixed results and make it difficult to determine optimal dosages and administration regimens. More studies are needed to clarify these factors and define what constitutes a positive response in pain management. Functional improvement should be a focus in these studies rather than just numerical pain intensity scale ratings. Ketamine has been the greatest considered NMDA receptor inhibitor and has the most scientific usefulness at this point. Dextromethorphan has the slightest clinical utility among the revised NMDA receptor inhibitors. Additionally, superior randomized controlled trials in distinct pain populaces are necessary to establish the character and safety of NMDA receptor antagonist before widespread use in pain management. With ongoing studies focused on NMDA receptor inhibitors for pain managing, it is expected that more clarity will be achieved in the coming years [102].

6. **NMDAR in epilepsy:**

   NMDAR inhibitor has shown narrow effectiveness for the management of seizure disorders, particularly in cases of focal seizures. Scientific studies have indicated that NMDA receptor antagonist may even worsen focal seizures, suggesting that they are unsuitable for this seizure type. However, NMDA receptor inhibitor may be more appropriate for conditions characterized by overexcited NMDA function, although identifying specific seizure types associated with this irregular NMDA receptor activity has proven to be challenging. Additionally, the AMPA receptor plays a more significant role in the generation of synchronized activity in seizure disorders compared to the NMDA receptor. AMPA receptors have been occupied in various stages of seizure development, and AMPA antagonists have shown efficacy in reducing hyper-synchronization. The broad spectrum effectiveness of the AMPA inhibitor, perampanel in different seizure forms may be attributed to its ability to reduce hyper-synchronization [103].

7. The involvement of the NMDA receptor in stroke:

Early attempts to block NMDARs with antagonists for neuroprotective therapies in ischemic stroke have been unsuccessful due to poor drug tolerance and a narrow treatment window. This is because NMDARs play a double character in neuronal signaling, both supporting survival and triggering cell death pathways. Blocking NMDARs completely can impair neuronal function and have adverse effects.

To overcome this challenge, researchers have focused on selectively blocking the NMDARs' pro-death effects while keeping pro-survival pathways intact. Important signaling complexes and downstream effectors implicated in neuronal death following excitotoxicity have been uncovered in recent investigations. With this understanding in mind, pharmacological inhibitors and interfering peptides have been created to precisely decouple NMDARs' neuronal death signaling while leaving their survival and functional signaling intact.

These novel therapies have the potential to offer a longer therapeutic window for intervention since they target the downstream pathways of NMDARs. To fully comprehend the significance and interplay of these signaling pathways in ischemic stroke, more investigation is necessary. Furthermore, combining many treatments that focus on various stroke pathways may enhance neuroprotection. The safety and effectiveness of these combined treatments in preventing ischemia injury should be the main focus of future research. It's important to remember that ischemic stroke and neurodegenerative illnesses frequently coexist in patients, indicating that there may be shared pathologies across neurological conditions. Consequently, knowledge of the signals involved in ischemia cell death and the discovered neuroprotective choices may aid in the making of treatments for many neurological disorders [11].

8. NMDAR in schizophrenia:

Research conclusions from clinical, genetic, and research on animals strongly recommend that the NMDARs hold a vital character in the pathological processes of schizophrenia. This is corroborated by the fact that NMDAR antagonists, such as PCP and ketamine, can cause indications and mental deficits like schizophrenia in healthy persons, while ketamine can exacerbate symptoms in stabilized schizophrenia patients. Furthermore, patients with NMDAR autoimmune encephalitis, characterized with production of antibodies against the GluN1 subunit of NMDAR, initially manifest psychiatric symptoms resembling those of schizophrenia. Furthermore, there is confirmation that a fraction of schizophrenia patients develop antibodies against the NMDAR. Furthermore, evidence from both patient studies and postmortem brain analyses suggests that NMDAR activity is reduced in schizophrenia. Additionally, large-scale genetic investigations have identified disruptions in genes related to glutamatergic signaling, the postsynaptic density, and dendritic spines, providing further support for the involvement of NMDAR dysfunction in schizophrenia. Animal models of NMDAR hypofunction, both pharmacological and genetic, have demonstrated various abnormalities in brain pathology, electrophysiology, and behavior that are consistent with schizophrenia.

However, it is important to recognize that schizophrenia is an extremely complex disease with multiple contributing factors. In addition to disturbances in the glutamatergic system and NMDAR function, there are also disruptions in other neurotransmitter systems, such as GABA, acetylcholine, and dopamine, in addition to abnormalities in astrocytes and white matter integrity. Additionally, there is compelling suggestion implicating the immunity and oxidative stress in the pathogenesis of schizophrenia. Given the etiological complexities of schizophrenia, it is unlikely that a solitary pharmacological method will completely treat the illness or be equally effective for all patients. A comprehensive understanding of the various genetic and environmental factors involved is necessary for developing personalized and targeted therapeutic strategies for schizophrenia. [104].

10.8 CONCLUSIONS AND FUTURE DIRECTIONS

Researchers have extensively studied the roles of glutamate and its receptors in neurodegenerative conditions, including dementia, synaptic plasticity, and AD. Recent findings indicate that glutamatergic neurotransmission through NMDARs can have divergent effects. While synaptic NMDAR signaling is necessary for neuron survival, the activation of extrasynaptic

NMDARs by glutamate released from astrocytes or presynaptic terminals can lead to excitotoxicity and neurodegeneration. Memantine, an FDA-approved inhibitor of NMDAR, has exhibited positive clinical outcomes from mild to severe AD. It's believed that memantine potentially acts by inhibiting extrasynaptic NMDAR transmission. Further research on memantine and related compounds will deepen our understanding on the molecular mechanisms that underpin the participation of glutamate and NMDAR in the progression of AD.

In summary, the therapeutic potential of NMDA receptor goes beyond AD. Its involvement in various neurological and psychiatric conditions suggests that modulating its activity may have beneficial effects in conditions, such as depression, schizophrenia, stroke, neuropathic pain, and substance abuse disorders. So, future research and clinical studies are needed to fully determine the protection and efficacy in targeting this receptor in these different therapeutic contexts. While these alternative therapeutic applications of NMDA receptor modulation hold promise, further study is required to abundantly understand their potential reimbursements and any associated risks or side effects.

KEYWORDS

- **Alzheimer's disease**
- **amyloid beta**
- **glutamate excitotoxicity**
- **memantine**
- **N-methyl-D-aspartate receptor**
- **NMDA GluN2B receptor**
- **synaptic plasticity**

REFERENCES

1. Ballard, C., Gauthier, S., Corbett, A., Brayne, C., Aarsland, D., & Jones, E. (2011). Alzheimer's disease. *Lancet (London, England), 377*(9770), 1019–1031. ISSN 1474–547X. Dostupné z: doi:10.1016/S0140-6736(10)61349-9
2. Melnikova, I. (2007). *Therapies for Alzheimer's disease.* květen. ISSN 1474–1776. Dostupné z: doi:10.1038/nrd2314

3. Mota, S. I., Ferreira, I. L., & Rego, A. C. (2014). Dysfunctional synapse in Alzheimer's disease - A focus on NMDA receptors. *Neuropharmacology, 76* Pt A, 16–26. ISSN 1873–7064. Dostupné z: doi:10.1016/j.neuropharm.2013.08.013
4. Traynelis, S. F., Wollmuth, L. P., Mcbain C. J., Menniti, F. S., Vance, K. M., Ogden, K. K., Hansen, K. B., Yuan, H., Myers, S. J., & Dingledine, R. (2010). Glutamate receptor ion channels: Structure, regulation, and function. *Pharmacological Reviews, 62*(3), 405–496. ISSN 1521–0081. Dostupné z: doi:10.1124/pr.109.002451
5. Rothman, S. M., & Olney, J. W. (1986). Glutamate and the pathophysiology of hypoxic--ischemic brain damage. *Annals of Neurology, 19*(2), 105–111. ISSN 0364–5134. Dostupné z: doi:10.1002/ana.410190202
6. Choi, D. W. (1988). Glutamate neurotoxicity and diseases of the nervous system. *Neuron, 1*(8), 623–634. ISSN 0896–6273. Dostupné z: doi:10.1016/0896-6273(88)90162-6
7. Lipton, S. A., & Rosenberg, P. A. (1994). Excitatory amino acids as a final common pathway for neurologic disorders. *The New England Journal of Medicine, 330*(9), 613–622. ISSN 0028–4793. Dostupné z: doi:10.1056/NEJM199403033300907
8. Choi, D. W. (1992). Excitotoxic cell death. *Journal of Neurobiology, 23*(9), 1261–1276. ISSN 0022–3034. Dostupné z: doi:10.1002/neu.480230915
9. Wang, R., & Reddy, P. H. (2017). Role of glutamate and NMDA receptors in Alzheimer's disease. *Journal of Alzheimer's disease: JAD, 57*(4), 1041–1048. ISSN 1875–8908. Dostupné z: doi:10.3233/JAD-160763
10. Choi, D. W. (1987). Ionic dependence of glutamate neurotoxicity. *The Journal of Neuroscience: The Official Journal of the Society for Neuroscience, 7*(2), 369–379. ISSN 0270–6474. Dostupné z: doi:10.1523/JNEUROSCI.07-02-00369.1987
11. Tymianski, M., Charlton, M. P., Carlen, P. L., & Tator, C. H. (1993). Source specificity of early calcium neurotoxicity in cultured embryonic spinal neurons. *The Journal of Neuroscience: The Official Journal of the Society for Neuroscience, 13*(5), 2085–2104. ISSN 0270–6474. Dostupné z: doi:10.1523/JNEUROSCI.13-05-02085.1993
12. Paoletti, P., Bellone, C., & Zhou, Q. (2013). NMDA receptor subunit diversity: Impact on receptor properties, synaptic plasticity and disease. *Nature Reviews. Neuroscience, 14*(6), 383–400. ISSN 1471–0048. Dostupné z: doi:10.1038/nrn3504
13. Kodis, E. J., Choi, S., Swanson, E., Ferreira, G., & Bloom, G. S. (2018). N-methyl-D-aspartate receptor-mediated calcium influx connects amyloid-β oligomers to ectopic neuronal cell cycle reentry in Alzheimer's disease. *Alzheimer's & Dementia: The Journal of the Alzheimer's Association, 14*(10), 1302–1312. ISSN 1552–5279. Dostupné z: doi:10.1016/j.jalz.2018.05.017
14. Danysz, W., & Parsons, C. G. (2003). The NMDA receptor antagonist memantine as a symptomatological and neuroprotective treatment for Alzheimer's disease: Preclinical evidence. *International Journal of Geriatric Psychiatry, 18*(Suppl 1), S23–32. ISSN 0885–6230. Dostupné z: doi:10.1002/gps.938
15. Wenk, G. L. (2006). Neuropathologic changes in Alzheimer's disease: Potential targets for treatment. *The Journal of Clinical Psychiatry, 67*(Suppl 3), 3–7; quiz 23. ISSN 0160–6689.
16. Fox, C., Crugel, M., Maidment, I., Auestad, B. H., Coulton, S., Treloar, A., Ballard, C., Boustani, M., Katona, C., & Livingston, G. (2012). Efficacy of memantine for agitation in Alzheimer's dementia: A randomised double-blind placebo controlled

trial. *PloS One*, 7(5), e35185. ISSN 1932–6203. Dostupné z: doi:10.1371/journal.pone.0035185

17. Wilkinson, D., Fox, N. C., Barkhof, F., Phul, R., Lemming, O., & Scheltens, P. (2012). Memantine and brain atrophy in Alzheimer's disease: A 1-year randomized controlled trial. *Journal of Alzheimer's Disease: JAD*, *29*(2), 459–469. ISSN 1875–8908. Dostupné z: doi:10.3233/JAD-2011-111616
18. Hardingham, G. E., & Bading, H. (2010). Synaptic versus extrasynaptic NMDA receptor signalling: implications for neurodegenerative disorders. *Nature Reviews. Neuroscience, 11*(10), 682–696. ISSN 1471–0048. Dostupné z: doi:10.1038/nrn2911
19. Amakiri, N., Kubosumi, A., Tran, J., & Reddy, P. H. (2019). Amyloid Beta and MicroRNAs in Alzheimer's Disease. *Frontiers in Neuroscience, 13*, 430. ISSN 1662–4548. Dostupné z: doi:10.3389/fnins.2019.00430
20. Silva, M. V. F., Loures, C. de M. G., Alves, L. C. V., de Souza, L. C., Borges, K. B. G., & Carvalho, M. das G. (2019). Alzheimer's disease: Risk factors and potentially protective measures. *Journal of Biomedical Science, 26*(1), 33. ISSN 1423–0127. Dostupné z: doi:10.1186/s12929-019-0524-y
21. Singh, A., Hasan, A., Tiwari, S., & Pandey, L. M. (2018). Therapeutic advancement in Alzheimer disease: New hopes on the horizon? *CNS & Neurological Disorders Drug Targets, 17*(8), 571–589. ISSN 1996–3181. Dostupné z: doi:10.2174/1871527317666180627122448
22. Sastre, M., Dewachter, I., Rossner, S., Bogdanovic, N., Rosen, E., Borghgraef, P., Evert, B. O., Dumitrescu-Ozimek, L., Thal, D. R., Landreth, G., Walter, J., Klockgether, T., Van Leuven, F., & Heneka, M. T. (2006). Nonsteroidal anti-inflammatory drugs repress beta-secretase gene promoter activity by the activation of PPARgamma. *Proceedings of the National Academy of Sciences of the United States of America, 103*(2), 443–448. ISSN 0027–8424. Dostupné z: doi:10.1073/pnas.0503839103
23. Kim, Jaewook, Yang, Y., Song, S. S., Na, J.-H., Oh, K. J., Jeong, C., Yu, Y. G., & Shin, Y.-K. (2014). Beta-amyloid oligomers activate apoptotic BAK pore for cytochrome c release. *Biophysical Journal, 107*(7), 1601–1608. ISSN 1542–0086. Dostupné z: doi:10.1016/j.bpj.2014.07.074
24. Kullmann, D. M., & Lamsa, K. P. (2007). Long-term synaptic plasticity in hippocampal interneurons. *Nature Reviews. Neuroscience*, *8*(9), 687–699. ISSN 1471–003X. Dostupné z: doi:10.1038/nrn2207
25. Nikseresht, Z., Ahangar, N., Badrikoohi, M., & Babaei, P. (2021). Synergistic enhancing-memory effect of D-serine and RU360, a mitochondrial calcium uniporter blocker in rat model of Alzheimer's disease. *Behavioural Brain Research, 409*, 113307. ISSN 1872–7549. Dostupné z: doi:10.1016/j.bbr.2021.113307
26. Zhang, Y., Li, P., Feng, J., & Wu, M. (2016). Dysfunction of NMDA receptors in Alzheimer's disease. *Neurological Sciences: Official Journal of the Italian Neurological Society and of the Italian Society of Clinical Neurophysiology, 37*(7), 1039–1047. ISSN 1590–3478. Dostupné z: doi:10.1007/s10072-016-2546-5
27. Bading, H. (2017). Therapeutic targeting of the pathological triad of extrasynaptic NMDA receptor signaling in neurodegenerations. *The Journal of Experimental Medicine, 214*(3), 569–578. ISSN 1540–9538. Dostupné z: doi:10.1084/jem.20161673

28. Lewerenz, J., & Maher, P. (2015). Chronic glutamate toxicity in neurodegenerative diseases-what is the evidence? *Frontiers in Neuroscience*, *9*, 469. ISSN 1662–4548. Dostupné z: doi:10.3389/fnins.2015.00469
29. Babaei, P. (2021). NMDA and AMPA receptors dysregulation in Alzheimer's disease. *European Journal of Pharmacology*, *908*, 174310. ISSN 1879–0712. Dostupné z: doi:10.1016/j.ejphar.2021.174310
30. Dore, K., Stein, I. S., Brock, J. A., Castillo, P. E., Zito, K., & Sjöström, P. J. (2017). Unconventional NMDA Receptor Signaling. *The Journal of Neuroscience: The Official Journal of the Society for Neuroscience*, *37*(45), 10800–10807. ISSN 1529–2401. Dostupné z: doi:10.1523/JNEUROSCI.1825-17.2017
31. Akashi, K., Kakizaki, T., Kamiya, H., Fukaya, M., Yamasaki, M., Abe, M., Natsume, R., Watanabe, M., & Sakimura, K. (2009). NMDA receptor GluN2B (GluR epsilon 2/NR2B) subunit is crucial for channel function, postsynaptic macromolecular organization, and actin cytoskeleton at hippocampal CA3 synapses. *The Journal of Neuroscience: The Official Journal of the Society for Neuroscience*, *29*(35), 10869–10882. ISSN 1529–2401. Dostupné z: doi:10.1523/JNEUROSCI.5531-08.2009
32. Suryavanshi, P. S., Ugale, R. R., Yilmazer-Hanke, D., Stairs, D. J., & Dravid, S. M. (2014). GluN2C/GluN2D subunit-selective NMDA receptor potentiator CIQ reverses MK-801-induced impairment in prepulse inhibition and working memory in Y-maze test in mice. *British Journal of Pharmacology*, *171*(3), 799–809. ISSN 1476–5381. Dostupné z: doi:10.1111/bph.12518
33. Chung, C. (2013). NMDA receptor as a newly identified member of the metabotropic glutamate receptor family: Clinical implications for neurodegenerative diseases. *Molecules and Cells*, *36*(2), 99–104. ISSN 0219–1032. Dostupné z: doi:10.1007/s10059-013-0113-y
34. Kamat, P. K., Kalani, A., Rai, S., Swarnkar, S., Tota, S., Nath, C., & Tyagi, N. (2016). Mechanism of oxidative stress and synapse dysfunction in the pathogenesis of Alzheimer's disease: Understanding the therapeutics strategies. *Molecular Neurobiology*, *53*(1), 648–661. ISSN 1559–1182. Dostupné z: doi:10.1007/s12035-014-9053-6
35. Saura, C. A., & Valero, J. (2011). The role of CREB signaling in Alzheimer's disease and other cognitive disorders. *Reviews in the Neurosciences*, *22*(2), 153–169. ISSN 0334–1763. Dostupné z: doi:10.1515/RNS.2011.018
36. Alberini, C. M. (2009). Transcription factors in long-term memory and synaptic plasticity. *Physiological Reviews*, *89*(1), 121–145. ISSN 0031–9333. Dostupné z: doi:10.1152/physrev.00017.2008
37. Du, Heng, Guo, L., Wu, X., Sosunov, A. A., Mckhann, G. M , Chen, J. X., & Yan, S. S. (2014). Cyclophilin D deficiency rescues Aβ-impaired PKA/CREB signaling and alleviates synaptic degeneration. *Biochimica et biophysica acta*, *1842*(12 Pt A), 2517–2527. ISSN 0006–3002. Dostupné z: doi:10.1016/j.bbadis.2013.03.004
38. Dieterich, D. C., Karpova, A., Mikhaylova, M., Zdobnova, I., König, I., Landwehr, M., Kreutz, M., Smalla, K.-H., Richter, K., Landgraf, P., Reissner, C., Boeckers, T. M., Zuschratter, W., Spilker, C., Seidenbecher, C. I., Garner, C. C., Gundelfinger, E. D., & Kreutz, M. R. (2008). Caldendrin-Jacob: A protein liaison that couples NMDA receptor signalling to the nucleus. *PLoS Biology*, *6*(2), e34. ISSN 1545–7885. Dostupné z: doi:10.1371/journal.pbio.0060034

39. Karpova, A., Mikhaylova, M., Bera, S., Bär, J., Reddy, P. P., Behnisch, T., Rankovic, V., Spilker, C., Bethge, P., Sahin, J., Kaushik, R., Zuschratter, W., Kähne, T., Naumann, M., Gundelfinger, E. D., & Kreutz, M. R. (2013). Encoding and transducing the synaptic or extrasynaptic origin of NMDA receptor signals to the nucleus. *Cell, 152*(5), 1119–1133. ISSN 1097–4172. Dostupné z: doi:10.1016/j.cell.2013.02.002
40. Qin, W., Zhao, W., Ho, L., Wang, J., Walsh, K., Gandy, S., & Pasinetti, G. M. (2008). Regulation of forkhead transcription factor FoxO3a contributes to calorie restriction-induced prevention of Alzheimer's disease-type amyloid neuropathology and spatial memory deterioration. *Annals of the New York Academy of Sciences, 1147*, 335–347. ISSN 1749–6632. Dostupné z: doi:10.1196/annals.1427.024
41. Dick, O., & Bading, H. (2010). Synaptic activity and nuclear calcium signaling protect hippocampal neurons from death signal-associated nuclear translocation of FoxO3a induced by extrasynaptic N-methyl-D-aspartate receptors. *The Journal of Biological Chemistry, 285*(25), 19354–19361. ISSN 1083–351X. Dostupné z: doi:10.1074/jbc.M110.127654
42. Xu, J., Kurup, P., Zhang, Y., Goebel-Goody, S. M., Wu, P. H., Hawasli, A. H., Baum, M. L., Bibb, J. A., & Lombroso, P. J . (2009). Extrasynaptic NMDA receptors couple preferentially to excitotoxicity via calpain-mediated cleavage of STEP. *The Journal of Neuroscience: The Official Journal of the Society for Neuroscience, 29*(29), 9330–9343. ISSN 1529–2401. Dostupné z: doi:10.1523/JNEUROSCI.2212-09.2009
43. Hetman, M., & Kharebava, G. (2006). Survival signaling pathways activated by NMDA receptors. *Current Topics in Medicinal Chemistry, 6*(8), 787–799. ISSN 1568–0266. Dostupné z: doi:10.2174/156802606777057553
44. Varga, E., Juhász, G., Bozsó, Z., Penke, B., Fülöp, L., & Szegedi, V. (2014). Abeta (1-42) enhances neuronal excitability in the CA1 via NR2B subunit-containing NMDA receptors. *Neural Plasticity, 2014*, 584314. ISSN 1687–5443. Dostupné z: doi:10.1155/2014/584314
45. Liu, J., Chang, L., Song, Y., Li, H., & Wu, Y. (2019). The role of NMDA receptors in Alzheimer's disease. *Frontiers in Neuroscience, 13*, 43. ISSN 1662–4548. Dostupné z: doi:10.3389/fnins.2019.00043
46. Shi, X.-D., Sun, K., Hu, R., Liu, X-Y., Hu, Q.-M., Sun, X.-Y., Yao, B., Sun, N., Hao, J.-R., Wei, P., Han, Y., & Gao, C. (2016). Blocking the interaction between EphB2 and ADDLs by a small peptide rescues impaired synaptic plasticity and memory deficits in a mouse model of Alzheimer's disease. *The Journal of Neuroscience: The Official Journal of the Society for Neuroscience, 36*(47), 11959–11973. ISSN 1529–2401. Dostupné z: doi:10.1523/JNEUROSCI.1327-16.2016
47. Jagust, W. (2018). Imaging the evolution and pathophysiology of Alzheimer disease. *Nature Reviews. Neuroscience, 19*(11), 687–700. ISSN 1471–0048. Dostupné z: doi:10.1038/s41583-018-0067-3
48. Bi, H., & Sze, C.-I. (2002). N-methyl-D-aspartate receptor subunit NR2A and NR2B messenger RNA levels are altered in the hippocampus and entorhinal cortex in Alzheimer's disease. *Journal of the Neurological Sciences, 200*(1–2), 11–18. ISSN 0022–510X. Dostupné z: doi:10.1016/s0022-510x(02)00087-4
49. Walsh, D. M., Tseng, B. P., Rydel, R. E., Podlisny, M. B., & Selkoe, D. J. (2000). The oligomerization of amyloid beta-protein begins intracellularly in cells derived

from human brain. *Biochemistry, 39*(35), 10831–10839. ISSN 0006–2960. Dostupné z: doi:10.1021/bi001048s

50. Palop, J. J., & Mucke, L. (2010). Amyloid-beta-induced neuronal dysfunction in Alzheimer's disease: From synapses toward neural networks. *Nature Neuroscience, 13*(7), 812–818. ISSN 1546–1726. Dostupné z: doi:10.1038/nn.2583
51. Manczak, M., Anekonda, T. S., Henson, E., Park, B. S., Quinn, J., & Reddy, P. H. (2006). Mitochondria are a direct site of A beta accumulation in Alzheimer's disease neurons: Implications for free radical generation and oxidative damage in disease progression. *Human Molecular Genetics, 15*(9), 1437–1449. ISSN 0964–6906. Dostupné z: doi:10.1093/hmg/ddl066
52. Hansson Petersen, C. A., Alikhani, N., Behbahani, H., Wiehager, B., Pavlov, P. F., Alafuzoff, I., Leinonen, V., Ito, A., Winblad, B., Glaser, E., & Ankarcrona, M. (2008). The amyloid beta-peptide is imported into mitochondria via the TOM import machinery and localized to mitochondrial cristae. *Proceedings of the National Academy of Sciences of the United States of America, 105*(35), 13145–13150. ISSN 1091–6490. Dostupné z: doi:10.1073/pnas.0806192105
53. Pagani, L., & Eckert, A. (2011). Amyloid-Beta interaction with mitochondria. *International Journal of Alzheimer's Disease, 2011*, 925050. ISSN 2090–0252. Dostupné z: doi:10.4061/2011/925050
54. Monyer, H., Burnashev, N., Laurie, D. J., Sakmann, B., & Seeburg, P. H. (1994). Developmental and regional expression in the rat brain and functional properties of four NMDA receptors. *Neuron, 12*(3), 529–540. ISSN 0896–6273. Dostupné z: doi:10.1016/0896-6273(94)90210-0
55. Takai, H., Katayama, K., Uetsuka, K., Nakayama, H., & Doi, K. (2003). Distribution of N-methyl-D-aspartate receptors (NMDARs) in the developing rat brain. *Experimental and Molecular Pathology, 75*(1), 89–94. ISSN 0014–4800. Dostupné z: doi:10.1016/s0014-4800(03)00030-3
56. Morris, G. P., Clark, I. A., & Vissel, B. (2018). Questions concerning the role of amyloid-β in the definition, aetiology and diagnosis of Alzheimer's disease. *Acta neuropathologica, 136*(5), 663–689. ISSN 1432–0533. Dostupné z: doi:10.1007/s00401-018-1918-8
57. Chen, Q., Zhou, Z., Zhang, L., Wang, Y., Zhang, Y., Zhong, M., Xu, S-c., Chen, C.-h., Li, L., & Yu, Z.-p. (2012). Tau protein is involved in morphological plasticity in hippocampal neurons in response to BDNF. *Neurochemistry International, 60*(3), 233–242. ISSN 1872–9754. Dostupné z: doi:10.1016/j.neuint.2011.12.013
58. Rong, Y., Lu, X., Bernard, A., Khrestchatisky, M., & Baudry, M. (2001). Tyrosine phosphorylation of ionotropic glutamate receptors by Fyn or Src differentially modulates their susceptibility to calpain and enhances their binding to spectrin and PSD-95. *Journal of Neurochemistry, 79*(2), 382–390. ISSN 0022–3042. Dostupné z: doi:10.1046/j.1471-4159.2001.00565.x
59. Groveman, B. R., Feng, S., Fang, X.-Q., Pflueger, M., Lin, S.-X., Bienkiewicz, E. A., & Yu, X. (2012). The regulation of N-methyl-D-aspartate receptors by Src kinase. *The FEBS Journal, 279*(1), 20–28. ISSN 1742–4658. Dostupné z: doi:10.1111/j.1742-4658.2011.08413.x
60. Zempel, H., Thies, E., Mandelkow, E., & Mandelkow, E.-M. (2010). Abeta oligomers cause localized Ca(2+) elevation, missorting of endogenous Tau into dendrites,

Tau phosphorylation, and destruction of microtubules and spines. *The Journal of Neuroscience: The Official Journal of the Society for Neuroscience, 30*(36), 11938–11950. ISSN 1529–2401. Dostupné z: doi:10.1523/JNEUROSCI.2357-10.2010

61. Roberson, E. D., Scearce-Levie, K., Palop, J. J., Yan, F., Cheng, I. H., Wu, T., Gerstein, H., Yu, G.-Q., & Mucke, L (2007). Reducing endogenous tau ameliorates amyloid beta-induced deficits in an Alzheimer's disease mouse model. *Science (New York, N.Y.), 316*(5825), 750–754. ISSN 1095–9203. Dostupné z: doi:10.1126/science.1141736
62. Hoover, B. R., Reed, M. N., Su, J., Penrod, R. D., Kotilinek, L. A., Grant, M. K., Pitstick, R., Carlson, G. A., Lanier, L. M., Yuan, L-L., Ashe, K. H., & Liao, D. (2010). Tau mislocalization to dendritic spines mediates synaptic dysfunction independently of neurodegeneration. *Neuron, 68*(6), 1067–1081. ISSN 1097–4199. Dostupné z: doi:10.1016/j.neuron.2010.11.030
63. Mondragón-Rodríguez, S., Trillaud-Doppia, E., Dudilot, A., Bourgeois, C., Lauzon, M., Leclerc, N., & Boehm, J. (2012). Interaction of endogenous tau protein with synaptic proteins is regulated by N-methyl-D-aspartate receptor-dependent tau phosphorylation. *The Journal of Biological Chemistry, 287*(38), 32040–32053. ISSN 1083–351X. Dostupné z: doi:10.1074/jbc.M112.401240
64. Amadoro, G., Ciotti, M. T., Costanzi, M., Cestari, V., Calissano, P., & Canu, N. (2006). NMDA receptor mediates tau-induced neurotoxicity by calpain and ERK/MAPK activation. *Proceedings of the National Academy of Sciences of the United States of America, 103*(8), 2892–2897. ISSN 0027–8424. Dostupné z: doi:10.1073/pnas.0511065103
65. De Montigny, A., Elhiri, I., Allyson, J., Cyr, M., & Massicotte, G. (2013). NMDA reduces Tau phosphorylation in rat hippocampal slices by targeting NR2A receptors, GSK3β, and PKC activities. *Neural Plasticity, 2013*, 261593. ISSN 1687–5443. Dostupné z: doi:10.1155/2013/261593
66. Reifert, J., Hartung-Cranston, D., & Feinstein, S. C. (2011). Amyloid beta-mediated cell death of cultured hippocampal neurons reveals extensive Tau fragmentation without increased full-length tau phosphorylation. *The Journal of Biological Chemistry, 286*(23), 20797–20811. ISSN 1083–351X. Dostupné z: doi:10.1074/jbc.M111.234674
67. Corlew, R., Wang, Y., Ghermazien, H., Erisir, A., & Philpot, B. D. (2007). Developmental switch in the contribution of presynaptic and postsynaptic NMDA receptorsto long-term depression. *The Journal of Neuroscience: The Official Journal of the Society for Neuroscience, 27*(37), 9835–9845. ISSN 1529–2401. Dostupné z: doi:10.1523/JNEUROSCI.5494-06.2007
68. Bouvier, G., Larsen, R. S., Rodríguez-Moreno, A., Paulsen, O., & Sjöström, P. J. (2018). Towards resolving the presynaptic NMDA receptor debate. *Current Opinion in Neurobiology, 51*, 1–7. ISSN 1873–6882. Dostupné z: doi:10.1016/j.conb.2017.12.020
69. Duguid, I. C. (2013). Presynaptic NMDA receptors: are they dendritic receptors in disguise? *Brain Research Bulletin, 93*, 4–9. ISSN 1873–2747. Dostupné z: doi:10.1016/j.brainresbull.2012.12.004
70. Mcguinness, L., Taylor, C., Taylor, R. D. T., Yau, C., Langenhan, T., Hart, M. L., Christian, H., Tynan, P. W., Donnelly, P., & Emptage, N. J. (2010). Presynaptic

NMDARs in the hippocampus facilitate transmitter release at theta frequency. *Neuron, 68*(6), 1109–1127. ISSN 1097–4199. Dostupné z: doi:10.1016/j.neuron.2010.11.023

71. Park, H., Popescu, A., & Poo, M. (2014). Essential role of presynaptic NMDA receptors in activity-dependent BDNF secretion and corticostriatal LTP. *Neuron, 84*(5), 1009–1022. ISSN 1097–4199. Dostupné z: doi:10.1016/j.neuron.2014.10.045
72. Bell, K. F. S., Bennett, D. A., & Cuello, A. C. (2007). Paradoxical upregulation of glutamatergic presynaptic boutons during mild cognitive impairment. *The Journal of Neuroscience: The Official Journal of the Society for Neuroscience, 27*(40), 10810–10817. ISSN 1529–2401. Dostupné z: doi:10.1523/JNEUROSCI.3269-07.2007
73. Hogan-Cann, A. D., & Anderson, C. M. (2016). Physiological roles of non-neuronal NMDA receptors. *Trends in Pharmacological Sciences, 37*(9), 750–767. ISSN 1873–3735. Dostupné z: doi:10.1016/j.tips.2016.05.012
74. Lalo, U., Pankratov, Y., Kirchhoff, F., North, R. A., & Verkhratsky, A. (2006). NMDA receptors mediate neuron-to-glia signaling in mouse cortical astrocytes. *The Journal of Neuroscience: The Official Journal of the Society for Neuroscience, 26*(10), 2673–2683. ISSN 1529–2401. Dostupné z: doi:10.1523/JNEUROSCI.4689-05.2006
75. Burzomato, V., Frugier, G., Pérez-Otaño, I., Kittler, J. T., & Attwel, D. L. (2010). The receptor subunits generating NMDA receptor mediated currents in oligodendrocytes. *The Journal of Physiology, 588*(Pt 18), 3403–3414. ISSN 1469–7793. Dostupné z: doi:10.1113/jphysiol.2010.195503
76. Sofroniew, M. V. (2009). Molecular dissection of reactive astrogliosis and glial scar formation. *Trends in Neurosciences, 32*(12), 638–647. ISSN 1878–108X. Dostupné z: doi:10.1016/j.tins.2009.08.002
77. Kato, H., Narita, M., Miyatake, M., Yajima, Y., & Suzuki, T. (2006). Role of neuronal NR2B subunit-containing NMDA receptor-mediated Ca2+ influx and astrocytic activation in cultured mouse cortical neurons and astrocytes. *Synapse (New York, N.Y.) 59*(1), 10–17. ISSN 0887–4476. Dostupné z: doi:10.1002/syn.20213
78. Li, Y., Chang, L., Song, Y., Gao, X., Roselli, F., Liu, J., Zhou, W., Fang, Y., Ling, W., Li, H., Almeida, O. F. X., & Wu, Y. (2016). Astrocytic GluN2A and GluN2B oppose the synaptotoxic effects of Amyloid-β1-40 in hippocampal cells. *Journal of Alzheimer's Disease: JAD, 54*(1), 135–148. ISSN 1875–8908. Dostupné z: doi:10.3233/JAD-160297
79. Káradóttir, R., Cavelier, P., Bergersen, L. H., & Attwell, D. (2005). NMDA receptors are expressed in oligodendrocytes and activated in ischaemia. *Nature, 438*(7071), 1162–1166. ISSN 1476–4687. Dostupné z: doi:10.1038/nature04302
80. Hamilton, N. B., Kolodziejczyk, K., Kougioumtzidou, E., & Attwell, D. (2016). Proton-gated Ca(2+)-permeable TRP channels damage myelin in conditions mimicking ischaemia. *Nature, 529*(7587), 523–527. ISSN 1476–4687. Dostupné z: doi:10.1038/nature16519
81. Wang, C., Pralong, W. F., Schulz, M. F., Rougon, G., Aubry, J. M., Pagliusi, S., Robert, A., & Kiss, J. Z. (1996). Functional N-methyl-D-aspartate receptors in O-2A glial precursor cells: A critical role in regulating polysialic acid-neural cell adhesion molecule expression and cell migration. *The Journal of Cell Biology, 135*(6 Pt 1), 1565–1581. ISSN 0021–9525. Dostupné z: doi:10.1083/jcb.135.6.1565
82. Kessels, H. W., Nabavi, S., & Malinow, R. (2013). Metabotropic NMDA receptor function is required for β-amyloid-induced synaptic depression. *Proceedings of the*

National Academy of Sciences of the United States of America, 110(10), 4033–4038. ISSN 1091–6490. Dostupné z: doi:10.1073/pnas.1219605110

83. Malenka, R. C. (1994). Synaptic plasticity in the hippocampus: LTP and LTD. *Cell, 78*(4), 535–538. ISSN 0092–8674. Dostupné z: doi:10.1016/0092-8674(94)90517-7
84. Perl, D. P. (2010). *Neuropathology of Alzheimer's disease.* Leden. ISSN 00272507. Dostupné z: doi:10.1002/msj.20157
85. Bleich, S., Römer, K., Wiltfang, J., & Kornhuber, J. (2003). Glutamate and the glutamate receptor system: A target for drug action. *International Journal of Geriatric Psychiatry, 18*(Suppl 1), S33–40. ISSN 0885–6230. Dostupné z: doi:10.1002/gps.933
86. Perl, D. P. (2010). Neuropathology of Alzheimer's disease. *The Mount Sinai Journal of Medicine, New York,* 77(1), 32–42. ISSN 1931–7581. Dostupné z: doi:10.1002/msj.20157
87. Alberdi, E., Sánchez-Gómez, M. V., Cavaliere, F., Pérez-Samartín, A., Zugaza, J. L., Trullas, R., Domercq, M., & Matute, C. (2010). Amyloid beta oligomers induce Ca2+ dysregulation and neuronal death through activation of ionotropic glutamate receptors. *Cell Calcium, 47*(3), 264–272. ISSN 1532–1991. Dostupné z: doi:10.1016/j.ceca.2009.12.010
88. Arias, C., Arrieta, I., & Tapia, R. (1995). beta-Amyloid peptide fragment 25-35 potentiates the calcium-dependent release of excitatory amino acids from depolarized hippocampal slices. *Journal of Neuroscience Research, 41*(4), 561–566. ISSN 0360–4012. Dostupné z: doi:10.1002/jnr.490410416
89. Farlow, M. R., Alva, G., Meng, X., & Olin, J. T. (2010). A 25-week, open-label trial investigating rivastigmine transdermal patches with concomitant memantine in mild-to-moderate Alzheimers disease: A post hoc analysis. *Current Medical Research and Opinion, 26*(2), 263–269. ISSN 03007995. Dostupné z: doi:10.1185/03007990903434914
90. Wu, L., Rosa-Neto, P., Hsiung, G.-Y. R., Sadovnick, A. D., Masellis, M., Black, S. E., Jia, J., & Gauthier, S. (2012). Early-onset familial Alzheimer's disease (EOFAD). *The Canadian Journal of Neurological Sciences. Le journal canadien des sciences neurologiques, 39*(4), 436–445. ISSN 0317–1671. Dostupné z: doi:10.1017/s0317167100013949
91. Choi, S. H., Park, K. W., Na, D. L., Han, H. J., Kim, E.-J., Shim, Y. S., & Lee, J.-H. (2011). Tolerability and efficacy of memantine add-on therapy to rivastigmine transdermal patches in mild to moderate Alzheimer's disease: A multicenter, randomized, open-label, parallel-group study. *Current Medical Research and Opinion, 27*(7), 1375–1383. ISSN 1473–4877. Dostupné z: doi:10.1185/03007995.2011.582484
92. Hardingham, G. E., & Bading, H. (2010). *Synaptic versus extrasynaptic NMDA receptor signalling: Implications for neurodegenerative disorders.* říjen. ISSN 1471003X. Dostupné z: doi:10.1038/nrn2911
93. Porsteinsson, A. P., Grossberg, G. T., Mintzer, J., & Olin, J. T. (2008). Memantine treatment in patients with mild to moderate Alzheimer's disease already receiving a cholinesterase inhibitor: a randomized, double-blind, placebo-controlled trial. *Current Alzheimer Research, 5*(1), 83–89. ISSN 1567–2050. Dostupné z: doi:10.2174/156720508783884576
94. Riepe, M. W., Adler, G., Ibach, B., Weinkauf, B., Tracik, F., & Gunay, I. (2007). Domain-specific improvement of cognition on memantine in patients with Alzheimer's

disease treated with rivastigmine. *Dementia and Geriatric Cognitive Disorders, 23*(5), 301–306. ISSN 1420–8008. Dostupné z: doi:10.1159/000100875

95. Lopez, O. L., Becker, J. T., Wahed, A. S., Saxton, J., Sweet, R. A., Wolk, D. A., Klunk, W., & Dekosky, S. T. (2009). Long-term effects of the concomitant use of memantine with cholinesterase inhibition in Alzheimer disease. *Journal of Neurology, Neurosurgery, and Psychiatry, 80*(6), 600–607. ISSN 1468–330X. Dostupné z: doi:10.1136/jnnp.2008.158964
96. Weiner, M. W., Sadowsky, C., Saxton, J., Hofbauer, R. K., Graham, S. M., Yu, S. Y., Li, S., Hsu, H-A., Suhy, J., Fridman, M., & Perhach, J. L . (2011). Magnetic resonance imaging and neuropsychological results from a trial of memantine in Alzheimer's disease. *Alzheimer's & Dementia: The Journal of the Alzheimer's Association,* 7(4), 425–435. ISSN 1552–5279. Dostupné z: doi:10.1016/j.jalz.2010.09.003
97. Farlow, M. R., Alva, G., Meng, X., & Olin, J. T. (2010). A 25-week, open-label trial investigating rivastigmine transdermal patches with concomitant memantine in mild-to-moderate Alzheimer's disease: A post hoc analysis. *Current Medical Research and Opinion, 26*(2), 263–269. ISSN 1473–4877. Dostupné z: doi:10.1185/03007990903434914
98. Amidfar, M., Woelfer, M., Réus, G. Z., Quevedo, J., Walter, M., & Kim, Y.-K. (2019). The role of NMDA receptor in neurobiology and treatment of major depressive disorder: Evidence from translational research. *Progress in Neuro-Psychopharmacology & Biological Psychiatry, 94*, 109668. ISSN 1878–4216. Dostupné z: doi:10.1016/j.pnpbp.2019.109668
99. Hovelsø, N., Sotty, F., Montezinho, L. P., Pinheiro, P. S., Herrik, K. F., & Mørk, A. (2012). Therapeutic potential of metabotropic glutamate receptor modulators. *Current Neuropharmacology, 10*(1), 12–48. ISSN 1875–6190. Dostupné z: doi:10.2174/157015912799362805
100. Lijffijt, M., Green, C. E., Balderston, N., Iqbal, T., Atkinson, M., Vo-Le, B., Vo-Le, B., O'Brien, B., Grillon, C., Swann, A. C., & Mathew, S. J. (2019). A proof-of-mechanism study to test effects of the NMDA receptor antagonist lanicemine on behavioral sensitization in individuals with symptoms of PTSD. *Frontiers in Psychiatry, 10*, 846. ISSN 1664–0640. Dostupné z: doi:10.3389/fpsyt.2019.00846
101. Greenamyre, J. T., & O'Brien, C. F. (1991). N-methyl-D-aspartate antagonists in the treatment of Parkinson's disease. *Archives of Neurology, 48*(9), 977–981. ISSN 0003–9942. Dostupné z: doi:10.1001/archneur.1991.00530210109030
102. Kreutzwiser, D., & Tawfic, Q. A. (2019). Expanding role of NMDA receptor antagonists in the management of pain. *CNS Drugs, 33*(4), 347–374. ISSN 1179–1934. Dostupné z: doi:10.1007/s40263-019-00618-2
103. Hanada, T. (2020). Ionotropic glutamate receptors in epilepsy: A review focusing on AMPA and NMDA receptors. *Biomolecules, 10*(3). ISSN 2218–273X. Dostupné z: doi:10.3390/biom10030464
104. Hanson, J. E., Yuan, H., Perszyk, R. E., Banke, T. G., Xing, H., Tsai, M-C., Menniti, F. S., & Traynelis, S. F. (2024). Therapeutic potential of N-methyl-D-aspartate receptor modulators in psychiatry. *Neuropsychopharmacology: Official Publication of the American College of Neuropsychopharmacology, 49*(1), 51–66. ISSN 1740–634X. Dostupné z: doi:10.1038/s41386-023-01614-3

CHAPTER 11

Unraveling the Insights of NMDA Receptors in Neurodegenerative Disorders

ANUP KUMAR CHAKRABORTY[1,*], PRASHANT TIWARI[2], SUNIL KUMAR KADIRI[2], and DEENANATH JAHDE[3]

[1]*IES Institute of Pharmacy, IES University, Bhopal, Madhya Pradesh, India*

[2]*College of Pharmaceutical Sciences, Dayananda Sagar University, Bengaluru, Karnataka, India*

[3]*School of Pharmacy, Mody University of Science and Technology, Sikar, Rajasthan, India*

*Corresponding author

ABSTRACT

Understanding the significance of NMDA receptors in neurodegenerative conditions is crucial for devising precise treatment approaches. These receptors, pivotal in excitatory neurotransmission, impact various brain functions like learning and memory. Yet, their malfunction contributes to neurodegeneration. To counter this, neuroprotective methods aim to maintain neuronal health, often by addressing factors causing damage and enhancing overall brain function. By comprehending the diverse roles of NMDA receptors in neurodegenerative disorders—such as regulating excitotoxicity, calcium levels, synaptic activity, and protein aggregation—tailored therapies can be developed to mitigate disease progression. Thus, exploring the involvement of NMDA receptors offers hope for targeted interventions against neurodegenerative diseases.

Cellular and Molecular Mechanisms in Alzheimer's Disease Therapeutics, Volume 2: Molecular Mechanisms. Prashant Tiwari, Sunil Kumar Kadiri, and Kunjbihari Sulakhiya (Eds)

DOI: 10.1201/9781779641816-11

11.1 INTRODUCTION

Neurodegeneration stands as a central pathological feature across a spectrum of brain-related disorders [1]. Despite relentless efforts in modern science, effective medical or surgical solutions remain elusive. Neurodegenerative disorders (NDs) like Alzheimer's and dementia persist as significant clinical concerns, particularly in aging populations [2, 3]. The formidable blood-brain barrier (BBB) poses a substantial obstruction to successful ND management. While surgeries and invasive methods have shown some effectiveness in treating NDs, their widespread use is restricted, arising from concerns about its long-term development efficacy and potential harm to the BBB [4]. As a solution, nanotherapeutics have made more commitments, capable of traversing the BBB without causing damage [5]. The adoption of nanotherapeutics is on the rise due to their various advantages compared to traditional dosage forms. However, further refinement of these nanotherapeutics is essential to ensure optimal results. This chapter offers an overview of the pathophysiology and current management strategies of major neurodegenerative diseases (ND), along with investigating the role of the BBB and other barriers to the delivery of drugs to the brain. Additionally, it explores the potential of nanotherapeutics in combating neurodegeneration. Finally, this chapter discusses recent advancements and ongoing research in nanotherapeutics for managing NDs, providing insights into their future applications.

11.1.1 NEURODEGENERATIVE DISEASES

Neurons are essential for the proper functioning of the human brain because they facilitate communication [6]. Although most neurons originate from the brain, they are found throughout the body [7]. Neural stem cells in childhood mainly generate neurons, but this production decreases significantly in adulthood [8]. While neurons aren't immortal, they perform crucial functions throughout life. However, the gradual loss of neurons, changes in their structure, and impaired functions, collectively known as neurodegeneration, are central to the pathophysiology of many brain disorders [9]. This presents a significant health challenge.

Neurodegeneration involves dysfunction at various levels, including the synapse, neural networks, and the accumulation of aberrant protein variants in the brain (see Figure 11.1) [10]. Diseases characterized by neurodegeneration as a prominent feature are collectively referred to as NDs [11].

Common examples of NDs encompass Alzheimer's disease (AD), Parkinson's disease (PD), prion disease, amyotrophic lateral sclerosis, motor neuron disease, Huntington's disease (HD), spinal muscular atrophy, and spinocerebellar ataxia [12]. These disorders not only impact the affected individuals but also pose significant challenges for healthcare systems worldwide.

NDs represent a significant global burden, affecting millions of individuals worlwide. While aging stands as the primary risk factor for all NDs, recent research highlights the intricate interplay between genetic predisposition and environmental factors in exacerbating this risk [13]. Additionally, while certain genes contribute to an individual's susceptibility to NDs, the onset and severity of neurodegeneration are heavily influenced by environmental cues. Recent findings suggest that multiple pathological mechanisms may converge within individual cases of NDs, resulting in varying degrees of severity [14]. Some cases may even pose life-threatening risks depending on the specific disease type and its stage of progression [15–18].

NDs significantly disrupt a range of human functions as the brain, vital for controlling bodily activities, is affected [19]. They hinder basic functions like speech, movement, and balance, along with more intricate tasks, such as bladder and bowel regulation, as well as cognitive abilities [20]. Most NDs exhibit a relentless progression without remission, although treatment approaches typically focus on managing symptoms, alleviating pain, and restoring functional capabilities [21, 22]. In the subsequent sections, we will provide concise overviews of some prevalent NDs [23, 24].

NDs encompass a range of debilitating conditions marked by progressive neuronal dysfunction and loss inside the CNS, while their causes are multifaceted, emerging research underscores the role of NMDA (N-methyl-D-aspartate) receptors, a type of glutamate receptor, in the development of these disorders [25].

NMDA receptors are integral to synaptic plasticity, learning, and memory. Comprising various subunits like GluN1, GluN2, and GluN3, they confer diverse functions to the receptor complex. Activation of NMDA receptors typically facilitates calcium ion influx (Ca^{2+}), vital for synaptic transmission and neuronal survival. However, aberrant NMDA receptor activity can induce excitotoxicity, oxidative stress, and neuroinflammation, contributing to neuronal damage and demise in neurodegenerative conditions [26–29].

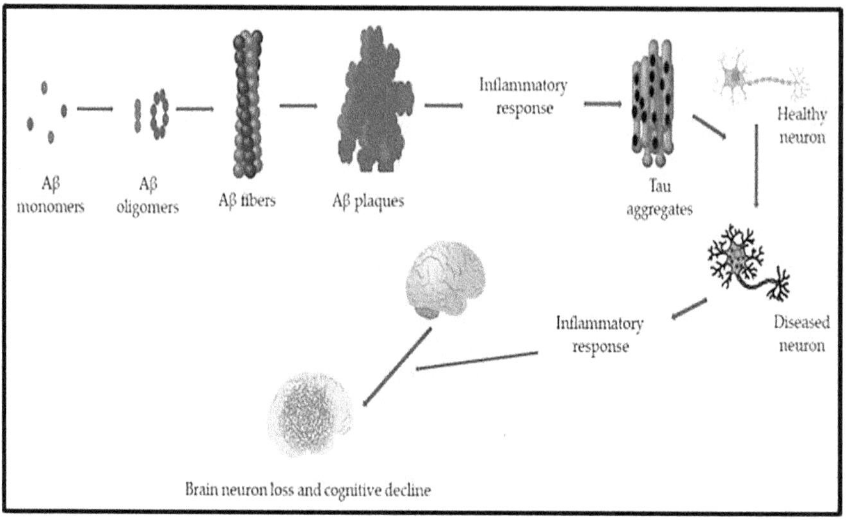

FIGURE 11.1 The path to cognitive decline in ND begins with the aggregation of amyloid-β (Aβ) monomers into oligomers of various structures. These oligomers then assemble into Aβ fibrils, which in turn dissociate into Aβ plaques. The formation of plaques leads to inflammatory reactions, including the accumulation of tau protein, which eventually causes healthy neurons to turn into diseased neurons. The increase in the number of neurons causes other pains, which leads to further damage to the neuron and decreased brain functions, ultimately leading to cognitive decline.

11.1.1.1 ALZHEIMER'S DISEASE (AD)

Recent research exploring the pathophysiology of AD has highlighted the significant involvement of Aβ and tau proteins in its development [30]. Hallmarks of AD include formation of Aβ-containing plaques in the brain and the presence of neurofibrillary tangles (NFTs) consisting of hyperphosphorylated tau [31]. These pathological changes disrupt hippocampal circuits and disrupt the integration of short-term circuits into long-term circuits [32]. AD is indicated by widespread neuronal loss, dysfunction, and damage to important neurotransmitters in the brain, especially memory. Therefore, selective memory impairment is an early symptom of AD. In addition, functions that rely on the hippocampus and the central nervous system (CNS), like mediating memory, particularly episodic memory, are frequently impacted by these disorders. In addition, impairment in functionality, decision-making, and problem solving are clinical findings that often occur early in the disease [33].

11.1.1.2 PARKINSON'S DISEASE (PD)

PD presents with symptoms like tremors, stiffness, balance issues, and coordination difficulties. Both genetic predisposition and environmental influences play roles in these disorders, including age, can contribute to its onset [34]. While there is a correlation between PD and factors, such as high caffeine consumption, smoking, and exposure to toxins, the exact mechanisms involved remain unclear [35]. The disease's pathophysiology involves changes, such as loss of pigmentation in particular areas of the brain due to the degeneration of dopamine-containing neurons, impacting dopamine levels and causing characteristic symptoms [36]. Cell loss in various brain areas also contributes to non-motor symptoms. Parkinson's progression involves several mechanisms, including protein misfolding, mitochondrial dysfunction, impaired protein clearance systems, and neuroinflammation [37]. Microscopically, the disease is defined by the existence of unusual protein buildups called Lewy bodies and dystrophic neurites in the brain [38]. These Lewy bodies can spread throughout the CNS, akin to AD pathology [39]. Additionally, protein misfolding in Parkinson's involves tau protein, leading to the creation of NFTs, resembling those observed in Alzheimer's patients [40].

11.1.1.3 HUNTINGTON'S DISEASE (HD)

HD is an inherited ND with an autosomal dominant pattern of inheritance condition characterized through involuntary movements and cognitive decline [41]. It stems from an expansion of the CAG trinucleotide repeat within the the Huntingtin (*HTT*) gene, which is situated on chromosome 4p16.3, resulting in an elongated polyglutamine sequence in the HTT protein. This genetic alteration triggers neurodegeneration and disrupts protein folding, leading to aggregation [42]. Typically, HD manifests in individuals aged 30 to 50, with earlier onset linked to longer CAG repeats. Adolescent HD refers to a condition that begins before the age of 20 and is characterized by academic difficulties and disruptive behavior in school. A correct diagnosis, presymptomatic testing can determine whether high-risk individuals have a genetic mutation [43]. There is currently no cure for HD, and the number of patients increases as the disease progresses. Treatment is designed to improve quality of life and manage problems. Pneumonia and suicide are common causes of death in HD patients [44, 45]. Targeting specific NMDA receptor subunits and downstream signaling pathways holds promise for treatment. A detailed understanding of the role of NMDA receptors in the neurodegenerative process is essential in devising therapeutic approaches.

11.2 THE FUNCTION OF NMDA RECEPTORS

The role of NMDA receptors in ND is complex and diverse. These receptors, belonging to the glutamate receptor family, are essential for synaptic plasticity, learning, and memory within the CNS. However, dysfunctions in NMDA receptor activity have been linked to the onset and progression of various neurodegenerative conditions, including AD, PD, and HD, among others [46–48].

11.2.1 EXCITOTOXICITY

Excitotoxicity, the process by which too many calcium ions (Ca^{2+}) enter neurons, can cause cell damage and death. This phenomenon is associated with neuronal loss seen in ND. In diseases such as AD, overstimulation of NMDA receptors can lead to excitotoxicity, causing synaptic impairment and neuronal damage [49, 50].

11.2.2 SYNAPTIC DYSFUNCTION

Is a progressive ND characterized by cognitive decline and memory impairment. For example, in AD, abnormal NMDA receptor signaling disrupts synaptic function and impairs memory and cognition [51–53].

11.2.3 NEUROINFLAMMATION

The reaction causes blood vessels to swell. Chronic neuroinflammation is one of many ND that exacerbates neuronal damage and causes disease [54, 55].

11.2.4 TAU PHOSPHORYLATION AND AB TOXICITY

NMDA receptor activation is associated with hyperphosphorylation of tau and production of Aβ peptide accumulation, a defining feature of AD pathology. Abnormal NMDA receptor activity can lead to deposition of tau tangles and beta-amyloid plaques, further enhancing neurodegenerative changes [56–58].

11.2.5 THE DEMISE OF NEURONS

In numerous neurodegenerative conditions, such as HD, NMDA receptor-mediated excitotoxicity can lead to neuron death. Excessive calcium ions from over-activated NMDA receptors can trigger the apoptotic pathway, resulting in neuron loss, neuronal dysfunction, and degenerative diseases [59].

Although careful consideration of the complex roles of NMDA receptors at different stages of the disease is necessary to increase therapeutic efficacy, modulation of NMDA receptor activity may have a potential nature to ameliorate these serious diseases [60].

11.3 GLUTAMATE EXCITOTOXICITY IN RELATION TO NDS

11.3.1 EXCITOTOXICITY IN RELATION TO ALZHEIMER'S DISEASE

AD is a progressive neurodegenerative condition marked by a decline in cognitive function and memory loss. Its molecular hallmarks include the buildup of Aβ in senile plaques, the presence of intraneuronal NFTs, reduced cholinergic activity, widespread neuronal loss, and altered production of neuronal peptides crucial for cognitive functions. The amyloid cascade hypothesis posits that AD initiates with excessive Aβ production and plaque formation, leading to various detrimental effects, such as excitotoxicity, oxidative stress, energy depletion, inflammation, and cell death. Recent research indicates that Aβ disrupts glutamate signaling by inhibiting synaptic glutamate receptors, contributing to early cognitive impairment [61].

Cellular events in AD primarily involve necrosis and apoptosis pathways. Specific peptides derived from Aβ's C-terminus can induce neuronal apoptosis, particularly in the hippocampal mossy fiber pathway, contributing to memory deficits. Neuronal autophagy is both induced and impaired in AD, resulting in intracellular Aβ accumulation and plaque formation. Mutations associated with AD exacerbate disruption of the autophagy-lysosomal pathway [62]. Microglia presence in senile plaques suggests an autoimmune response, leading to neuronal damage through free radical production. Oxidative stress and lipid peroxidation, exacerbated by glutamate and Aβ, contribute to neurotoxic effects [63].

Evidence suggests that disruption of glutamate-related systems, particularly excitatory amino acid L-glutamate systems, is pivotal in AD pathogenesis [64]. Mitochondrial dysfunction, characterized by the loss of membrane potential and the subsequent release of apoptotic proteins, further exacerbates cell damage in AD. NMDA receptor hyperactivation-induced excitotoxicity may increase neuronal sensitivity in affected brain regions, aligning with AD neuropathology [65]. Consequently, NMDA receptor antagonists hold promise for alleviating AD symptoms. Studies investigating the relationship between oxidative stress, Aβ binding, and NMDAR activity highlight their crucial roles in AD development [66].

11.3.2 EXCITOTOXICITY IN RELATION TO PARKINSON'S DISEASE

PD is a neurological disorder characterized by the degeneration of dopaminergic neurons in the substantia nigra region of the brain, leading to reduced dopamine levels in the striatum [67]. Despite considerable research efforts, the underlying mechanisms of PD remain not fully elucidated. Evidence suggests a significant role of oxidative stress in initiating and hastening this degenerative process. Oxidative stress triggers the release of reactive oxygen species (ROS) and free radicals that cause harm to cellular proteins, lipids, and DNA components. In PD, the presynaptic protein, α-synuclein, is notably present in Lewy bodies, which are distinctive neuronal inclusions [68] associated with the condition. Certain studies suggest that damaged neurons in the substantia nigra of PD brains might accumulate aggregated α-synuclein. This extracellular α-synuclein can activate microglia to produce ROS and worsen dopaminergic neurodegeneration, thereby playing a role in the loss of substantia nigra neurons in PD.

Studies suggest that parkin, encoded by the *PARK2* gene associated with PD, has a significant impact on regulating the activity and stability of excitatory glutamatergic synapses. When parkin is expressed postsynaptically, it reduces excitatory synaptic transmission and leads to the depletion of excitatory synapses on hippocampal neurons [69]. Conversely, decreasing parkin expression levels or introducing PD-related changes to synapses enhances synaptic efficiency and promotes the development of glutamatergic synapses, thereby increasing susceptibility to synaptic excitotoxicity. Excessive glutamatergic activity resulting from this synaptic density increase can lead to excitotoxicity in the substantia nigra, where NMDA receptor activation elevates intracellular calcium levels [70].

The preservation of dopaminergic neurons that express the calcium-binding protein calbindin in PD suggests the involvement of elevated calcium levels in the cellular demise process [71].

The identification of 1-methyl-4-phenyl-1,2,3,6-tetrahydropyridine (MPTP) has offered valuable insights into the environmental factors contributing to PD. MPTP serves as a critical tool in PD research, facilitating the development of animal models to study pathogenesis and explore potential treatments for mitigating dopaminergic neurodegeneration. The active metabolite of MPTP, MPP+, inhibits complex I in the mitochondrial electron transport chain, contributing to the onset of PD. Additionally, the noncompetitive NMDA receptor antagonist MK-801 (dizocilpine) has demonstrated protective effects against MPTP-induced excitotoxicity in the dopaminergic system [72]. Emerging evidence suggests that MPTP may induce neuronal demise through dopamine vesicle release, ensuing oxidative damage, iron accumulation, and other mechanisms, such as cytoplasmic degradation and intraneuronal calcium dysregulation. While PD has not always been classified as a mitochondrial disorder, recent research underscores the impact of PD-associated proteins on mitochondrial function or their interaction with the mitochondrial functional hub [73].

11.3.3 EXCITOTOXICITY IN RELATION TO HUNTINGTON'S DISEASE

HD is a ND marked by cognitive, motor, and emotional disturbances, usually due to the failure or loss of certain neuron cells in the brain. Among these, GABAergic projections of medium-sized spiny neurons (MSNs) in the neostriatum are particularly affected. Interestingly, studies have shown that improving GABA receptor function may promote neuronal survival after ischemia by reducing overall cellular excitability [74]. The HD gene and its protein, HTT, were discovered more than a decade ago, leading to extensive research into the mechanisms of the disease. Researchers have made significant progress in uncovering the molecular pathways underlying the pathogenesis of HD [75].

This process involves alterations in electrical characteristics and sensitivity of glutamate receptors in cells expressing polyglutamine-expanded HTT [76]. A notable instance is HD, where the mutant HTT protein is thought to aggregate into insoluble structures within the nucleus, contributing to neuronal cell death. Moreover, research indicates that mutant HTT expression elevates NMDA receptor activity and disrupts calcium signaling.

Some studies propose a potential connection between HTT and NMDA receptor function through the postsynaptic density protein PSD-95. PSD-95 is known to interact with the C-terminal tail of the NR2 subunit, regulating NMDA receptors on the surface of cells. Data from experiments conducted on transplanted cells and human cortical tissue suggest an association between HTT, PSD-95, and NMDA receptor subunits NR1, NR2A, and NR2B [77].

These findings imply that HTT may interact with NMDA receptors via PSD-95. Recent research indicates that expanded polyglutamine sequences in HTT are linked to heightened NR1A/NR2B-mediated cell death due to excitotoxicity. Mutant HTT boosts NR1A/NR2B expression, which are the primary NMDA receptor subtypes in neostriatal spiny neurons [78]. This correlation may elucidate why these neurons are particularly susceptible to apoptosis, particularly in response to NMDA-induced excitotoxicity. Stimulation has been demonstrated to trigger apoptosis in projection neurons and animal models [79]. However, certain antiapoptotic proteins present in interneurons may offer neuroprotection against excitotoxic challenges. This could account for the enhanced survival of striatal interneurons observed in patients with HD and animal models of excitotoxicity [80]. Reports indicate altered NMDA receptor function at corticostriatal synapses, characterized by increased NMDA receptor-mediated currents and/or toxicity in striatal neurons across various HD mouse models and mutant brains [81]. NMDA receptor stimulation is closely related to mitochondrial activity, and studies have shown that the addition of NMDA receptor antagonists can reverse damage caused by mitochondrial toxins. Future studies should focus on uncovering the molecular pathways between HTT and NMDA receptors, as well as the mechanism by which mutant HTT increases NMDA receptor activity. Understanding these pathways and mechanisms may provide better insights into the pathogenesis of HD and potentially identify targets for therapeutic intervention.

11.4 CALCIUM DYSREGULATION

Calcium has a crucial function in brain processes and is important for maintaining the health and function of neurons. In the extracellular space of brain cells, the calcium level is generally between 1.1 and 1.4 mm, while in neurons the cytosolic level is tightly controlled and hovers around 100 nm [82]. This counterbalancing calcium gradient is essential for numerous brain functions, such as gene transcription, synaptic transmission, memory formation, apoptosis, and other cellular processes. Calcium primarily enters neurons through plasma membrane channels and is then regulated by buffers, calcium-binding proteins, and cellular organelles like the endoplasmic reticulum (ER), mitochondria, and lysosomes. Increased calcium levels can occur through the activation of plasma membrane channels or the release of internal calcium stores near the cell membrane, initiating various cellular processes. The progression of various cellular processes relies on factors, like cellular localization, calcium concentration, and metabolic status. When intracellular calcium utilization and homeostasis are disrupted, it can promote the development of ND, such as AD, HD, and PD. Extensive studies have shed light on how disturbances in neuronal calcium regulation contribute to motor and cognitive impairments in these disorders. Recognizing calcium dysregulation as an early or progressive mechanism in ND has prompted research into targeting calcium specifically as a therapeutic strategy. As understanding of the role of calcium dysregulation in these conditions improves, therapeutic approaches focusing on calcium channels are being explored [83].

11.4.1 CALCIUM DYSREGULATION AND ALZHEIMER'S DISEASE

AD is characterized histopathologically by two main features: the accumulation of extracellular plaques containing Aβ, and the development of intracellular NFTs composed of hyperphosphorylated microtubule-associated proteins. While certain genetic risk factors have been identified, the cause of the majority (>95%) of AD cases, which are sporadic, remains unclear. The most significant factor associated with AD is the presence of the E4 allele of apolipoprotein E (ApoE); having one allele increases the risk of AD threefold, while possessing both alleles elevates the risk by 12-fold. ApoE exists in three variants, with ApoE4 expression linked to AD risk, ApoE3 considered the "normal" allele, and ApoE2 is believed to provide protection against AD. ApoE plays a role in lipid homeostasis and cholesterol transport. ApoE4 is thought to have a higher affinity for the pathogenic form of Aβ, leading to increased amyloid accumulation in the CNS [84].

11.4.2 CALCIUM DYSREGULATION AND PARKINSON'S DISEASE

PD manifests with symptoms like slurred speech, cognitive decline, and psychiatric issues. Physical signs include slowed or diminished voluntary movements (bradykinesia), tremors, muscle stiffness (rigidity), and impaired posture. In advanced stages, PD may involve dementia and deficits in motor and visuospatial cognition [85]. These symptoms arise from the specific degeneration of dopamine neurons in the substantia nigra pars compacta (SNpc) in the midbrain, resulting in decreased dopamine levels in crucial brain regions that regulate movement, such as the striatum. Levodopa, a dopamine precursor, is a primary short-term treatment for PD, alleviating motor symptoms [86]. However, prolonged use may lead to dyskinesia or motor fluctuations. PD typically presents between ages 60 and 70, with an average survival duration ranging from 6 to 22 years. Like AD, PD occurs in both familial (FPD) and idiopathic (IPD) forms [87]. FPD is characterized by mutations inherited in an autosomal dominant pattern, whereas IPD accounts for over 80% of cases and lacks a known genetic cause [88]. Triploidy with duplications and mutations of the *SNCA* gene, formerly known as *Park1*, has been detected. SNCA encodes α-synuclein, a protein extensively expressed in the presynaptic compartment. It holds significance in synaptic function by orchestrating and assembling synaptic vesicles and controlling their release. Mutations in SNCA are linked to diminished nigrostriatal dopamine (DA) transmission in PD [89]. In PD, a key histological characteristic is the buildup of abnormal forms of α-synuclein protein, forming structures known as Lewy bodies. These Lewy bodies are linked to the loss of connections between neurons (synaptic loss) and disturbances in calcium regulation. Specifically, calcium channels, including Cav1.2 and Cav1.3, are implicated in this dysregulation [90]. DA neurons in the SNpc rely on calcium channels for regulating DA release. In PD, there's a shift in the relative importance of these channels, with Cav1.3 becoming more prominent compared to Cav1.2. This shift results in elevated levels of cytosolic calcium, leading to stress on mitochondria and eventual neuronal death. Moreover, abnormal calcium flow from the ER through ryanodine receptors (RyRs) has been observed in PD. Genome-wide association studies (GWAS) have identified genes related to calcium regulation in PD, suggesting that disruptions in cellular calcium handling and voltage-dependent calcium entry may act together to induce mitochondrial stress, contributing to the disease's progression [91].

11.4.3 CALCIUM DYSREGULATION AND HUNTINGTON'S DISEASE

HD is characterized by choreiform motor movements, which involve sudden, uncontrollable jerking motions of the face, body, and limbs.

This loss of motor control progresses quickly, eventually impairing walking and balance. As the disease advances, it can lead to breathing difficulties and ultimately death. HD can also involve cognitive and psychological issues, reflecting damage to the fronto-striatal motor control pathways. Onset of HD usually occurs around the age of 40, though this can vary based on the number of CAG repeats in the affected gene. Currently, there are no cures or effective treatments for HD, and individuals diagnosed with it typically have a life expectancy of 10 to 20 years post-diagnosis [92].

Unlike many other ND, HD has well-understood genetic roots. The gene responsible, *HTT*, is located on chromosome 4. It encodes a protein that interacts with over 200 other proteins, crucial for functions, such as development, cellular transportation, and apoptosis. The pathogenicity of mutant HTT arises from an excessive number of CAG repeats, leading to unusually long polyglutamine chains at the protein's amino-terminal end. This places HD among the polyglutamine disorders, like spinocerebellar ataxia. Mutant HTT undergoes proteolytic cleavage, resulting in amino-terminal fragments that aggregate to form inclusion bodies within neurons. The age of onset and severity of HD are largely determined by the number of CAG repeats in the *HTT* gene [93]. Individuals typically have 10–35 repeats of the CAG sequence in the *HTT* gene, while those with 40 or more repeats (referred to as mHTT) show signs of HD. Intriguingly, although the HTT protein is expressed throughout the body, the GABAergic medium spiny neurons (MSNs) in the neostriatum are notably vulnerable to the effects of the disease.

11.5 SYNAPTIC DYSFUNCTION AND PLASTICITY

Synaptic dysfunction and plasticity are fundamental concepts in neuroscience that underlie learning, memory, and various neurological disorders. Here's an overview of these concepts:

1. **Synaptic dysfunction:** Synapses are the junctions between neurons where communication occurs. Synaptic dysfunction refers to abnormalities in the function of these synapses, leading to impaired neuronal communication. This dysfunction can arise from various factors, including alterations in neurotransmitter release, changes in receptor sensitivity, disruptions in synaptic structure, and dysfunction of supporting cells, such as astrocytes [94].

2. **Plasticity:** Synaptic plasticity is the capacity of synapses to alter their strength or effectiveness in response to activity. This phenomenon is essential for processes, such as learning and memory, as well as the brain's capability to adjust to new experiences. Various types of synaptic plasticity exist, including long-term potentiation (LTP) and long-term depression (LTD), which entail the enhancement and reduction of synaptic connections, respectively [95].
3. **Mechanisms of synaptic plasticity:** Synaptic plasticity involves a range of cellular and molecular mechanisms. Central to this process is the activation of glutamate receptors, notably NMDA receptors and AMPA receptors. Adjustments in the quantity and operation of these receptors, along with modifications in intracellular signaling pathways, play vital roles in both initiating and sustaining synaptic plasticity [96].
4. **Role in neurological disorders:** The dysregulation of synaptic plasticity has been linked to various neurological disorders, including AD, PD, schizophrenia, and autism spectrum disorders. In these conditions, abnormal synaptic plasticity may contribute to cognitive impairments, motor dysfunction, and other symptoms. Understanding the underlying mechanisms of synaptic dysfunction and plasticity is crucial for developing new therapeutic strategies for these disorders [97].

Overall, synaptic dysfunction and plasticity are core concepts in neuroscience that are pivotal for both normal brain function and the development of neurological disorders. Ongoing research in this field is vital for unraveling the intricate mechanisms of synaptic plasticity and devising effective treatments for neurological conditions.

11.6 TAU AND AMYLOID PATHOLOGY

The presence of tau and amyloid pathology in AD is closely associated with the activity of NMDA receptors, a subtype of glutamate receptors located at excitatory synapses in the brain. Here's a detailed explanation of the connection between tau and amyloid pathology and NMDA receptors:

1. **Tau pathology and NMDA receptors:** Tau pathology is characterized by the accumulation of hyperphosphorylated tau protein

in intracellular NFTs, which may affect NMDA receptor function. This disrupts the neuron's intracellular transport system, leading to poor organ function and neuronal degeneration. This effect may directly affect NMDA receptor traffic and localization at synapses. Hyperphosphorylated tau can bind to the NR2B subunit of the NMDA receptor, resulting in increased receptor activity and excitotoxicity, resulting in impaired brain function and cell death in AD.

2. **Amyloid pathology and NMDA receptors:** Amyloid pathology involves the buildup of Aβ peptide in extracellular plaques and impacts NMDA receptor function. Specifically, Aβ oligomers have been found to heighten NMDA receptor-mediated excitatory synaptic transmission, resulting in calcium overload and excitotoxicity. This alteration in NMDA receptor-dependent signaling pathways disrupts synaptic plasticity, leading to impairments, such as LTP and LTD. These synaptic dysfunctions, coupled with neuronal hyperexcitability, contribute to neurodegeneration in AD [98].

3. **Implications for synaptic dysfunction and neurodegeneration:** Dysfunction of NMDA receptors due to tau and amyloid pathology leads to functional impairment, immune dysfunction, and excitotoxicity, all of which contribute to neurodegeneration in AD. Calcium influx causes oxidative stress, mitochondrial dysfunction, and cell death, ultimately leading to neuron loss. Various NMDA receptor antagonists have been studied for their neuroprotective effects in clinical trials, although their results are still controversial. Interactions disrupt synaptic function, induce excitotoxicity, and cause neurodegenerative changes. Understanding the mechanisms of these interactions is important for the development of effective services for AD.

11.7 THERAPEUTIC STRATEGIES

Therapeutic strategies targeting NMDA receptors in NDs aim to modulate NMDA receptor activity to alleviate synaptic dysfunction, excitotoxicity, and neuronal degeneration. Here are some insights into these strategies.

11.7.1 NMDA RECEPTOR ANTAGONISTS

NMDA receptor antagonists work by inhibiting the activity of NMDA receptors, thus reducing excitatory neurotransmission and preventing excitotoxicity. Memantine, an example of such antagonist, is approved for treating moderate to severe AD. Memantine acts as a low-affinity, voltage-dependent, noncompetitive antagonist, specifically aiming at overactive NMDA receptors while preserving normal synaptic function [99].

Ketamine and dextromethorphan are among the other NMDA receptor antagonists studied for their potential neuroprotective effects in ND. However, their effectiveness in clinical settings is still being investigated.

11.7.2 POSITIVE ALLOSTERIC MODULATORS

Positive allosteric modulators work by augmenting the activity of NMDA receptors through binding to allosteric sites on the receptor complex. This action boosts the probability of channel opening without directly activating the receptor itself [100].

There's promising potential for these modulators to improve synaptic plasticity and cognitive function in neurodegenerative conditions. Nevertheless, careful consideration is essential to prevent over activation of NMDA receptors and the associated excitotoxicity.

11.7.3 SUBTYPE-SELECTIVE MODULATORS

Subtype-selective modulators target specific subunits or splice variants of NMDA receptors to achieve more precise modulation of receptor function.

For example, NR2B-selective antagonists, such as ifenprodil and Ro 25-6981, have been investigated for their potential neuroprotective effects in NDs. These compounds preferentially block NR2B-containing NMDA receptors, which are implicated in excitotoxicity and synaptic dysfunction.

11.7.4 SYNAPTIC STABILIZERS

Synaptic stabilizers aim to restore synaptic function and prevent excitotoxicity by modulating NMDA receptor activity indirectly.

These compounds may target downstream signaling pathways involved in synaptic plasticity and excitability regulation.

Examples include glycine transporter inhibitors, which enhance NMDA receptor function by increasing synaptic glycine levels and AMPA receptor potentiators, which promote synaptic transmission and plasticity.

11.7.5 NOVEL THERAPEUTIC TARGETS

Emerging research is uncovering novel therapeutic targets within the NMDA receptor signaling pathway, such as intracellular signaling molecules and regulatory proteins.

Targeting these molecules may offer new opportunities for therapeutic intervention with improved efficacy and reduced side effects [101].

In summary, therapeutic strategies targeting NMDA receptors in NDs aim to restore synaptic function, prevent excitotoxicity, and promote neuronal survival. While NMDA receptor modulation holds promise as a therapeutic approach, further research is needed to optimize drug efficacy, safety, and specificity for clinical use.

11.8 GENETIC AND EPIGENETIC FACTORS

Genetic and epigenetic factors play pivotal roles in regulating NMDA receptors in NDs. Following is an overview of how these factors contribute to NMDA receptor dysfunction in such conditions:

1. **Genetic factors:**
    a. **Gene mutations:** Mutations in genes encoding NMDA receptor subunits or associated proteins directly impact NMDA receptor function. For instance, mutations in NR1, NR2A, NR2B, and NR2C subunit genes have been implicated in various NDs.
    b. **Single nucleotide polymorphisms (SNPs):** Genetic variations, including SNPs, in genes encoding NMDA receptor subunits or regulatory proteins influence receptor expression, function, and susceptibility to ND. SNPs in the *GRIN2A* gene, encoding the NR2A subunit, are associated with increased AD risk.
    c. **Gene expression regulation:** Regulatory elements and transcription factors control NMDA receptor subunit gene expression. Dysregulation, such as altered promoter methylation

or histone modifications, leads to aberrant NMDA receptor expression and function in NDs.

2. **Epigenetic factors:**
    a. **DNA methylation:** DNA methylation, altering chromatin structure, regulates gene expression. Changes in DNA methylation patterns at NMDA receptor gene promoters occur in ND, affecting receptor expression levels.
    b. **Histone modifications:** Post-translational histone modifications influence chromatin remodeling and gene transcription. Dysregulation at NMDA receptor gene loci contributes to altered receptor expression and function in NDs.
    c. **Non-coding RNAs:** MicroRNAs (miRNAs) and long non-coding RNAs (lncRNAs) play roles in regulating gene expression after transcription. Dysregulation of these molecules, particularly in targeting NMDA receptor subunit genes or regulatory factors, is associated with ND.
3. **Interaction with environmental factors:** Stress, diet, and toxin exposure interact with genetic and epigenetic mechanisms, affecting NMDA receptor function and contributing to neurodegenerative processes. Environmental stressors induce epigenetic modifications at NMDA receptor gene loci, altering receptor expression and increasing susceptibility to neurodegeneration.

In summary, genetic and epigenetic factors are critical in regulating NMDA receptors in NDs, leading to dysfunctional NMDA receptor activity, synaptic impairment, and neuronal degeneration. Understanding the intricate interplay among genetic, epigenetic, and environmental factors is vital for developing targeted therapeutic strategies for AD, PD, HD, and other devastating NDs.

11.9 NEUROPROTECTIVE STRATEGIES

Neuroprotective strategies targeting NMDA receptors in NDs aim to mitigate excitotoxicity, synaptic dysfunction, and neuronal degeneration. Following are some neuroprotective approaches along with references supporting their potential therapeutic efficacy.

1. **NMDA receptor antagonists:** Memantine, treating moderate to severe AD, acts as an NMDA receptor antagonist. By targeting excessive NMDA receptor activity while preserving normal neurotransmission, it reduces excitotoxicity and sustains synaptic function [102].
2. **Positive allosteric modulators:** D-cycloserine acts as a partial agonist at the glycine modulatory site of the NMDA receptor, thereby augmenting its activity. It has exhibited neuroprotective effects in animal models of NDs by promoting synaptic plasticity and improving cognitive function [103].
3. **Subtype-selective modulators:** Ifenprodil is an NR2B-selective antagonist that exhibits neuroprotective properties in animal models of neurodegenerative conditions like PD and HD. It specifically inhibits NR2B-containing NMDA receptors, thereby mitigating excitotoxicity and neuronal loss [104].
4. **Synaptic stabilizers:** Glycine transporter inhibitors: Glycine transporter inhibitors, such as sarcosine, enhance NMDA receptor function by increasing synaptic glycine levels. Preclinical studies have shown that sarcosine attenuates excitotoxicity and improves cognitive function in animal models of ND [105].
5. **Novel therapeutic targets:** Tau protein modulators: Dysregulation of tau protein is implicated in NDs. Targeting tau protein aggregation and phosphorylation, which can influence NMDA receptor function, is an emerging neuroprotective strategy [106].
6. **Combination therapies:** Combining memantine with cholinesterase inhibitors like donepezil has demonstrated synergistic neuroprotective benefits in AD. This approach targets various pathways involved in neurodegeneration, including excitotoxicity mediated by NMDA receptors and cholinergic dysfunction [107].

11.10 EMERGING RESEARCH ON NMDA RECEPTORS IN NDS

Research has provided valuable insights into the involvement of NMDA receptors in various NDs, including AD, PD, HD, and amyotrophic lateral sclerosis (ALS). The following are some notable findings from these emerging research endeavors:

1. **Alzheimer's disease (AD):** Research suggests that dysfunctional NMDA receptors contribute to synaptic dysfunction and cognitive decline observed in AD. Furthermore, there is increasing evidence indicating that NMDA receptor-mediated excitotoxicity contributes to neurodegeneration in AD by causing excessive neuronal calcium levels and oxidative stress. Studies have highlighted the role of Aβ oligomers in disrupting NMDA receptor function, leading to synaptic dysfunction and neuronal damage [108].
2. **Parkinson's disease (PD):** NMDA receptor-mediated excitotoxicity is associated with the degeneration of dopaminergic neurons in PD. Studies indicate that abnormal NMDA receptor activation exacerbates neuronal susceptibility to mitochondrial dysfunction and oxidative stress, thereby contributing to the development of PD. Modulating NMDA receptor activity has demonstrated neuroprotective effects in preclinical PD models, indicating the potential therapeutic value of NMDA receptor-targeted interventions.
3. **Huntington's disease (HD):** Dysregulation of NMDA receptor function relates to the pathogenesis of HD, a ND characterized by impaired motor function and cognitive decline. Studies have demonstrated increased NMDA receptor activity in HD, leading to calcium dysregulation, excitotoxicity, and neuronal death. Targeting NMDA receptors has emerged as a potential therapeutic strategy for ameliorating excitotoxicity and mitigating neurodegeneration in HD.
4. **Amyotrophic Lateral Sclerosis (ALS):** Changes in the expression and activity of NMDA receptors have been noted in ALS, a progressive motor neuron ailment. Excitotoxic mechanisms involving NMDA receptors are believed to contribute to the preferential susceptibility of motor neurons in ALS. Preclinical investigations into NMDA receptor modulation have shown promising results in delaying disease progression and preserving motor function in animal models of ALS. In conclusion, emerging research underscores the significance of NMDA receptors in the pathogenesis of neurodegenerative conditions and suggests their potential as therapeutic targets for modifying disease trajectory. Further exploration is necessary to fully comprehend the exact mechanisms underlying NMDA receptor dysfunction in various NDs and to develop effective treatments aimed at halting or slowing disease progression [109, 110].

TABLE 11.1 Endogenous and Pharmacological Compounds that Target NMDARs.

Endogenous Molecules	Pharmacological Molecules
Channel blockers and modulators: Mg^{2+}	**Channel blockers and modulators:**
Neurosteroids	MK-801
	Aptiganel
	Dextromethorphan
	Amantadine
	Memantine
	Ketamine
	Methadone
	Dimebon
Glycine/Glutamate site antagonists:	**Glycine/Glutamate site antagonists:**
Polyamines	Selfotel
Zn^{2+}; Cu^{2+}	Rapastinel
	NYX-2925
	Apimostinel
	Gavestinel
	AV-101
	Phencyclidine
PAMs and NAMs:	**PAMs and NAMs:**
Spermines and spermidines Cholesterol	SAGE-718
Pregnelonone sulfate	NP10679
ATP	Neu2000
Downstream antagonist:	**Downstream antagonists:**
Uric acid	Nerinetide
	AVLX-144
	ARG-007
	ZL006
	Tramiprosate
	Edaravone dexborneol
	Tat-CAPON and ZLc-002
	Bpv and Tat-K13
Modulators:	**Antibody:**
Amyloid-$ (AB)	Glunomab
tPA	

11.11 CONCLUSION

In summary, unraveling the insights of NMDA receptors in NDs has deepened our comprehension of disease development and treatment strategies, leading to new opportunities for therapeutic intervention. By addressing the complex interplay between NMDA receptors and neurodegeneration, researchers aim to pave the way for innovative treatments that alleviate the burden of these disorders on patients and caregivers.

KEYWORDS

- **calcium dysregulation**
- **excitotoxicity**
- **neurodegenerative disorders**
- **nanotherapeutics**
- **NMDA receptors**
- **neuronal death**
- **synaptic plasticity**

REFERENCES

1. Merelli, A., Czornyj, L., & Lazarowski, A. (2013). Erythropoietin: A neuroprotective agent in cerebral hypoxia, neurodegeneration, and epilepsy. *Current Pharmaceutical Design, 19*, 6791–6801.
2. Choonara, Y. E., Pillay, V., Du Toit, L. C., Modi, G., Naidoo, D., Ndesendo, V. M., & Sibambo, S. R. (2009). Trends in the molecular pathogenesis and clinical therapeutics of common neurodegenerative disorders. *International Journal of Molecular Sciences, 10*, 2510–2557.
3. Rapp, T., Chauvin, P., Costa, N., & Molinier, L. (2015). Health economic considerations in neurodegenerative disorders. *Imaging in Neurodegenerative Disorders, 42*(3), 1851.
4. Harilal, S., Jose, J., Parambi, D. G. T., Kumar, R., Mathew, G. E., Uddin, M. S., Kim, H., & Mathew, B. (2019). Advancements in nanotherapeutics for Alzheimer's disease: Current perspectives. *Journal of Pharmacy and Pharmacology, 71*, 1370–1383.
5. Montazersaheb, S., Ahmadian, E., Maleki Dizaj, S., Jahanbani, Y., Davaran, S., Huseynova, I., Zhdanov, R., Keskin, C., Khalilov, R., & Eftekhari, A. (2021). Emerging nanotherapeutic strategies in Alzheimer's disease. In *Frontiers in clinical drug research-dementia* (Vol. 2, p. 173). Singapore: Bentham Science Publishers Private Limited.

6. NIH. (2002). Brain Basics: The Life and Death of a Neuron. Bethesda, Maryland, USA: Office of Communications and Public Liaison, National Institute of Neurological Disorders and Stroke.
7. Van den Heuvel, M. P., & Sporns, O. (2013). Network hubs in the human brain. *Trends in Cognitive Sciences, 17*, 683–696.
8. Kempermann, G. (2006). *Adult neurogenesis: Stem cells and neuronal development in the adult brain.* New York, USA: Oxford University Press.
9. Ganat, Y. M., Silbereis, J., Cave, C., Ngu, H., Anderson, G. M., Ohkubo, Y., Ment, L. R., & Vaccarino, F. M. (2006). Early postnatal astroglial cells produce multilineage precursors and neural stem cells in vivo. *The Journal of Neuroscience, 26*, 8609–8862.
10. Przedborski, S., Vila, M., & Jackson-Lewis, V. (2003). Series introduction: neurodegeneration: What is it and where are we? *Journal of Clinical Investigation, 111*, 3–10.
11. Hoover, B. R., Reed, M. N., Su, J., Penrod, R. D., Kotilinek, L. A., Grant, M. K., Pitstick, R., Carlson, G. A., Lanier, L. M., & Yuan, L. L. (2010). Tau mislocalization to dendritic spines mediates synaptic dysfunction independently of neurodegeneration. *Neuron, 68*, 1067–1081.
12. Martin, J. B. (1999). Molecular basis of the neurodegenerative disorders. *The New England Journal of Medicine, 340*, 1970–1980.
13. Hague, S., Klaffke, S., & Bandmann, O. (2005). Neurodegenerative disorders: Parkinson's disease and Huntington's disease. *Journal of Neurology, Neurosurgery and Psychiatry, 76*, 1058–1063.
14. Mehta, P., Raymond, J., Han, M., Larson, T., Kaye, W., & Paganoni, S. (2021). Amyotrophic lateral sclerosis (ALS) prevalence, incidence, and demographic characteristics: Findings from the National ALS Registry, 2015–2018. *Morbidity and Mortality Weekly Report (MMWR), 70*(29), 1013–1018.
15. Gaugler, J., James, B., Johnson, T., Marin, A., & Weuve, J. (2020). 2020 Alzheimer's disease facts and figures. *Alzheimer's & Dementia, 16*(3), 391–460.
16. Bekris, L. M., Yu, C. E., Bird, T. D., & Tsuang, D. W. (2010). Genetics of Alzheimer disease. *Journal of Geriatric Psychiatry and Neurology, 23*(4), 213–227.
17. Patterson, C. (2018). World Alzheimer Report 2018. The state of the art of dementia research: New frontiers. *Alzheimer's Disease International (ADI)*. https://www.alzint.org/resource/world-alzheimer-report-2018/.
18. Dugger, B. N., & Dickson, D. W. (2017). Pathology of neurodegenerative diseases. *Cold Spring Harbor Perspectives in Biology, 9*(7), a028035.
19. DeTure, M. A., & Dickson, D. W. (2019). The neuropathological diagnosis of Alzheimer's disease. *Molecular Neurodegeneration, 14*(1), 32.
20. Perez Ortiz, J. M., & Swerdlow, R. H. (2023). Mitochondrial dysfunction in Alzheimer's disease: Role in pathogenesis and novel therapeutic opportunities. *British Journal of Pharmacology, 180*(9), 1963–1975. https://doi.org/10.1111/bph.15859.
21. Cummings, J., Lee, G., Nahed, M., Kullar, R., & Zhong, K. (2023). Alzheimer's disease drug development pipeline: 2023. *Alzheimer's & Dementia: Translational Research & Clinical Interventions, 9*(1), e12385. https://doi.org/10.1002/trc2.12385.
22. Mancuso, M., Orsucci, D., LoGerfo, A., Calsolaro, V., Siciliano, G., & Murri, L. (2023). Diagnostic approach to neurodegenerative disorders: Alzheimer's disease

and Parkinson's disease. *Current Medicinal Chemistry, 30*(3), 305–317. https://doi.org/10.2174/0929867329666190227121115.

23. Ballard, C., Gauthier, S., Corbett, A., Brayne, C., Aarsland, D., & Jones, E. (2023). Alzheimer's disease. *Lancet, 397*(10284), 1577–1590. https://doi.org/10.1016/S0140-6736(21)00270-8.
24. Swerdlow, R. H., & Khan, S. M. (2023). Mitochondrial therapeutics in Alzheimer's disease. *Neurotherapeutics, 20*(3), 516–533. https://doi.org/10.1007/s13311-022-01185-5.
25. Wyss-Coray, T. (2023). Ageing, neurodegeneration and brain rejuvenation. *Nature, 591*(7851), 343–355.
26. Hardingham, G. E., & Do, K. Q. (2016). Linking early-life NMDAR hypofunction and oxidative stress in schizophrenia pathogenesis. *Nature Reviews Neuroscience, 17*(2), 125–134.
27. Paoletti, P., Bellone, C., & Zhou, Q. (2013). NMDA receptor subunit diversity: Impact on receptor properties, synaptic plasticity and disease. *Nature Reviews Neuroscience, 14*(6), 383–400.
28. Traynelis, S. F., Wollmuth, L. P., McBain, C. J., Menniti, F. S., Vance, K. M., Ogden, K. K., Hansen, K. B., Yuan, H., Myers, S. J., & Dingledine, R. (2010). Glutamate receptor ion channels: structure, regulation, and function. *Pharmacological Reviews, 62*(3), 405–496.
29. Lau, C. G., & Zukin, R. S. (2007). NMDA receptor trafficking in synaptic plasticity and neuropsychiatric disorders. *Nature Reviews Neuroscience, 8*(6), 413–426.
30. Papouin, T., Ladépêche, L., Ruel, J., Sacchi, S., Labasque, M., Hanini, M., Groc, L., Pollegioni, L., Mothet, J. P., & Oliet, S. H. R. (2012). Synaptic and extrasynaptic NMDA receptors are gated by different endogenous coagonists. *Cell, 150*(3), 633–646.
31. Parsons, M. P., & Raymond, L. A. (2014). Extrasynaptic NMDA receptor involvement in central nervous system disorders. *Neuron, 82*(2), 279–293.
32. Bayer, T. A., & Wirths, O. (2010). Intracellular accumulation of amyloid-Beta-a predictor for synaptic dysfunction and neuron loss in Alzheimer's disease. *Frontiers in Aging Neuroscience*, *2*, 8.
33. Butterfield, D. A., Griffin, S., Munch, G., & Pasinetti, G. M. (2002). Amyloid β-peptide and amyloid pathology are central to the oxidative stress and inflammatory cascades under which Alzheimer's disease brain exists. *Journal of Alzheimer's Disease*, *4*, 193–201.
34. Spires-Jones, T. L., & Hyman B. T. (2014). The intersection of amyloid beta and tau at synapses in Alzheimer's disease. *Neuron, 82*, 756–771.
35. Akhondzadeh, S., & Noroozian, M. (2002). Alzheimer's disease: Pathophysiology and pharmacotherapy. *IDrugs Investigation Drugs Journal*, *5*, 1062–1069.
36. Kouli, A., Torsney, K. M., & Kuan, W. L. (2018). Parkinson's disease: Etiology, neuropathology, and pathogenesis. In T. B. Stoker & J. C. Greenland (Eds.), *Parkinson's disease: Etiology, neuropathology, and pathogenesis* (pp. 3–26). Brisbane (AU): Codon Publications.
37. Priyadarshi, A., Khuder S. A., Schaub, E. A., & Priyadarshi, S. S. (2001). Environmental risk factors and Parkinson's disease: A metaanalysis. *Environmental Research, 86*, 122–127.

38. Bartels, A. L., & Leenders, K. L. (2009). Parkinson's disease: The syndrome, the pathogenesis and pathophysiology. *Cortex*, *45*, 915–921.
39. Chaudhuri, K. R., & Schapira, A. H. (2009). Non-motor symptoms of Parkinson's disease: Dopaminergic pathophysiology and treatment. *The Lancet Neurology*, *8*, 464–474.
40. Moore, D. J., West, A. B., Dawson, V. L., & Dawson, T. M. (2005). Molecular pathophysiology of Parkinson's disease. *Annual Review of Neuroscience*, *28*, 57–87.
41. Schulz-Schaeffer, W. J. (2010). The synaptic pathology of α-synuclein aggregation in dementia with Lewy bodies, Parkinson's disease and Parkinson's disease dementia. *Acta Neuropathologica*, *120*, 131–143.
42. NINDS. (2023). *Huntington's disease*. National Institute of Neurological Disorders and Stroke.
43. Ross, C. A., & Tabrizi, S. J. (2011). Huntington's disease: From molecular pathogenesis to clinical treatment. *Lancet Neurology, 10*(1), 83–98.
44. Walker, F. O. (2007). Huntington's disease. *Lancet, 369*(9557), 218–228.
45. Gendron, T. F., & Petrucelli, L. (2009). The role of tau in neurodegeneration. *Molecular Neurodegeneration*, *4*, 1–19.
46. Roos, R. A. (2010). Huntington's disease: A clinical review. *Orphanet Journal of Rare Diseases*, *5*, 40.
47. Liu, Y., Wong, T. P., Aarts, M., & et al. (2007). NMDA receptor subunits have differential roles in mediating excitotoxic neuronal death both in vitro and in vivo. *The Journal of Neuroscience*, *27*(11), 2846–2857.
48. Lipton, S. A. (2006). Paradigm shift in neuroprotection by NMDA receptor blockade: Memantine and beyond. *Nature Reviews Drug Discovery*, *5*(2), 160–170.
49. Parsons, M. P., Raymond, L. A. (2014). Extrasynaptic NMDA receptor involvement in central nervous system disorders. *Neuron*, *82*(2), 279–293.
50. Mattson, M. P. (2003). Excitotoxic and excitoprotective mechanisms: Abundant targets for the prevention and treatment of neurodegenerative disorders. *Neuromolecular Medicine*, *3*(2), 65–94.
51. Choi, D. W. (1985). Glutamate neurotoxicity in cortical cell culture is calcium dependent. *Neuroscience Letters*, *58*(3), 293–297.
52. Selkoe, D. J. (1997). Alzheimer's disease: Genotypes, phenotype, and treatments. *Science*, *275*(5300), 630–631.
53. Shankar, G. M., & Walsh, D. M. (2009). Alzheimer's disease: Synaptic dysfunction and Aβ. *Molecular Neurodegeneration*, *4*, 48.
54. Terry, R. D., Masliah, E., Salmon, D. P., & et al. (1991). Physical basis of cognitive alterations in Alzheimer's disease: Synapse loss is the major correlate of cognitive impairment. *Annals of Neurology*, *30*(4), 572–580.
55. Wyss-Coray, T., & Mucke, L. (2002). Inflammation in neurodegenerative disease—A double-edged sword. Neuron, *35*(3), 419–432.
56. Glass, C. K., Saijo, K., Winner, B., Marchetto, M. C., & Gage, F. H. (2010). Mechanisms underlying inflammation in neurodegeneration. *Cell*, *140*(6), 918–934.
57. Li, S., Hong, S., Shepardson, N. E., & et al. (2009). Soluble oligomers of amyloid Beta protein facilitate hippocampal long-term depression by disrupting neuronal glutamate uptake. *Neuron*, *62*(6), 788–801.

58. Ittner, L. M., & Götz, J. (2011). Amyloid-β and tau--A toxic pas de deux in Alzheimer's disease. *Nature Reviews Neuroscience*, *12*(2), 67–72.
59. Shankar, G. M., Li, S., Mehta, T. H., & et al. (2008). Amyloid-beta protein dimers isolated directly from Alzheimer's brains impair synaptic plasticity and memory. *Nature Medicine*, *14*(8), 837–842.
60. Zeron, M. M., Hansson, O., Chen, N., & et al. (2002). Increased sensitivity to N-methyl-D-aspartate receptor-mediated excitotoxicity in a mouse model of Huntington's disease. *Neuron*, *33*(6), 849–860.
61. Cepeda, C., Wu, N., André, V. M., Cummings, D. M., & Levine, M. S. (2007). The corticostriatal pathway in Huntington's disease. *Progress in Neurobiology*, *81*(5–6):253–271.
62. Finkbeiner, S. (2011). Huntington's disease. *Cold Spring Harbor Perspectives in Biology*, *3*(6), a007476.
63. Busciglio, J., & Yankner, B. A. (1995). Apoptosis and increased generation of reactive oxygen species in Down's syndrome neurons in vitro. *Nature*, *378*, 776–779.
64. Parameshwaran, K., Dhanasekaran, M., & Suppiramaniam, V. (2008). Amyloid beta peptides and glutamatergic synaptic dysregulation. *Experimental Neurology*, *210*, 7–13.
65. Nixon, R. A. (2007). Autophagy, amyloidogenesis and Alzheimer disease. *Journal of Cell Science*, *120*, 4081–4091.
66. Hynd, M. R., Scott, H. L., & Dodd, P. R. (2004). Glutamate mediated excitotoxicity and neurodegeneration in Alzheimer's disease. *Neurochemistry*, *45*, 583–595.
67. Kuo, C. T., Hsua, M. J., Chen, B. C., Chen, C. C., Teng, C. M., Pan, S. L., & et al. (2008). Denbinobin induces apoptosis in human lung adenocarcinoma cells via Akt inactivation, Bad activation, and mitochondrial dysfunction. *Toxicology Letters*, *177*, 48–58.
68. Miguel-Hidalgo, J. J., Alvarez, X. A., Cacabelos, R., & Quack, G. (2002). Neuroprotection by memantine against neurodegeneration induced by ß-amyloid1-40. *Brain Research*, *958*, 210–221.
69. De Felice, F. G., Velasco, P. T., Lambert, M. P., Viola, K., Fernandez, S. J., Ferreira, S. T., & et al. (2007). Aß oligomers induce neuronal oxidative stress through an N-methyl-D-aspartate receptor-dependent mechanism that is blocked by the Alzheimer drug memantine. *Journal of Biological Chemistry*, *282*, 11590–11601.
70. Norris, E. H., Giasson, B. I., Hodara, R., & et al. (2005). Reversible inhibition of alpha-synuclein fibrillization by dopaminochrome-mediated conformational alterations. *Journal of Biological Chemistry*, *280*(22), 21212–21219. doi:10.1074/jbc.m501825200
71. Zhu, X. R., Maskri, L., Herold, C., & et al. (2007). Non-motor behavioural impairments in parkin-deficient mice. *European Journal of Neuroscience*, *26*(7), 1902–1911.
72. Obeso, J. A., Olanou, C. W., & Nutt, J. G. (2000). Levodopa motor complications in Parkinson's disease. *Trends in Neurosciences*, *23*, 2–7.
73. Zhang, W., Wang, T. G., Pei, Z., Miller, D. S., Wu, X. F., Block, M. L., & et al. (2005). Aggregated-synuclein activates microglia: A process leading to disease progression in Parkinson's disease. FASEB Journal, *19*, 533–542.

74. Chinta, S. J., & Andersen, J. K. (2005). Dopaminergic neurons. *The International Journal of Biochemistry and Cell Biology*, *37*(5), 942–946.
75. Vonsattel, J. P., & DiFiglia, M. (1998). Huntington disease. *Journal of Neuropathology and Experimental Neurology*, *57*(5), 369–384.
76. Reiner, A., Albin, R. L., Anderson, K. D., & et al. (1988). Differential loss of striatal projection neurons in Huntington disease. *Proceedings of the National Academy of Sciences of the United States of America*, *85*(15), 5733–5737.
77. Lang, A. E., & Lozano, A. M. (1998). Parkinson's disease-first of two parts. *Medical Progression*, *339*, 1044–1053.
78. Moring, J., Niego, L. A., Ganley, L. M., Trumbore, M. W., & Herbette, L. G. (1994). Interaction of the NMDA receptor noncompetitive antagonist mk-801 with model and native membranes. *Biophysical Journal*, *67*, 2376–2386.
79. Li, S. H., & Li, X. J. (2004). Huntingtin-protein interactions and the pathogenesis of Huntington's disease. *Trends in Genetics*, *20*(3), 146–154.
80. Milnerwood, A. J., & Raymond, L. A. (2010). Early synaptic pathophysiology in neurodegeneration: Insights from Huntington's disease. *Trends in Neurosciences*, *33*(11), 513–523.
81. Cha, J. H. (2007). Transcriptional signatures in Huntington's disease. *Progress in Neurobiology*, *83*(4), 228–248.
82. Van Raamsdonk, J. M., Pearson, J., Rogers, D. A., Bissada, N., Vogl, A. W., Hayden, M. R., & et al. (2005). Loss of wild-type huntingtin influences motor dysfunction and survival in the YAC128 mouse model of Huntington's disease. *Human Molecular Genetics*, *14*, 1379–1392.
83. Sun, Y., Savanenin, A., Reddy, P. H., & Liu, Y. F. (2001). Polyglutamine-expanded huntingtin promotes sensitization of N-methyl-D-aspartate receptors via post-synaptic density 95. *Journal of Biological Chemistry*, *276*, 24713–24718.
84. Liang, Z. Q., Wang, X. X., Wang, Y. M., Chang, D. M., DiFiglia, M., Chase, T. N., & et al. (2005). Susceptibility of striatal neurons to excitotoxic injury correlates with basal levels of Bcl-2 and the induction of P53 and c-Myc immunoreactivity. *Neurobiology of Disease*, *20*, 562–573.
85. Berridge, M. J., Lipp, P., & Bootman, M. D. (2000). The versatility and universality of calcium signalling. *Nature Reviews Molecular Cell Biology*, *1*, 11–21.
86. Jankovic, J. (2008). Parkinson's disease: Clinical features and diagnosis. *Journal of Neurology, Neurosurgery and Psychiatry*, *79*(4), 368–376. doi:10.1136/jnnp.2007.131045.
87. Kalia, L. V., & Lang, A. E. (2015). Parkinson's disease. Lancet, *386*(9996), 896–912. doi:10.1016/s0140-6736(14)61393-3.
88. Fahn, S., Oakes, D., Shoulson, I., & et al. (2004). Levodopa and the progression of Parkinson's disease. *The New England Journal of Medicine*, *351*(24), 2498–2508.
89. Oules, B., Del Prete, D., Greco, B., Zhang, X., Lauritzen, I., Sevalle, J., Moreno, S., Paterlini-Brechot, P., Trebak, M., Checler, F., & et al. (2012). Ryanodine receptor blockade reduces amyloid-load and memory impairments in Tg2576 mouse model of Alzheimer disease. *The Journal of Neuroscience*, *32*, 11820–11834.
90. Carter, D. B. (2005). The interaction of amyloid-β with APoE. *Subcell Biochem*, *38*, 255–272.

91. Schiesling, C., Kieper, N., Seidel, K., & Krüger, R. (2008). Review: Familial Parkinson's disease—Genetics, clinical phenotype and neuropathology in relation to the common sporadic form of the disease. *Neuropathology and Applied Neurobiology, 34*, 255–271.
92. Surmeier, D. J., Guzman, J. N., Sanchez-Padilla, J., & Goldberg, J. A. (2010). What causes the death of dopaminergic neurons in Parkinson's disease? *Progress in Brain Research, 183*, 59–77.
93. Vargas, K. J., Schrod, N., Davis, T., Fernandez-Busnadiego, R., Taguchi, Y. V., Laugks, U., Lucic, V., & Chandra, S. S. (2017). Synucleins have multiple effects on presynaptic architecture. *Cell Reports, 18*, 161–173.
94. Surmeier, D. J., Obeso, J. A., & Halliday, G. M. (2017). Selective neuronal vulnerability in Parkinson disease. *Nature Reviews Neuroscience, 18*, 101–113.
95. Südhof, T. C. (2004). The synaptic vesicle cycle. *Annual Review of Neuroscience, 27*, 509–547.
96. Bliss, T. V., & Collingridge, G. L. (1993). A synaptic model of memory: Long-term potentiation in the hippocampus. *Nature, 361*(6407), 31–39.
97. Malenka, R. C., & Bear, M. F. (2004). LTP and LTD: An embarrassment of riches. *Neuron, 44*(1), 5–21.
98. Citri, A., & Malenka, R. C. (2008). Synaptic plasticity: Multiple forms, functions, and mechanisms. *Neuropsychopharmacology, 33*(1), 18–41.
99. Ittner, L. M., & Götz, J. (2011). Amyloid-β and tau—A toxic pas de deux in Alzheimer's disease. *Nature Reviews Neuroscience, 12*(2), 67–72.
100. Olivares, D., Deshpande, V. K., Shi, Y., Lahiri, D. K., Greig, N. H., Rogers, J. T., & Huang, X. (2012). N-methyl D-aspartate (NMDA) receptor antagonists and memantine treatment for Alzheimer's disease, vascular dementia and Parkinson's disease. *Current Alzheimer Research, 9*(6), 746–758.
101. Traynelis, S. F., Wollmuth, L. P., McBain, C. J., Menniti, F. S., Vance, K. M., Ogden, K. K., Hansen, K. B., Yuan, H., Myers, S. J., & Dingledine, R. (2010). Glutamate receptor ion channels: structure, regulation, and function. *Pharmacological Reviews, 62*(3), 405–496.
102. Hansen, K. B., Yi, F., Perszyk, R. E., & Furukawa, H. (2018). Structure, function, and allosteric modulation of NMDA receptors. *Journal of General Physiology, 150*(8), 1081–1105.
103. Suchowersky, O. (2019). *Huntington disease: Clinical features and diagnosis.*
104. Keum, J. W., Shin, A., Gillis, T., Mysore, J. S., Elneel, K. A., Lucente, D., Hadzi, T., Homans, P., Jones, L., Orth, M., & et al. (2016). The HTT CAG-expansion mutation determines age at death but not disease duration in Huntington disease. *The American Journal of Human Genetics, 98*, 287–298.
105. Wenk, G. L., Parsons, C. G., & Danysz, W. (2006). Potential role of N-methyl-D-aspartate receptors as executors of neurodegeneration resulting from diverse insults: Focus on memantine. *Behavioural Pharmacology, 17*(5–6), 411–424.
106. Wang, R., & Reddy, P. H. (2017). Role of glutamate and NMDA receptors in Alzheimer's disease. *Journal of Alzheimer's Disease, 57*(4), 1041–1048.
107. Dingledine, R., Borges, K., Bowie, D., & Traynelis, S. F. (1999). The glutamate receptor ion channels. *Pharmacological Reviews, 51*(1), 7–61.

108. Lane, H. Y., Lin, C. H., Huang, Y. J., Liao, C. H., Chang, Y. C., & Tsai GE. (2013). Sarcosine (N-methylglycine) treatment for acute schizophrenia: A randomized, double-blind study. *Biological Psychiatry*, *73*(1), 9–15.
109. Koss, D. J., Robinson, L., Drever, B. D., Plucińska, K., Stoppelkamp, S., Veselcic, P., Riedel, G., & Platt, B. (2016). Mutant Tau knock-in mice display frontotemporal dementia relevant behaviour and histopathology. *Neurobiology of Disease*, *91*, 105–123.
110. Tariot, P. N., Farlow, M. R., Grossberg, G. T., Graham, S. M., McDonald, S., Gergel, I.; & Memantine Study Group. (2004). Memantine treatment in patients with moderate to severe Alzheimer disease already receiving donepezil: A randomized controlled trial. *Journal of the American Medical Association*, *291*(3), 317–324.

CHAPTER 12

Pharmacological Modulation of Neurodegenerative Disorders: Therapeutic Potential and Persisting Obstacles

RADHA GOEL[1], ROSALINE MISHRA[2,*], NIHARIKA[2], NITIN KUMAR[3], PRAVEEN KUMAR GAUR[2], MOHD. YAASIR[4], and RAMZA RAHAT HASHMI[2]

[1]*Lloyd Institute of Management and Technology, Greater Noida, Uttar Pradesh, India*

[2]*Metro College of Health Sciences and Research, Greater Noida, Uttar Pradesh, India*

[3]*Department of Pharmacy, Meerut Institute of Technology, Meerut, Uttar Pradesh, India*

[4]*Arsi University, Asella, Oromia, Ethiopia*

*Corresponding author

ABSTRACT

Millions of people worldwide are afflicted by neurodegenerative disabilities characterized by the decline in functioning axons. Substantial evidence argues that liaison uniting autophagy dysfunction could result in protein aggregation and further progress demonstrativeness diseases. Genetic defects that alter biological processes and genomic products at multiple levels are another potential reason for the development of demonstrativeness diseases. The cyclic nucleotide, cyclic guanosine

Cellular and Molecular Mechanisms in Alzheimer's Disease Therapeutics, Volume 2: Molecular Mechanisms. Prashant Tiwari, Sunil Kumar Kadiri, and Kunjbihari Sulakhiya (Eds)

DOI: 10.1201/9781779641816-12

monophosphate (cGMP), and cyclic adenosine monophosphate (cAMP) are involved in deterioration. Intracellular cAMP signaling is regulated by phosphodiesterase 7 (PDE) and is implicated in many vital physiological and pathological processes. Normalizing cyclic nucleotide signaling via PDE inhibition remains attractive for demonstrativeness disease therapy. In this chapter, we will study the involvement of PDE7s in demonstrativeness illnesses, notably Alzheimer's disease (AD). This chapter outlines the available data about the etiology of dysfunction and discusses how autophagy controls the breakdown of AD-related proteins and the inflammation of axons and neurogliacyte. Therefore, the purpose of this chapter is to explore the therapeutic potential of autophagy modulators that have been used to treat AD models, as well as the challenges and possibilities for the novel autophagy activators to be used clinically for AD therapy.

12.1 INTRODUCTION TO NEURODEGENRATIVE DISEASES

"Neurodegeneration" is a term commonly used, presumed to have universal understanding. It signifies the central pathophysiological change in most brain-related disorders. Etymologically, it comprises the prefix "neuro-" denoting nerve cells (neurons) and "degeneration," indicating the loss of structure or function in tissues or organs [1]. Therefore, strictly speaking, neurodegeneration encompasses any pathological condition primarily impacting neurons. Despite ongoing efforts in modern science to develop medical or surgical interventions, favorable outcomes have remained elusive. The loss of these neurons affects walking and balances [2]. Amyotrophic lateral sclerosis (ALS), Parkinson's disease, Alzheimer's disease (AD), Huntington's disease, and other dementia conditions are examples of demonstrativeness disorders. Even though they are just as fatal, many of the less well-known or prevalent neurological disorders have largely gone unnoticed [3].

Demonstrativeness disorders have a wide range of consequences on human functioning since the nervous system governs many different physical systems. These effects include impairments to more sophisticated activities, such as urination and digestion and mental processes, as well as basic capabilities, including communication, moving, stability, and coordination [4]. The progression of most demonstrativeness diseases is relentless, with little to no remission. However, in certain cases, treatments aim to alleviate symptoms, manage pain if present, and restore balance and mobility. In demonstrativeness disorder, the most reliable possibility is increasing the duration of life, particularly in AD or Parkinson's disease. In industrialized nations, the population of those aged 65 and older has grown at a rate significantly faster than the overall population during the last century. Therefore, it is expected that the number of be elderly citizens would double over the next several generations, and possibly along with it, the proportion of those who suffer from a neurological disease [5, 6].

AD is thought to affect approximately 65–75% of people around the world. Recent studies on AD pathophysiology have highlighted the pivotal part of beta-amyloid (Aβ) and tau proteins in disease progression. The accumulation of Aβ and initiation of tau-containing neurofibrillary tangles (NFTs) are recognized as key pathological features of AD. These processes disrupt hippocampal circuitry, impairing the consolidation of short-term memories of extended trace [4]. Consequently, AD is characterized by vast neuronal loss, synaptic dysfunction, and impairment to crucial neurotransmitter systems vital for memory and other brain functions. Memory impairment, particularly in the early stages, is a very typical clinical sign of AD. In addition to selective memory deficits, AD often affects hippocampal and medial temporal lobe-dependent processes, such as declarative episodic memory. Moreover, impairment of executive functions, judgement issues, and difficulties in problem-solving typically manifest early in the disease progression [7]. This chapter aims to highlight the current pharmacological modulation of demonstrativeness disorders.

12.2 AUTOPHAGY DYSREGULATION IN AD

Both ecological and inherited factors can contribute to the development of AD. Of all these risk factors, aging is the most significant. The probability of getting AD increases with age, from about 3% at age 65 to over 30% by age 85 [8]. Although the prevalence of AD among those under 65 is poorly known, it is thought to account for about 3% of all cases in this age group. Age-specific incidence rates appear to be falling in a lot of nations, despite the fact that the number of cases is rising due to the aging population. The restoration of cyclic nucleotide signaling via PDE inhibition is still a fascinating possibility for the treatment of demonstrativeness diseases. We will specifically look at PDE7s' involvement in demonstrativeness illnesses in this chapter.

There are many types of dementia groups based on when it first manifests and whether if it runs in families. Over 95% of dementia-related instances are late-onset Alzheimer's disease (LOAD). Before the age of 65, AD symptoms can develop in the majority of cases. Familial AD has mendelian inheritance, sometimes with a dominant pattern, whereas sporadic AD lacks a clear family relationship [9, 10].

Numerous investigations conducted globally have revealed that possessing a particular variant of the *APOE* gene, namely *APOE4*, heightens the likelihood of developing AD at some stage in one's life. However, the connection is not unequivocal. Certain individuals carrying *APOE4* do not develop Alzheimer's, while others manifest the disease despite lacking the *APOE4* variant in their genetic makeup [11]. Essentially, although the *APOE* gene undeniably impacts the risk of Alzheimer's, its presence or absence does not consistently predict the onset of the disease. Early-onset Alzheimer's disease (EOAD) stems from infrequent and dominantly inherited mutations in the *APP*, *PSEN1*, and *PSEN2* genes. On the other hand, LOAD, also known as sporadic AD, exhibits a substantial genetic component. Approximately 60–80% of LOAD cases are heritable, although the interplay of inherited and ecological factors significantly influences disease onset, progression, and severity [12]. The *APOE* gene's ε4 allele is the main risk factor in the pathophysiology of LOAD.

Utilizing groundbreaking technologies like genome-wide association studies (GWAS), researchers have pinpointed several genomic regions that may elevate an individual's susceptibility to LOAD to varying extents. Furthermore, developments in next-generation sequencing (NGS) have revealed uncommon variants that alter vulnerability within a number of AD-related genes [13]. Nonetheless, other genes contributing to causation and risk in AD remain to be identified to fully comprehend its etiology. This offers an overview of the genetic underpinnings of both EOAD and LOAD, shedding light on the association between threatening chromatin and the development the AD. GWAS along with genotyping, have identified more than 20 risk loci connected to rare genetic disorders, yet many cases of these disorders still lack an identifiable genetic cause [14].

Linkage analysis represented a pivotal breakthrough in deciphering the genetic foundations of AD, particularly in families exhibiting autosomal predominant patterns. This study aimed to pinpoint the inheritable part linked to the disease, rather than pinpointing individual genes or mutations associated with it [15]. Within affected families, genetic linkage investigations facilitated the discovery of dominant chromosomal variation situated in the *APP* gene on chromosome 21q, the *PSEN1* gene on 14q, and the *PSEN2* gene on 1q, all of which are linked to EOAD [3].

Additionally, the discovery of the *APOE ε4* allele gene, recognized as a key inherited factor for LOAD, was also achieved through genetic linkage studies. However, while linkage mapping proved highly effective after determining the monoclonal features of EOAD, its success in identifying risk factors in LOAD has been limited, likely due to the complex nature of the trait and the presence of unidentified variants [16].

12.3 SOURCE OF OXIDATIVE STRESS IN AD

Individuals with AD exhibit a notable degree of oxidative damage to their brains, which is linked to the abnormal buildup of Amyloid-β and neurofibrillary filament development. There is abundant evidence that the brain is susceptible to oxidative stress tissue that AD patients experience as the disease progresses [17]. Table 12.1 indicates that the primary cause of age-associated declines in cognition and neurodegeneration is believed to be a variance of producing the ROS, or irritable oxygenated category, and antioxidant defense, or oxidative stress. High amounts of protein carbonylation, advanced products of glycation, lipid peroxidation end products, production of hazardous species, and oxidative changes in DNA are indicative of oxidative damage in AD. Decreases in brain and plasma antioxidant defense systems, such as reduction of the antioxidant glutathione, which scavenges and gives electrons to reactive oxygen species (ROS), are linked to age-related memory impairments. Protein-based carbonyls or hydroxyl groups are introduced when side-chains of proteins are oxidized by ROS-mediated processes [18].

A number of mechanisms, such as the immediate oxidation of residues of amino acids, peptide bond cleavage, and interactions with additional molecules like lipids, DNA, and sugars, can lead to this process. Measuring the carbonylation of proteins aids in determining the degree of oxidative damage brought on by aging, AD, and oxidative stress. In AD brains, Aβ causes lipoperoxidation of membranes and the resulting products are associated with antioxidant enzymes, lipid peroxides, amyloid plaques, and NFTs [19]. In AD brains, a number of oxidative stress breakdown products have been detected, including 4-hydroxy-2,3-nonenal (HNE). HNE has the ability to alter proteins, block neuronal transporters, trigger kinases, and start an apoptotic sequence. Greater membrane damage is seen in dystrophic neurites and NFTs [20].

Biometals like iron, zinc, and copper play a crucial role in Aβ and neurodegeneration. Aβ and the preceding *APP* have abundant sites of

TABLE 12.1 List of Risk Factors for Alzheimer's Disease.

Risk Factor		
Age	AD affects 40% of people at 90 years of age and older, and the prevalence increases every five years after age 65.	[2, 21]
	Aging and neuroinflammation are linked, with aging impairing brain-immune interactions, leading to heightened inflammatory cytokine responses and severe cognitive deficits.	
Sex	A considerable component for AD is sex: men are half as likely as women to have dementia in their lifetime.	[2, 12]
	Studies show that due to variations in the neuroimmune system, such as microglial activation, generation of pro-inflammatory cytokines, and dysinformational synaptic plasticity, women are more susceptible to the deleterious consequences of inflammation.	
	Additionally, more than 80% of all diagnoses of autoimmune diseases, including scleroderma, myasthenia gravis, Hashimoto thyroiditis, systemic lupus erythematosus, Sjogren syndrome, and Graves' disease, are made in women. These sex-based neuroimmune differences may explain the associated variations in AD frequency and severity between sexes.	
Arterial hypertension	Systemic arterial hypertension has been connected to a decline in cognitive function in research pertaining to AD, making hypertension a higher risk factor for the disease.	[2, 4]
	While there are more disputed correlations among older groups, middle excessive blood pressure (BP) rise and cognitive deterioration in AD are positively associated. A complex relationship exists between AD and hypertension, with vascular and neuroinflammatory mechanisms being the main contributors.	
	Increased neuroinflammatory reactions and perivascular inflammation are consequences of chronic hypertension.	
Smoking	Research indicates that smoking tobacco products raises dementia and AD risk. Cigarette smoking is also associated with neuropathology. Smoking also contributes to immune-based inflammation, enhancing inflammatory responses and reducing immune defenses. Higher levels of neural inflammation genes in cerebrospinal fluid are indicative of cigarette smoking's association with a higher level of susceptible biomarkers for AD.	[3, 15]

TABLE 12.1 *(Continued)*

Risk Factor		
Hypercholesterolemia	AD along with additional disorders is at risk due to dysregulated metabolism and production of cholesterol.	[2, 12]
	In rats and mice, AD-like illness and cognitive decline are brought on by a high-cholesterol diet (HCD). According to epidemiological research, there may be a connection between AD and hypercholesterolemia, with higher cholesterol being linked to more severe pathology.	
	There are several contributing aspects to the complex relationship between hypercholesterolemia and AD, but two main ones are vascular and neuroinflammatory. One risk factor for AD is excessive cholesterol buildup, particularly in middle age.	
Obesity	Due to its association with immunological dysregulation and neuroinflammation, obesity is a known risk factor for AD. Increased amounts of circulating fats, agents of inflammation, and systemic inflammation are brought on by obesity and heavy fat consumption. These factors then cause local neurological inflammation and neurodegeneration.	[2, 15]
	This early suppression of immune system activity leads to AD and cognitive impairment.	
Physical inactivity	AD is associated with a higher risk of physical inactivity, and improved activity levels can reduce the risk.	[2,12]
	Exercise improves cognitive function and behavioral indicators of AD. Exercise can also prevent AD by influencing hippocampus neurogenesis and controlling immune responses in the brain.	
	Studies show that exercise enhances neuroinflammation-mediated microglia activation and downregulates pro-inflammatory cytokine release.	
Dietary factors	Studies show that diet, particularly an emphasis on aging, has an impact on cognitive abilities and increases the risk of AD. Western diet patterns are a major contributing factor to AD, while Mediterranean, ketogenic, and omega-3 fatty acid-rich diets are certainly neuroprotective. Healthy dietary regimens like the Mediterranean diet and MIND diet have cognitive benefits and decreased AD risk. Dietary changes also affect the microbiota, changing the homeostatic axis between the digestive tract and brain, and releasing pro-inflammatory bacterial metabolites.	[2, 12]

TABLE 12.1 *(Continued)*

Risk Factor		
Diabetes mellitus	Compared to non-diabetic controls, diabetics, particularly those with Type II Insulin resistance, have a 65% increased chance of suffering from AD. The prevalence of diabetes is closer to 35% in AD patients, which is greater than anticipated. Moreover, insulin resistance, an inflammatory predictor of diabetes, affects 46% of AD patients. Diabetes and AD have a complicated relationship that involves several factors, such as malfunctioning mitochondria, tau hyperphosphorylation, Aβ misfolding, and neuroinflammation. Diabetes causes certain reactions, one of which is the NLRP3 inflammasome.	[32]
Insomnia	Insomnia and other sleep disturbances provide a serious risk for AD. Memory retention and consolidation depend heavily on sleep, and the molecular pathophysiology of AD is impacted by sleep deprivation. Prolonged sleeplessness causes Aβ to build up in brain tissue, which advances the condition. A key physiological and molecular link between insomnia and AD is neuroinflammation, which is caused by disrupted sleep architecture that raises pro-inflammatory cytokines and activates microglia, which affects memory and learning.	[2, 13]
Depression	There is a continuum between depression and dementia, with depression causing dementia and dementia causing depression. An increased risk of AD and kindred dementias is linked to early-life depression. The primary mechanism tying depression to dementia is neuroinflammation, which is accelerated by long-term innate immune activation, which raises inflammatory cytokine levels and central inflammation. These cytokines affect hormonally induced mood changes via influencing neurogenesis, interneuronal cross-talk, and mood-related brain areas.	[2, 14]

binding for copper and zinc. The brains of people with AD experience oxidative stress due to copper's strong hydroxyl radical mediatory effects. Elevated levels of zinc have been linked to harmful Aβ aggregation and oxidative stress in memory and cognitive areas. Cytotoxicity results from these metals building up uncontrollably [18].

Due to its composition of fatty acids that are polyunsaturated, the brain membrane is susceptible to free radical damage, which can result in elevated lipid peroxidation and deterioration associated with AD. Enzymes essential for the proper functioning of neurons and glia can be impacted by protein oxidation, namely glutamine synthetase and creatine kinase, among others. Proteins that aggregate pathologically can form fibrils and become insoluble, which can result in NFTs. In addition to causing base modification, sister chromatid exchange, and strand breaks in DNA, oxidation can also produce advanced glycation end products [19].

Therefore, oxidative stress caused by an excessive amount of ROS may be detrimental and a primary cause of the degeneration of the structures of cells, which in turn leads to aging and a range of disease conditions. But antioxidant treatments have demonstrated that AD is connected to oxidative stress, suggesting that the illness is more complex [21].

12.4 MECHANISM OF PROTEIN AGGREGATION AND AD

Amyloid plaques containing Aβ and NFTs inferred from Aβ are considered as neurotic signs of AD [22–24]. These two proteins show few biochemical properties. The basic components of amyloid plaques are fibrils, namely fibrillary tangles with Aβ fibrils (β fibrils; NFTs). The two proteins have been considered through post-translational alterations like phosphorylation, cleavage, or potential pyroglutamate development in AD. These adjustments speed up *in vitro* conglomeration. Moreover, Aβ has been shown to advance β conglomeration [25, 26], indicating that a major cause of the disease's onset and/or *APP* is protein aggregation. Other proteins, including receptors for N-methyl-D-aspartate (NMDA), which function as molecular switches to cause neuronal malfunction and neurodegeneration, may be necessary for Aβ to interact with τ effectively.

The primary neuropathological characteristic of AD is its restrained process, as seen by the dispersion of NFTs and Aβ plaques throughout the brain [27, 28]. According to recent studies, these phases are similarly

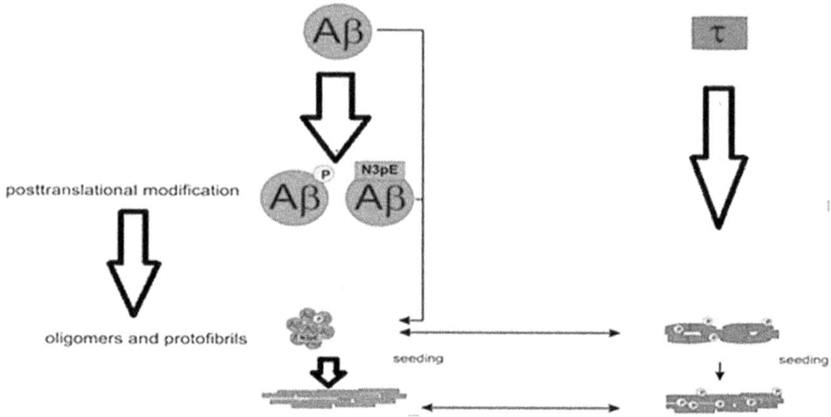

FIGURE 12.1 Aβ accumulation, post-translational alterations, and the presence of interstitials and fibril forms in the brain's physiochemical and biochemical systems.

mirrored in Aβ biochemical alterations, particularly in post-translational modifications with the formation of AD from experimental to clinical forms. The researchers discuss the cellular pathways via which Aβ carries out its harmful effects and explore potential methods of blocking it with antibodies and other medications.

12.4.1 EFFECTS OF AB AGGREGATION ON NFT, β AND NMDA RECEPTORS, AND Fyn PROTEIN KINASE

Protein aggregation, post-translational modification of proteins, and interactions between these aggregates and other proteins are the causes of AD. The negative effect of Aβ and τ is mostly associated with toxic intermediates (oligomers and protofibrils) at the cellular biology level. The effects of toxic intermediates can be amplified by Aβ and τ aggregates by displacing functional cytoplasm within cells (τ-containing NFTs) or generating mechanical barriers to budding, cell migration in the brain parenchyma (Aβ fibrils), and diffusion. Given that the production of τ-derived NFTs and the aggregation and deposition of Aβ share similar biophysical mechanisms of protein aggregation, one may be tempted to hypothesize that AD is brought on by an event that can set off a τ-protein biophysical process.

12.5 GENETIC DEFECTS IN DEVELOPMENT OF AD

Hereditary, biochemical, and neuropathological information demonstrate that Aβ collection is integral to the improvement of *APP* pathogenesis. Aβ is a portion of the *APP*, or amyloid precursor protein that is enzymatic and is framed to consequently separate β and γ secretases. In neurons and associated processes, the accumulation of NFTs follows the accumulation of extracellular amyloid plaques [29]. NFT's are made of hyperphosphorylated tau. The presence of neurofibrillary tangles in the nervous system of an Alzheimer's brain is strongly associated with neuronal dysfunction and disease progression. The amyloid outpouring speculation recommends that modifications in *APP* as well as Aβ homeostasis lead to Aβ conglomeration and plaque gathering, and that these occasions remember the collection of tau proteins for NFTs, an obsessive and clinical component related with Alzheimer's illness [30]. Inherited mutations in the β-amyloid precursor protein (APP), presenilin 1 (*PSEN1*), and presenilin 2 (*PSEN2*) genes cause beginning stage AD. These qualities and *APOE* were distinguished through family hereditary linkage studies.

12.5.1 β-AMYLOID PRECURSOR PROTEIN (APP)

APP, often referred to as type 1 transmembrane protein, is located on chromosome 21q21. APP, on the other hand, was combined to form APP695, APP751, and APP770. The predominant APP isoform is APP695 communicated in axons, and APP751 isoform is profoundly communicated in astroglia [31]. Proteolytic handling of APP prompts the development of pieces through non-pathogenic and amyloidogenic pathways. Most proteases, including α- and γ-secretases, cleave amyloid precursor protein (APP), resulting in the release of amyloid-β (Aβ) peptides into the extracellular space. The outcome is non-pathogenic pieces: sAPPα and α-C-terminal section (CTF). Then again, APP is separated by successive β-and γ-secretases' proteolytic cleavage to shape β-CTF, Aβ peptides, and sAPPβ. Associated with cell, APP is assimilated, permitting Aβ development by means of the endocytic pathway and emission into the extracellular space. Prevailing changes in APP represent around 14% of instances of beginning stage autosomal predominant AD [32]. Most changes in APP group are situated in the locale contiguous or inside the Aβ space; in any case, early hereditary examinations just sequenced APP exons encoding the Aβ succession, leaving open likelihood of variations somewhere else in APP that reason or increment AD risk. APP transformations have uncovered numerous significant parts of the atomic instruments, basic for AD pathogenesis. The Swedish transformation (KM670/671NL) increments plasma Aβ levels 2-3-overlay by modifying the cleavage productivity of β-secretase [33]. APP and encompassing succession covers are likewise connected with beginning stage of Alzheimer's. Families transferring these duplications have exemplary AD and cerebral amyloid angiopathy (CAA). What's more, people with down condition because of trisomy 21, generate AD neuropathology. People with an incomplete trisomy of chromosome 21 that doesn't contain the APP quality, don't generate neuropathology [34].

12.5.2 PSEN1 AND PSEN2

PSEN2 is located on chromosomes 1q31–q42, while PSEN1 is located on chromosome 14q24.3. PSEN1 and PSEN2, which have nine transmembrane gaps and a hydrophilic intracellular circular district, are essentially equivalent fundamental film proteins. PSEN1 and PSEN2 are basic parts inside the complex of γ-secretase that cuts *APP* into Aβ sections. PSEN1 and PSEN2 restrict to the endoplasmic reticulum and golgi mechanical assembly, where they assume a significant part in protein handling. Up to 185 predominant pathogenic changes have been distinguished in PSEN, representing roughly 80% of beginning stage familial AD cases [35].

Changes in PSEN1 and PSEN2 are disseminated all through the protein, with some bunching in transmembrane spaces. Overwhelmingly acquired changes in PSEN1 and APP have likewise been distinguished in late-beginning of AD cases. These families might have other hereditary variations that postpone the onset of the change. Different variations have been depicted for PSEN1 and PSEN2 that are not pathogenic or whose pathogenicity stays indistinct. Ongoing evidence demonstrates that a portion of these variations, like PSEN1 E318G and PSEN2 R62H, may act as genetic factors for AD [36].

12.5.3 ADAM10

It was recently shown that seven LOAD families are segregated by rare coding variations of ADAM10. The primary α-secretase responsible for cleaving ADAM10 risk variants Q170H and R181G in the APP ectodomain and raising Aβ levels *in vitro* is ADAM10. ADAM10 Q170H and R181G increase plaque load in Tg2576 mice via disrupting α-secretase function, shifting the analysis of APP, which leads to amyloidogenic fragmentation. Processing APPs and producing Aβ is enough, as shown by these genetic and biochemical studies [37].

12.5.4 APOE

Encoding a pleiotropic glycoprotein, apolipoprotein E (APOE), located on chromosome 19q13.2, plays a crucial role in the transport, composition, and redistribution of cholesterol and lipids within hepatocytes, brain cells, and macrophages, influencing lipid metabolism and neurodegenerative disease susceptibility. APOE is additionally engaged with nerve cell development and repair, nerve recovery, insusceptible reaction, and actuation of lipolytic proteins. APOE is found in three possible isoforms, APOEe2, APOEe3, and APOEe4, which differ by two amino corrosive deposits (112 and 158). The most prevalent subtype of APOE, APOEε3, is found in around 72% of the population worldwide [38]. Family-based techniques initially recognized a hereditary connection between AD and the region of chromosome 19 containing the *APOE* gene. The APOEε4 allele significantly increases the risk of both familial and sporadic AD, with a 3-overlay expansion in heterozygous transporters and 8-to 10 times expansion in homozygous transporters. APOEε4 likewise has a portion subordinate impact at an early age. Strangely, APOEε2 lessens the risk and postpones the onset of late-beginning of AD. Uncommon coding variations influencing AD may likewise happen in APOE. However, extensive analysis of the APOE level across big datasets hasn't been carried out. Furthermore, administrative locations may have APOE variants that affect APOE articulation, which in turn may affect plaque aggregation. APOE articulation levels influence the accumulation of amyloid plaque in animal models [39].

12.6 GENOMIC ASSOCIATION STUDIES CASE/CONTROL STUDIES

Just half of Alzheimer patients transfer the APOEε4 allele, recommending that other hereditary elements should add to illness risk. Before 2005, a typical methodology for distinguishing new risk qualities was to dissect at least one polymorphism of a competitor quality. The primary justification for this is that other normal hereditary risk factors most likely affect APOE. Accordingly, these examinations coming up short on the ability to recognize new risk qualities in light of the fact that the sample sizes utilized in past investigations were very few. Starting around 2005, a few innovative advances have reformed the investigation of the hereditary qualities of perplexing characteristics [40].

12.7 CURRENT CHALLENGES AND POSSIBILITY OF NOVEL AUTOPHAGY ACTIVATOR FOR CLINICAL TREATMENT OF AD

The confirmation that autophagy regulation is a promising technique for treating demonstrative disorders highlights its potential to clear misfolded proteins, reduce neuroinflammation, and restore cellular homeostasis, offering therapeutic benefits in neurodegenerative disease management and progression control. Since 2008, a number of research studies have looked into the possibility for the creation of autophagy-targeting therapeutics for the intervention of AD pathology through the use of chemical autophagy modulators. The majority of autophagy regulators that exhibit anti-AD action are mTOR-dependent autophagy inducers. Autophagy modulators, which are small-molecule drugs, have been shown to modify AD-related diseases, such as amyloid plaque and NFT formation, neuroinflammation, oxidative damage, and neuronal and synaptic loss [41, 42].

12.7.1 AUTOPHAGY ACTIVATORS

Several recent publications have discussed how increasing autophagy may have therapeutic benefits in a variety of disorders. Autophagy activators are a prominent area of focus in new research because of their potential clinical benefit. The anti-AD effects of a few well-known autophagy modulator properties are mentioned in Table 12.2.

12.7.2 AUTOPHAGY INHIBITORS

Autophagy may increase the rate of the autophagic flux. Many pharmacologic autophagy mechanisms lower the amount of Aβ, improve neuronal loss, and avoid cognitive impairments (Table 12.2). Therefore, pharmacological

TABLE 12.2 Autophagy Modulators and their Anti-AD Properties.

Activator Name	Mechanism	Phenotype Effect	References
Rapamycin	mTOR inhibition	Reduces Aβ, APP, and tau levels	[43, 44]
	Wnt/GSK3b/b-catenin signaling pathway	Reduces neurodegeneration and neuroinflammation	
		Increases cell viability	
		Improves memory and learning impairment	
Trehalose	mTOR independent activator and signaling pathway	Reduces Aβ, APP, and tau concentrations	[45, 46]
		Boosts the vitality of cells	
		Improves memory and learning impairment	
Resveratrol	AMPK activation and inhibition of mTOR as AMPK targets	Reduces the amount of Aβ and encourages the release of synaptic biomarkers and neurotrophins	[47, 48]
		Boosts the vitality of cells	
		Reduces inflammation, apoptosis, and oxidative stress	
		Reduces the harm caused by mitochondria	
Temsirolimus	mTOR inhibition	Lowers tau and Aβ levels	[49]
		Reduces the death of cells	
		Enhances memory impairment and learning	
Latrepirdine	mTOR inhibition	Lowers the amounts of Aβ and CTF-b/a. Stops cognitive impairments	[50]
Arctigenin	mTOR inhibition	Reduces Aβ level	[51]
		Improves memory impairment	
b-A sarone	Parkin- and PINK1/ mediated pathway	Decreases Aβ level	[52]
		Enhances memory impairment	
Berberine	PtdIns3K/BECN-1 pathway	Lowers the amounts of tau and Aβ Ameliorates memory loss	[53]
Carbamazepine	m-TOR-indepenedent	Reduces Aβ levels	[54]
		Improves memory and learning impairment	
Curcumin	TFEB nuclear translocation is promoted by GSK-3b and mTOR inhibition	Lowers the levels of APP and Aβ Lowers the quantity of ROS to enhance memory impairment	[55, 56]

TABLE 12.2 *(Continued)*

Activator Name	Mechanism	Phenotype Effect	References
Lithium	IMPase inhibitor	Lowers the levels of tau and p-tau	[57]
		Avoids any motor disturbances	
Methylene blue	mTOR inhibition	Lowers the levels of tau and p-tau	[58]
Fistein	mTOR inhibition	Reduces p-tau levels	[59]
	Promotes TFEB/Nrf2 nuclear translocation		
Selenomethionine	mTOR inhibition	Boosts the vitality of cells	[60]
		Avoids mental impairments	
Thioperamide	CREB-1-mediated autophagy	Lowers the amount of Aβ, improves neuronal loss, and avoids cognitive impairments	[61]

constraints are not completely selective, and the outcomes produced by these substances, particularly in terms of dose and incubation duration, should be interpreted with caution [62–64].

12.8 CHALLENGES

Autophagy enhancers, both genetic and pharmacological, have shown therapeutic potential in a variety of neurodegenerative disease models, including AD [65, 66]. Chemical autophagy modulators have been identified as a new class of anti-AD drugs [67]. Although autophagy modulators are becoming more prevalent, a number of obstacles have prevented their successful clinical development [68–70].

The primary issue is the lack of precision: Many substances that are used to either promote or inhibit autophagy don't show any pharmacological selectivity for the targets of autophagy. For instance, sustained rapamycin delivery results in mTORC2 disassembly, while acute rapamycin administration has a high selectivity for inhibiting mTORC1 [71, 72]. In addition to promoting autophagy, rapamycin also suppresses mTORC1-mediated translation and cell division [73]. Targeting these proteins to either activate or inhibit autophagy will have an impact on their functions outside of autophagy, as many autophagy machinery components have autophagy-independent functions.

Another barrier is that it is unclear how current autophagy modulators work to target autophagy in certain nervous system cell types. Every cell uses the same fundamental autophagy machinery, making autophagy a highly conserved process. Autophagy modulators may generate an adverse reaction in tissues by inducing undesired autophagy responses due to their limited cell type selectivity.

Another challenge is the difficulty of tracking autophagy flux *in vivo*. Since autophagy is so complex, there isn't a single reliable indicator of autophagy activity; autophagy flux must always be determined using a mix of tests. Autophagy flux at different phases of AD is difficult to assess without adequate assays and biomarkers for tracking its presence in human tissues. It is also difficult to define the optimal time point for inducing autophagy and optimizing the dosage of autophagy modulators in patients.

The targeted destruction of disease-associated damaged organelles and toxic proteins is hampered by the lack of substrate-selective autophagy modulators. Extremely harsh conditions, including hunger, oxidative stress, or cytotoxicity greatly activate unselective autophagy, a costly process that results in the loss of bulk cellular content. Through the selective degradation of substrates including aggregated proteins and damaged organelles, autophagy can preserve cellular homeostasis. Still, most selective autophagy modulators are not truly "selective"; rather, they have the ability to trigger global autophagy.

With over half of all known autophagy inducers being mTOR-dependent, it is evident from the discussion above that these modulators are still the most studied. There has been constant reporting on the anti-AD potential of mTOR inhibition, and the influence of mTOR inhibition on lifetime extension adds to mTOR inhibitors' therapeutic advantages in the management of AD [74].

KEYWORDS

- **Alzheimer's disease**
- **autophagy**
- **autophagy modulator**
- **dysfunction**
- **genetic modulation**
- **neurodegenerative disease**
- **therapeutic potential**

REFERENCES

1. Agrawal, M., Ajazuddin., Tripathi, D. K., Saraf, S., Antimisiaris, S., Mourtas, S., Hammarlund-Udenaes, M., & Alexander, A. (2017). Recent advancements in liposomes targeting strategies to cross blood-brain barrier (BBB) for the treatment of Alzheimer disease. *Journal of Controlled Release: Official Journal of the Controlled Release Society*, *260*, 61–77. https://doi.org/10.1016/j.jconrel.2017.05.019.
2. Cacàbelos, R. (2020). Pharmacogenomics of Alzheimer's and Parkinson's diseases. *Neuroscience Letters*, *726*, 133807.
3. Alexander, A., & Ajazuddin, K. J. (2014). Formulation and evaluation of chitosan-based long-acting injectable hydrogel for PEGylated melphalan conjugate. *The Journal of Pharmacy and Pharmacology*, *66*(9), 1240–1250.
4. Gobom, J., & et al. (2022). Validation of the LUMIPULSE automated immunoassay for the measurement of core AD biomarkers in cerebrospinal fluid. *Clinical Chemistry and Laboratory Medicine, 60*, 207–219.
5. Budd Haeberlein, S., & et al. (2022). Two randomized phase 3 studies of aducanumab in early Alzheimer's disease. *The Journal of Prevention of Alzheimer's Disease*, *9*, 197–210.
6. Patel, M., Souto, E. B., & Singh, K. K. (2013). Advances in brain drug targeting and delivery: Limitations and challenges of solid lipid nanoparticles. *Expert Opinion on Drug Delivery*, *10*(7), 889–905.
7. Mukherjee, S., Ray, S., & Thakur, R. S. (2009). Solid lipid nanoparticles: A modern formulation approach in drug delivery system. *Indian Journal of Pharmaceutical Sciences*, *71*(4), 349–358
8. Bernardi, A., Frozza, R. L., Meneghetti, A., Hoppe, J. B., Battastini, A. M. O., Pohlmann, A. R., Guterres, S. S., & Salbego, C. G. (2012). Indomethacin-loaded lipid-core nanocapsules reduce the damage triggered by Aβ1-42 in Alzheimer's disease models. *The International Journal of Nanomedicine, 7*, 4927–4942
9. Brambilla, D., Le Droumaguet, B., Nicolas, J., Hashemi, S. H., Wu, L. P., Moghimi, S. M., & et al. (2011). Nanotechnologies for Alzheimer's disease: Diagnosis, therapy, and safety issues. *Nanomedicine*, *7*, 521–540.
10. Cacabelos, R. (2007). Donepezil in Alzheimer disease: From conventional trials to pharmacogenetics. *Neuropsychiatric Disease and Treatment*, *3*(3), 303–333.
11. Cummings, J., Aisen, P., Lemere, C., Atri, A., Sabbagh, M., & Salloway, S. (2021). Aducanumab produced a clinically meaningful benefit in association with amyloid lowering. *Alzheimer's Research and Therapy*, *13*(1), 98.
12. Barrett, T., Stangis, K. A., Saito, T., Saido, T., & Park, K. H. J. (2021). Neuronal cell cycle re-entry enhances neuropathological features in AppNLF knock-in mice. *The Journal of Alzheimer's Disease*, *82*, 1683–1702. https://doi.org/10.3233/jad-210091.
13. Basun, H., Bogdanovic, N., Ingelsson, M., Almkvist, O., Naslund, J., Axelman, K., & et al. (2008). Clinical and neuropathological features of the arctic APP gene mutation causing early-onset Alzheimer disease. *Archives of Neurology*, *65*, 499–505. https://doi.org/10.1001/archneur.65.4.499.
14. Gonzalez, A., Singh, S. K., Churruca, M., & Maccioni, R. B. (2022). Alzheimer's disease and tau self-assembly: In the search of the missing link. *International Journal of Molecular Sciences*, *23*, 4192. https://doi.org/10.3390/ijms23084192.

15. Bellenguez, C., Kucukali, F., Jansen, I. E., Kleineidam, L., Moreno-Grau, S., & et al. (2022). New insights into the genetic etiology of Alzheimer's disease and related dementias. *Nature Genetics*, *54*, 412–436. https://doi.org/10.1038/s41588-022-01024-z.
16. Safieh, M., Korczyn, A. D., & Michaelson, D. M. (2019). ApoE4: An emerging therapeutic target for Alzheimer's disease. *BMC Medicine*, *17*, 64.
17. Reddy, A. P., Ravichandran J., & Carkaci-Salli, N. (2020). Neural regeneration therapies for Alzheimer's and Parkinson's disease-related disorders. *Biochimica et Biophysica Acta - Molecular Basis of Disease*, *1866*(4), 165506.
18. Budd Haeberlein, S., & et al. (2022). Two randomized phase 3 studies of aducanumab in early Alzheimer's disease. *Journal of Prevention of Alzheimer's Disease*, *9*, 197–210.
19. Mattsson-Carlgren, N., & et al. (2022). Cerebrospinal fluid biomarkers in autopsy-confirmed Alzheimer disease and frontotemporal lobar degeneration. *Neurology*, *98*, 1137–1150.
20. Gobom, J., & et al. (2022). Validation of the LUMIPULSE automated immunoassay for the measurement of core AD biomarkers in cerebrospinal fluid. *Clinical Chemistry and Laboratory Medicine*, *60*, 207–219.
21. Baldeiras, I., Santana, I., Leitão, M. J., Ribeiro, M. H., Pascoal, R., Duro, D., Lemos, R., Santiago, B., Almeida, M. R., & Oliveira, C. R. (2015). Cerebrospinal fluid Aβ40 is similarly reduced in patients with Frontotemporal Lobar Degeneration and Alzheimer's Disease. *Journal of the Neurological Sciences, 358*(1–2), 308–316.
22. Alzheimer, A., Stelzmann, R. A., Schnitzlein, H. N., & Murtagh, F. R. (1995). An English translation of Alzheimer's 1907 paper, "Uber eine eigenartige Erkankung der Hirnrinde". *Clinical Anatomy*, *8*(6), 429–431.
23. Grundke-Iqbal, I., Iqbal, K., Quinla, M., Tung, Y. C., Zaidi, M. S., & Wisniewski, H. M. (1986). Microtubule-associated protein tau. A component of Alzheimer paired helical filaments. *The Journal of Biological Chemistry*, *261*(13), 6084–6089.
24. Masters, C. L., Simms, G., Weinman, N. A., Multhaup, G., McDonald, B. L., & Beyreuther K. (1985). Amyloid plaque core protein in Alzheimer disease and Down syndrome. *Proceedings of the National Academy of Sciences*, *82*, 4245–4249. https://doi.org/10.1073/pnas.82.12.4245.
25. Götz, J., Chen, F., van Dorpe, J., & Nitsch, R. M. (2001). Formation of neurofibrillary tangles in P301 l tau transgenic mice induced by Abeta 42 fibrils. *Science*, *293*, 1491–1495. https://doi.org/10.1126/science.1062097
26. Lasagna-Reeves, C. A., Castillo-Carranza, D. L., Guerrero-Muoz, M. J., Jackson, G. R., & Kayed, R. (2010). Preparation and characterization of neurotoxic tau oligomers. *Biochemistry*, *49*, 10039–10041. https://doi.org/10.1021/bi1016233.
27. Braak, H., & Braak, E. (1991). Neuropathological stageing of Alzheimer-related changes. *Acta Neuropathologica*, *82*, 239–259. https://doi.org/10.1007/BF00308809.
28. Thal, D. R, Rüb, U., Orantes, M., & Braak, H. (2002). Phases of Abeta-deposition in the human brain and its relevance for the development of AD. *Neurology*, *58*, 1791–1800. https://doi.org/10.1212/wnl.58.12.1791
29. Thinakaran., G., & Koo., E. H. (2008). Amyloid precursor protein trafficking, processing, and function. *The Journal of Biological Chemistry*, *283*, 29615–29619. https://doi.org/10.1074/jbc.R800019200.

30. Yoshikai, S., Sasaki, H., Doh-ura, K., Furuya, H., & Sakaki, Y. (1990). Genomic organization of the human amyloid beta-protein precursor gene. *Gene*, *87*, 257–263. https://doi.org/10.1016/0378-1119(90)90310-n.
31. Guerreiro, R. J., Baquero, M., Blesa, R., Boada, M., Brás, J. M., Bullido, M. J., Calado, A., Crook, R., Ferreira, C., Frank, A., & et al. (2010). Genetic screening of Alzheimer's disease genes in Iberian and African samples yields novel mutations in presenilins and APP. *Neurobiology Aging*, *31*(5), 725–731. https://doi.org/10.1016/j.neurobiolaging.2008.06.012.
32. Mullan, M., Crawford, F., Axelman, K., Houlden, H., Lilius, L., Winblad, B., & Lannfelt, L. (1992). A pathogenic mutation for probable Alzheimer's disease in the APP gene at the N-terminus of beta-amyloid. *Nature Genetics*, *1*(5), 345–347. https://doi.org/10.1038/ng0892-345.
33. Bergmans, B. A., & De Strooper, B. (2010). Gamma-secretases: From cell biology to therapeutic strategies. *The Lancet Neurology*, *9*(2), 215–226. https://doi.org/10.1016/S1474-4422(09)70332-1.
34. Nilsberth, C., Westlind-Danielsson, A., Eckman, C. B., Condron, M. M., Axelman, K., Forsell, C., Stenh, C., Luthman, J., Teplow, D. B., Younkin, S. G., & et al. (2001). The 'Arctic' APP mutation (E693G) causes Alzheimer's disease by enhanced Abeta protofibril formation. *Nature Neuroscience*, *4*(9), 887–893. https://doi.org/10.1038/nn0901-887.
35. De Strooper, B. (2003). Aph-1, Pen-2, and nicastrin with presenilin generate an active gamma-secretase complex. *Neuron*, *38*(1), 9–12. https://doi.org/10.1016/s0896-6273(03)00205-8.
36. Kovacs, D. M., Fausett, H. J., Page, K. J., Kim, T. W., Moir, R. D., Merriam, D. E., Hollister, R. D., Hallmark, O. G., Mancini, R., Felsenstein, K. M., & et al. (1996). Alzheimer-associated presenilins 1 and 2: Neuronal expression in brain and localization to intracellular membranes in mammalian cells. *Nature Medicine*, *2*(2), 224–229. https://doi.org/10.1038/nm0296-224.
37. Benitez, B. A., Karch, C. M., Cai, Y., Jin, S. C., Cooper, B., Carrell, D., Bertelsen, S., Chibnik, L., Schneider, J. A., Bennett, D. A. & et al. (2013). Alzheimer's disease neuroimaging initiative; Genetic and environmental risk for Alzheimer's disease consortium GERAD. The PSEN1, p.E318G variant increases the risk of Alzheimer's disease in APOE-ε4 carriers. *PLoS Genetics*, *9*, e1003685. https://doi.org/10.1371/journal.pgen.1003685.
38. Cruchaga, C., Kauwe, J. S. K., Mayo, K., Spiegel, N., Bertelsen, S., Nowotny, P., Shah, A. R., Abraham, R., Hollingworth, P., Harold, D., & et al. (2010). SNPs associated with cerebrospinal fluid phospho-tau levels influence rate of decline in Alzheimer's disease. *PLoS Genetics*, *6*(9), e1001101. https://doi.org/10.1371/journal.pgen.1001101.
39. Wakabayashi, T., & De Strooper, B. (2008). Presenilins: Members of the gamma-secretase quartets, but part-time soloists too. *Physiology (Bethesda)*, *23*, 194–204. https://doi.org/10.1152/physiol.00009.
40. Lambert, J. C., Heath, S., Even, G., Campion, D., Sleegers, K., Hiltunen, M., Combarros, O., Zelenika, D., Bullido, M. J., Tavernier, B., & et al. (2009). Genome-wide association study identifies variants at CLU and CR1 associated with Alzheimer's disease. *Nature Genetics*, *41*(10), 1094–1099.

41. Olive, D. M., & Reddy, P. H. (2019). Small molecules as therapeutic drugs for Alzheimer's disease. *Molecular and Cellular Neuroscience, 96*, 47–62. https://doi.org/10.1016/j.mcn.2019.03.001
42. Suresh, S., Chakravorty, A., Giridharan, M., Garimella, L., & Manjithaya, R. (2020). Pharmacological tools to modulate autophagy in neurodegenerative diseases. *The Journal of Molecular Biology, 432*(8), 2822–2842. https://doi.org/10.1016/j.jmb.2020.02.023.
43. Siman, R., Cocca, R., & Dong, Y. (2015). The mTOR inhibitor rapamycin mitigates perforant pathway neurodegeneration and synapse loss in a mouse model of early-stage Alzheimer-type tauopathy. *PLoS One, 10*(11), e0142340. https://doi.org/10.1371/journal.pone.0142340.
44. Chen, J., Long, Z., Li, Y., Luo, M., Luo, S., & He, G. (2019). Alteration of the Wnt/GSK3b/b-catenin signalling pathway by rapamycin ameliorates pathology in an Alzheimer's disease model. *International Journal of Molecular Medicine, 44*(1), 313–323. https://doi.org/10.3892/ijmm.2019.4198.
45. Rodríguez-Navarro, J. A., Rodríguez, L., Casarejos, M. J., & Solano, R. M. (2010). Trehalose ameliorates dopaminergic and tau pathology in parkin deleted/tau overexpressing mice through autophagy activation. *Neurobiology of Disease, 39*(3), 423–438. https://doi.org/10.1016/j.nbd.2010.05.014.
46. Krüger, U., Wang, Y., Kumar, S., & Mandelkow, E. M. (2012). Autophagic degradation of tau in primary neurons and its enhancement by trehalose. *Neurobiology of Aging, 33*(10), 2291–2305. https://doi.org/10.1016/j.neurobiolaging.2011.11.009.
47. Deng, H., & Mi, M. T. (2016). Resveratrol attenuates Ab25e35 caused neurotoxicity by inducing autophagy through the TYrRSePARP1eSIRT1 signaling pathway. *Neurochemical Research, 41*(9), 2367–2379. https://doi.org/10.1007/s11064-016-1950-9.
48. Broderick, T. L., Rasool, S., Li, R., Zhang, Y., & Anderson, M. (2020). Neuroprotective effects of chronic resveratrol treatment and exercise training in the 3xTg-AD mouse model of Alzheimer's disease. *The International Journal of Molecular Sciences, 21*(19), 7337. https://doi.org/10.3390/ijms21197337.
49. Jiang, T., Yu, J. T., Zhu, X. C., & Zhang, Q. Q. (2014). Temsirolimus attenuates tauopathy *in vitro* and *in vivo* by targeting tau hyperphosphorylation and autophagic clearance. *Neuropharmacology, 85*, 121–130. https://doi.org/10.1016/j.neuropharm.2014.05.032.
50. Bharadwaj, P. R., Verdile, G., Barr, R. K., Gupta, V., & Steele, J. W. (2012). Latrepirdine (dimebon) enhances autophagy and reduces intracellular GFP-Ab42 levels in yeast. *The Journal of Alzheimer's Disease, 32*(4), 949–967. https://doi.org/10.3233/JAD-2012-120178.
51. Zhu, Z., Yan, J., Jiang, W., Yao, X. G., & Chen, J. (2013). Arctigenin effectively ameliorates memory impairment in Alzheimer's disease model mice targeting both β-amyloid production and clearance. *The Journal of Neuroscience, 33*(32), 13138–13149. https://doi.org/10.1523/JNEUROSCI.4790-12.2013.
52. Wang, N., Wang, H., Li, L., Li, Y., & Zhang, R. (2020). β-Asarone inhibits amyloid-β by promoting autophagy in a cell model of Alzheimer's disease. *Frontiers in Pharmacology, 10*, 1529. https://doi.org/10.3389/fphar.2019.01529.

53. Chen, Y., Chen, Y., Liang, Y., Chen, H., Ji, X., & Huang, M. (2020). Berberine mitigates cognitive decline in an Alzheimer's disease mouse model by targeting both tau hyperphosphorylation and autophagic clearance. *Biomedicine and Pharmacotherapy*, *121*, 109670. https://doi.org/10.1016/j.biopha.2019.109670.
54. Li, L., Zhang, S., Zhang, X., Li, T., Tang, Y., & Liu, H. (2013). Autophagy enhancer carbamazepine alleviates memory deficits and cerebral amyloid- β pathology in a mouse model of Alzheimer's disease. *Current Alzheimer Research*, *10*(4), 433–441. https://doi.org/10.2174/1567205011310040008.
55. Wang, C., Zhang, X., Teng, Z., Zhang, T., & Li, Y. (2014). Downregulation of PI3K/Akt/mTOR signaling pathway in curcumin-induced autophagy in APP/PS1 double transgenic mice. *The European Journal of Pharmacology*, *740*, 312–320. https://doi.org/10.1016/j.ejphar.2014.06.051.
56. Song, H. C., Chen, Y., Chen, Y., Park, J., & Zheng, M. (2020). GSK-3b inhibition by curcumin mitigates amyloid genesis via TFEB activation and anti-oxidative activity in human neuroblastoma cells. *Free Radical Research*, *54*(11–12), 918–930. https://doi.org/10.1080/10715762.2020.1791843.
57. Uddin, M. N., Elahi, M., Shimonaka, S., Kakuta, S., & Ishiguro, K. (2021). Strain-specific clearance of seed-dependent tau aggregation by lithium-induced autophagy. *Biochemical and Biophysical Research Communications*, *543*, 65–71. https://doi.org/10.1016/j.bbrc.2020.12.113.
58. Congdon, E. E., Wu, J. W., Myeku, N., Figueroa, Y. H., Herman, M., & Marinec, P. S. (2012). Methylthioninium chloride (methylene blue) induces autophagy and attenuates tauopathy in vitro and in vivo. *Autophagy*, *8*(4), 609–622. https://doi.org/10.4161/auto.19048.
59. Kim, S., Choi, K. J., Cho, S. J., & Yun, S. M. (2016). Fisetin stimulates autophagic degradation of phosphorylated tau via the activation of TFEB and Nrf2 transcription factors. *Scientific Reports*, *6*, 1–13. https://doi.org/10.1038/srep24933.
60. Zhang, Z. H., Wu, Q. Y., Zheng, R., Chen, C., & Chen, Y. (2017). Selenmethionine mitigates cognitive decline by targeting both tau hyperphosphorylation and autophagic clearance in an Alzheimer's disease mouse model. *The Journal of Neuroscience*, *37*(9), 2449–2462. https://doi.org/10.1523/JNEUROSCI.3229-16.2017.
61. Wang, J., Liu, B., Xu, Y., & Yang, M. (2021). Activation of CREB-mediated autophagy by thioperamide ameliorates β-amyloid pathology and cognition in Alzheimer's disease. *Aging Cell*, *20*(3), e13333. https://doi.org/10.1111/acel.13333.
62. Takatsuka, C., Inoue, Y., Matsuoka, K., & Moriyasu, Y. (2004). 3-methyladenine inhibits autophagy in tobacco culture cells under sucrose starvation conditions. *Plant and Cell Physiology*, *45*(3), 265–274. https://doi.org/10.1093/pcp/pch031.
63. Christian, F., Anthony, D. F., Vadrevu, S., & Riddell, T. (2010). p62 (SQSTM1) and cyclic AMP phosphodiesterase-4A4 (PDE4A4) locate to a novel, reversible protein aggregate with links to autophagy and proteasome degradation pathways. *Cell Signal*, *22*(10), 1576–1596. https://doi.org/10.1016/j.cellsig.2010.06.003.
64. Oliva, O., Rez, G., Palfia, Z., & Fellinger, E. (1992). Dynamics of vinblastine-induced autophagocytosis in murine pancreatic acinar cells: Influence of cycloheximide post-treatments. *Experimental and Molecular Pathology*, *56*, 76–86. https://doi.org/10.1016/0014-4800(92)90025-7.

65. Wu, Y. C., Wu, W. K., Li, Y., & Yu, L. (2009). Inhibition of macroautophagy by bafilomycin A1 lowers proliferation and induces apoptosis in colon cancer cells. *Biochemical and Biophysical Research Communications*, *382*(2), 451–456.
66. Harhaji-Trajkovic, L., Arsikin, K., Kravic-Stevovic, T., Petricevic, S., & Tovilovic, G. (2012). Chloroquine-mediated lysosomal dysfunction enhances the anticancer effect of nutrient deprivation. *Pharmaceutical Research*, *29*(8), 2249–2263. https://doi.org/10.1007/s11095-012-0753-1.
67. McAfee, Q., Zhang, Z., Samanta, A., Levi, S. M., & Ma, X. H. (2012). Autophagy inhibitor Lys05 has single-agent antitumor activity and reproduces the phenotype of a genetic autophagy deficiency. *Proceedings of the National Academy of Sciences of the United States of America*, *109*(21), 8253–8258. https://doi.org/10.1073/pnas.1118193109.
68. Moriyasu, Y., & Inoue, Y. (2008). Use of protease inhibitors for detecting autophagy in plants. *Methods in Enzymology*, *451*, 557–580. https://doi.org/10.1016/S0076-6879(08)03232-1.
69. Dong, X. X., Wang, Y. R., Qin, S., & Liang, Z. Q. (2012). p53 Mediates autophagy activation and mitochondria dysfunction in kainic acid-induced excitotoxicity in primary striatal neurons. *Neuroscience*, *207*, 52–64. https://doi.org/10.1016/j.neuroscience.2012.01.018.
70. Tanida, I., Minematsu-Ikeguchi, N., Ueno, T., & Kominami, E. (2005). Lysosomal turnover, but not a cellular level, of endogenous LC3 is a marker for autophagy. *Autophagy*, *1*(2), 84–91. https://doi.org/10.4161/auto.1.2.1697.
71. Uddin, M. S., Rahman, M. A., Kabir, M. T., Behl, T., & Mathew, B. (2020). Multifarious roles of mTOR signaling in cognitive aging and cerebrovascular dysfunction of Alzheimer's disease. *IUBMB Life*, *72*(9), 1843–1855. https://doi.org/10.1002/iub.2324.
72. Schreiber, K. H., Ortiz, D., Academia, E. C., Anies, A. C., Liao, C. Y., & Kennedy, B. K. (2015). Rapamycin-mediated mTORC2 inhibition is determined by the relative expression of FK506-binding proteins. *Aging Cell*, *14*(2), 265–273. https://doi.org/10.1111/acel.12313.
73. Galluzzi, L., Bravo-San Pedro, J. M., Levine, B., Green, D. R., & Kroemer, G. (2017). Pharmacological modulation of autophagy: therapeutic potential and persisting obstacles. *Nature Reviews Drug Discovery*, *16*(7), 487–511. https://doi.org/10.1038/nrd.2017.22.
74. Kataura, T., Tashiro, E., Nishikawa, S., Shibahara, K., & Muraoka, Y. (2021). A chemical genomics-aggrephagy integrated method studying functional analysis of autophagy inducers. *Autophagy*, *17*(8), 1856–1872.

Index

M

O

P

R

S

T

V

For Product Safety Concerns and Information please contact our EU representative GPSR@taylorandfrancis.com
Taylor & Francis Verlag GmbH, Kaufingerstraße 24, 80331 München, Germany

www.ingramcontent.com/pod-product-compliance
Lightning Source LLC
LaVergne TN
LVHW022359021026
849161LV00021B/655
* 9 7 8 1 7 7 9 6 4 1 8 0 9 *